HUMAN ASPECTS OF LIBRARY AUTOMATION

The Library Association Information Services

HUMAN ASPECTS OF LIBRARY AUTOMATION

SECOND EDITION

ANNE MORRIS AND HILARY DYER

Gower

First edition published in 1990 by Gower Publishing Company Ltd

Published by
Gower Publishing Limited
Gower House
Croft Road
Aldershot
Hampshire GU11 3HR
England

Gower
Old Post Road
Brookfield
Vermont 05036
USA

Anne Morris and Hilary Dyer have asserted their right under the Copyright, Designs and Patents Act 1988 to be identified as the authors of this work.

British Library Cataloguing in Publication Data
Morris, Anne, 1950–
Human aspects of library automation. – 2nd ed.
1. Libraries – Automation 2. Information storae and retrieval systems
I. Title II. Dyer, Hilary
025'.00258

ISBN 0 566 07504 0

Library of Congress Cataloging-in-Publication Data
Morris, Anne, 1950–
Human aspects of library automation / Anne Morris and Hilary Dyer. —2nd ed.
p. cm.
Dyer's name appears first on the earlier edition.
Includes bibliographical references and index.
ISBN 0–566–07504–0 (hardback)
1. Libraries—Automation. 2. Libraries—Data processing. 3. Human engineering. I. Dyer, Hilary, 1954– . II. Title.
Z678.9.D85 1998 97–28407
025'.00285'4019—dc21 CIP

Typeset in Century Schoolbook by Bournemouth Colour Press, Parkstone and printed in Great Britain by Biddles Limited, Guildford

CONTENTS

List of Figures		ix
List of Tables		xi
Acknowledgements		xiii
Introduction		xv
PART I	**HUMAN CHARACTERISTICS**	**1**
	1 Human Characteristics Introduction – Human information processing – Factors affecting human information processing – Humans versus machines – Physical characteristics – Summary	3
PART II	**HEALTH AND SAFETY CHARACTERISTICS**	**27**
	2 The Importance of Health and Safety Why health and safety matters – Legal aspects – The alleged health hazards of VDU work	29
	3 Musculoskeletal and Postural Problems Introduction – Forward-leaning posture – Static posture – Repeated work movement – Intensity of work – Psychosocial factors – Individual aspects – How to minimize postural problems	34
	4 Work-related Upper Limb Disorders Introduction – Range of conditions – The causes of WRULDs – Who suffers from WRULDs – Treatment of WRULDs – WRULDs and the law – Management of WRULDs – Prevention of WRULDs	40

5 Eye and Vision Problems 48
Introduction – Eyesight of the user – The VDU as a cause of visual discomfort – The type of work performed – Summary

6 Stress 66
Introduction – What is stress? – Stress and health – The causes of stress – The need to reduce stress – Stress and the law – How to monitor stress levels – How to reduce work-related stress – Summary

7 Reproductive Hazards 78
Introduction – Reports of adverse pregnancy outcomes – Alleged causes of reproductive hazards – Minimizing risk to pregnant (VDU) workers

8 Other Alleged Health Hazards 97
Cataracts – Facial rashes – Static electric shocks – Photosensitive epilepsy – Cabling – Fire – Laser printers – Recommendations

PART III WORKPLACE DESIGN: WORKSTATION CONSIDERATIONS 103

9 Basic Considerations 105
Introduction – Users of the workstation – The task(s) to be performed – Sitting posture – Upright posture – Forward-tilting posture – Backward-leaning posture – The ideal posture?

10 Workstation Components 112
Chairs – Desks – Recommended dimensions for workstation chair and desk – The VDU screen – The keyboard – Forearm rests – Touch screens – Mice – Printers – Barcode readers – Summary

PART IV WORKPLACE DESIGN: ENVIRONMENTAL CONSIDERATIONS 139

11 Lighting 141
Introduction – Units – Recommended illumination levels – Natural daylight – Artificial light – Light sources – Lighting colour – Lights as glare sources – Floors, walls and ceilings – Emergency lighting – Recommendations

12 Noise 155
Introduction – Units – Guidelines on noise levels – The causes of noise – Problems caused by noise – Recommendations

13 Heating and Ventilation 164
Introduction – Body temperature regulation – Air temperature – Radiant heat – Relative humidity – Air movement – Type of activity – Clothing – Thermal indices – Sick building syndrome – Recommendations

PART V WORKPLACE DESIGN: WORKSTATION DESIGN 173

14 Workstation Layout and Design 175
Introduction – Task analysis – Equipment considerations – Operator limitations – Workstation design – Effect on productivity

15 Workstation Layouts for Library Functions 195
Issue counter – Cataloguing – Serials control – Online searching – Recommendations

PART VI THE HUMAN – COMPUTER INTERFACE 205

16 The Software Interface 207
Introduction to software interface – Software interface

17 User Needs 210
Types of user – Accommodating variations and changes in user types – Short-term memory and overload – Psychological closure – Perception – Long-term memory and mental models – Designers' conceptual models: training and the user of analogy and metaphor – Consistency and transferable skills – Summary

18 Dialogue Design 222
Principles of interface design – Input methods – Dialogue styles – Summary

19 Screen Design 241
Information content – Formatting – Coding – The use of colour – Screen colour – Summary

20 Interaction features affecting usability 256
Consistency – Initiative – Control – Help – Feedback – Error handling – Recovery – Response time – Rules, guidelines and standards – Summary

21 Software Evaluation 270
The types of library systems – Evaluation methods – Custom software, or what to do if you can't find a suitable system! – Summary

PART VII JOB DESIGN 281

22 Job Design 283
Introduction – Job design and job satisfaction – Techniques for improving job satisfaction – The impact of automation on staff structures

PART VIII PLANNING AUTOMATED SYSTEMS 295

23 Technological Change 297
Reactions to technological change – People affected by technological change – Steps in the management of change – Summary

24 Training 325
The staff training programme – Support mechanisms – Further training – Training the user

Useful WWW Sites and Addresses 341

References 365

Subject Index 387

Index of Authors Cited 395

LIST OF FIGURES

1.1	Basic information processing model	4
1.2	The Yerkes–Dodson law on arousal	7
1.3	Variation in size of library staff	21
1.4	Normal distribution	23
5.1	Simplified section of the eye	49
5.2	Simplified section of VDU screen	53
5.3	Two dot matrices used to create the letter 'R'	57
7.1	Radiation patterns, indicating the inverse relationship between frequency and wavelength	90
7.2	Types of radiation and alleged hazards	90
9.1	The kneeling chair	110
10.1	Cluttered library workstations	115
10.2	The effect of working at low desks: lack of space and knee room at workstation	116
10.3	Desk and chair dimensions	118
10.4	Use of VDU stand to alter height of screen	120
10.5	The Sholes (QWERTY) keyboard	122
10.6	The Dvorak keyboard	123
10.7	Keyboard layouts used by Norman and Fisher (1982)	125
10.8	Kathleen Gardner's keyboard layout	126
10.9	The Maltron keyboard	127
10.10	Kinesis Ergonomic keyboard	128
10.11	Bat Personal keyboard	129
10.12	DataHand	130
10.13	Preferred cursor control layout	133
10.14	Wrist support for keyboarding	134
11.1	Different light types used in one work area	147
11.2	The effect of glare from lights on the VDU screen	150
11.3	The effect of reflected light from windows on the VDU screen	151
11.4	The effect of diffuse lighting	152
12.1	The dB(A) scale	156
14.1	Different workstation layouts	178
14.2	Self-issue, renew and return system	181

14.3	Maximum and optimum lengths of horizontal reach	183
14.4	Reach problems at the issue counter	184
14.5	Operator viewing angles	186
14.6	The increased work surface offered by modular furniture and a range of angles between adjacent surfaces	189
14.7	The use of modular furniture to extend the working surface within reach of the operator	190
14.8	How system units on the floor can be difficult to access and cause cabling problems	191
14.9	Trailing wires and the potential hazards they cause	192
15.1	Original counter	196
15.2	Redesigned counter	196
15.3	Low desks and the problems they cause clients and staff	198
15.4	A workstation layout for cataloguing	201
15.5	A workstation layout for serials check-in	202
18.1	Cursor-based broad menu with optional selection using letters	233
18.2	The use of palette menus on screens	235
19.1	Screen displays showing logical groupings	248, 249
22.1	Percentage of workers reporting selected health symptoms 'almost daily' by hours per day worked on VDUs	290
22.2	Percentage of workers reporting selected health symptoms 'almost daily' by longest time worked on VDUs without a break	290

LIST OF TABLES

1.1	Anthropometric measurements, application and correction factors	22
1.2	Anthropometric estimates for various categories of library staff and users	24
6.1	Social readjustment rating scale	72
7.1	Research into adverse pregnancy and VDU use	80
7.2	Permitted radiation levels	96
11.1	Recommended lighting levels for different work areas	143
11.2	Recommended lighting levels for different kinds of VDU work	143
12.1	Maximum noise levels for various activities	158
13.1	Estimated metabolic heat generated for different tasks	168
13.2	The relationship between PMV values and PPD values	169
14.1	5th percentile measurements relating to reach	182
17.1	The knowledge domains of different user types	211
19.1	The visible spectrum of colour	253

ACKNOWLEDGEMENTS

This book could not have been completed without the help of a number of people: librarians and information officers, typists, proofreaders, and assistants. Our grateful thanks are extended to (in alphabetical order): Claire Bradshaw, Graham Dyer, Sally Fensome, Cliff McKnight, Sally Maynard, Janine Mascia, Alexia Waldron to name but a few.

AM
HD

INTRODUCTION

This is the second edition of *Human Aspects of Library Automation*. Since the first edition of this book, published in 1990, the introduction of computer technology in libraries has become commonplace, developments in technology has continued apace, and several new European rules and regulations have appeared. The aim of the second edition is to reflect these changes and summarize guidelines and the latest research into the various aspects of human–computer interaction which is applicable to libraries whether implementing new systems for the first time, second time or even third time. However, as research is continually being undertaken and books have a long gestation period, it is not possible to ensure that the latest report to hit the headlines will be covered. Every effort has been made to give a thorough examination of the topic, but no guarantee of completeness can be given.

The implementation of computer-based systems has received much attention in the literature. Focus has, however, centred primarily on the technical and practical aspects of the process, with relatively little mention being made of the effect on the people who will operate such systems or be served by them. Over a decade ago Fine (1982, p. 209) expressed it thus:

> The greatest marvel of technology is that if it breaks down, we can fix it; if it has flaws, we can debug it; if it doesn't work at all, we can ignore it; and if it works well, we can make it work better. No one has as yet figured out a way to debug the human factor. It is the most complicated aspect of any technological system, yet it's the one that gets the least attention, is least discussed, the least researched, and perhaps the least understood.

This still holds true today.

As many libraries have found, there are advantages to be gained by introducing new or upgrading old computer systems, such as increased productivity, improved reader services, faster movement of acquisitions between their delivery and appearance on the shelves, improved stock control and a better quality of working life for all concerned. Nevertheless, there are many pitfalls on the road to successful computerization, which can counterbalance several of these advantages. It is the problems with the human aspect of library automation that this book seeks to address.

The importance of the human factor must not be underestimated. If people respond badly to the introduction of a new system, the anticipated effectiveness will not be achieved. Poor workstation and job design can result in poor health and can induce, or increase, stress. Poorly designed software interfaces result in under-used systems and decreased accuracy. Thus, in order to be successful, attention needs to be paid to the human factor.

Part I examines how people process information and how factors such as attention, arousal, motivation, mental capacity, perception and memory affect the way people respond to events. It is extremely important to be aware of people's limitations in the processing of information, in order to make allowances for these in systems design.

Parts II to V concentrate on the health and safety factors and workplace design. Rumours about the possible risks to health abound in the general and professional press and it is hoped that the review of the research in this field will clarify some of the issues and allay some of the fears. One of the primary causes of health problems among visual display unit (VDU) operators is poor workstation and/or job design. The latter part of the book therefore concentrates on methods of improving the operator's workplace and job design. It is appreciated that lack of funds and inadequate buildings are in part to blame, but it is believed that if librarians become more aware of the alternatives available, they may be better able to argue for an improvement in these conditions.

The software interface (Part VI) has no small part to play in the overall level of operator satisfaction. With the relatively small choice of library systems available, it is perhaps difficult enough to choose a system which offers appropriate functionality without complicating the issue by assessing the quality of the user interface. It is considered important, however, to outline those factors which contribute to or hinder ease of use, if only to heighten the awareness of potential software purchasers and so to stimulate and make more articulate the demand for a more usable interface.

Parts VII and VIII examine the implications of introducing, adding or upgrading computer systems on the operator's work. Changes here may, in turn, affect the design of the workstation to incorporate variations in job procedures. Having studied the detail of what is affected by automating processes, attention is paid to how new systems should be implemented. Consideration is given to the management of change, the planning and implementation process and the training of both staff and users.

Where possible, reference has been made not only to British standards and laws but also to those of other countries. The World Wide Web sites and addresses at the end of the book are primarily targeted at British readers, but addresses of human factor and ergonomic societies around the world have been included to enable other readers to obtain details about contacts in their own country. The general advice on ergonomic, health and safety issues is valuable to all librarians no matter where they live.

PART

I

HUMAN CHARACTERISTICS

CHAPTER 1

HUMAN CHARACTERISTICS

INTRODUCTION

The design of an information system is rarely undertaken by a single person. Rather, a design *team* consisting of people with a variety of skills and backgrounds will be involved. If this team is to produce a successful design – one which can be used easily and effectively by the target users – then it is important that they understand several aspects of the socio-technical system in which the information system will play a part.

The present chapter will consider the general psychological characteristics which a design team might reasonably be expected to take into account. People's mental attributes affect the way in which they react to the world around them. In terms of an automated library system, the surrounding environment will contain several factors which impinge on a person's private world. For example, a change of system will generate stress to a greater or lesser degree depending on the attitude of the person towards the change; the need to learn and remember new procedures will make demands on the memory process; and the need for concentration or attention may alter.

Physically, people will also vary. Rather than forcing them to adapt to a given arrangement of furniture, possibly inducing harmful postures, it is desirable to redesign the workplace to take into account the human as well as the computer. Tables 1.1 and 1.2 at the end of this chapter outline the main physical characteristics which need to be taken into account.

HUMAN INFORMATION PROCESSING

People interact with computers by giving and receiving information. The quality of the communication between human and machine will depend upon an accurate exchange of data. For this reason, it is important to consider how people process information and the factors which affect their performance.

Information is fed into the brain via one or more senses. Once there, it

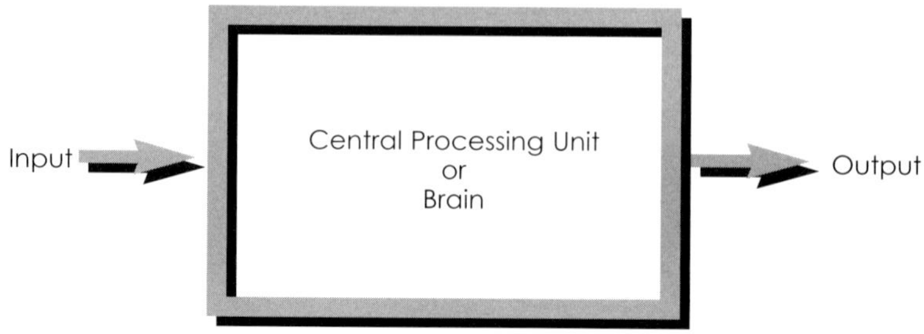

Figure 1.1 Basic information processing model

is operated on and then the manipulated data are either stored for future use or transmitted via various output devices (for example, via speech or actions). The basic information processing model can therefore be represented at its highest level of abstraction as comprising *input*, *processing* and *output* (see Figure 1.1). Like any abstraction this is necessarily over-simplified – it might be taken wrongly to imply, for example, that all human output is driven by external data input – but serves as a useful orienting device.

DATA INPUT

Information is received by a person via the many body receptors. These take in data from the outside world, via one or more of the five senses: sound, sight, touch, smell and taste; from inside the body, for example, they inform the brain when the body is tired or hungry or thirsty; and from the neuro-muscular system which gives information to the brain about body position. Once the information has been received by the receptors, it is passed to the brain via the nervous system.

Human information processing is necessarily complex – to paraphrase Beale and Jackson (1990, p. 109), if the brain was simple to understand we would be too simple to understand it! A wide variety of processes can act on the received information with the result that it is interpreted by the brain in an idiosyncratic manner. Such factors as perception, attention, arousal, motivation, emotion, past experience and environmental features can affect singularly or interactively the analysing process, sometimes leading the brain to misinterpret the information. Many of the factors affecting human information processing will be discussed further later in this chapter.

CENTRAL PROCESSING FACILITIES

The central processing facilities in the brain may be seen as being divided into various mental activities or processes. Information or stimuli are first received and held in a very short-term sensory store. The information then passes to the brain, where it is 'perceived'. The process of perception is extremely complex and clearly involves much more than simply matching the sensory information with a 'template' or similar representation which might be held in a long-term memory store. Consider, for example, the perceptual processes operating as you read the words on this page. If you had to identify characters individually, build them up into words, identify each word and retrieve its meaning from memory, retaining this meaning while you continued to build up the rest of the sentence, reading would be a much slower process than it is. Although the process is data driven (or 'bottom-up') to some extent, there is a large part which is concept driven (or 'top-down') processing. You bring to the page all sorts of experiences which not only help you to recognize words but, for example, provide common syntactic structures and even give you expectations of what words typically follow each other.

Once the information is identified (assuming it is), a decision has to be made as to what action to take. A response can be evoked immediately and/or the information can be committed to working memory. Once in the working memory, information can be simply forgotten, in a relatively short time, committed to long-term memory, that is, learned, or used to generate a response. Responses, either before or after committing information to the working memory, involve some sort of controlled muscular movement as stated above. The consequences of each response are then perceived as feedback (that is, the output produces an effect which provides further input), and the whole process starts again. Obviously, this cycle of events takes place rapidly, but it is important to note that there are limits on both the quantity of information that can be processed in a certain time and the amount of 'mental capacity' available when undertaking information processing tasks.

DATA OUTPUT

There is very little to say about data output except that it generally takes the form of some sort of controlled muscular movement. This includes not only the movement of limbs but also the mouth (for speech), face (facial expression) and eyes. Output can be affected by constraints imposed by the working environment, an operator's physical well-being and other factors.

FACTORS AFFECTING HUMAN INFORMATION PROCESSING

ATTENTION

The concept of attention refers to a 'global limitation on the resources available to time-share activities and to process information in parallel' (Wickens, 1987).

People are capable of attending to only a limited number of stimuli at any one time. In fact, some theorists propose that a person is only capable of attending to one source of information at a time (the single channel theory). Others believe that this theory can be extended. They believe that, although people might only be able to process or attend to one item at a time, they are nevertheless capable of rapidly 'switching' attention across several tasks thereby giving the appearance of processing information in parallel. A number of other theorists contend that people can perform two or more tasks simultaneously, particularly if one of these can be performed without conscious attention in an 'automatic' mode of processing, that is, where stimuli produce a response but are not consciously perceived. For example, car drivers often report that they suddenly realized they had no recollection of the last five minutes of driving, yet information was almost certainly being processed and acted upon during this time otherwise they would have crashed. However, for managerial and system design purposes it is safer to assume that people can pay attention to only one source of information at a time.

Certain kinds of stimuli or information are more successful in gaining our attention than others. These include stimuli that are of high intensity (a loud noise has the advantage over a low murmur and, similarly, a bright light over a dim one), stimuli that suddenly change (the fan on the computer may go unnoticed until it stops working), stimuli that differ or contrast significantly with their surroundings, stimuli which involve movement (such as flashing error messages) and stimuli which are repeated (the repeated high pitched bleep of an activated library security system, for example, invokes more attention and a better response than if there was only one bleep). Stimuli which have some sort of emotional importance (such as the occurrence of a person's own name) also attract attention.

The general level of attention is also determined by internal factors. These include:

- individual habits of attention and inattention (people can learn to attend to certain inputs and disregard others);
- level of interest (the more interesting the stimuli, the more likely the person is to respond);
- level of motivation (highly motivated people will attend longer, see 'Motivation' below);
- the perceived relevance or importance of the input (relevant and important stimuli are more likely to be filtered through and attended to than less relevant and less important material);
- the general level of arousal (see 'Arousal' below);
- to some extent a person's mental capacity (see 'Mental capacity' below).

Notwithstanding the level of motivation mentioned above, people also differ in their attention span, that is the amount of time they can concentrate on a single task without being distracted. Attention factors must therefore be considered not only by the system design team but also by the library manager as job designer. Inattentive staff can at one extreme be mildly irritating to both managers and clients and, at the other, incur vast unnecessary expense, particularly if unheeded warnings lead to computer breakdowns.

AROUSAL

To arouse is to 'stir into activity' (*Concise Oxford Dictionary*, 1990), the concept of arousal being associated with the amount of brain activity in a human being. People talk about being aroused from sleep, that is experiencing a change of state from where brain activity is low to one of increased activity or arousal. Research has shown that arousal is correlated to performance levels. Too little arousal (caused by insufficient external and internal stimulation) and too much arousal (excessive external and internal stimulation), for example, cause performance to deteriorate. Optimum performance can only be obtained if the arousal conditions are perfect for a particular operator and task. This relationship is often called the Yerkes–Dodson law and is illustrated in Figure 1.2.

The external factors include such things as the number of customers using the library (arousal may be very low if there are only few readers or, conversely, too high if there are queues of customers); noise (too much and too little can cause over- and under-stimulation respectively); and lighting (glare on the VDU screen may over-stimulate the eyes). In addition, there are numerous physical, psychological, social and personal internal factors

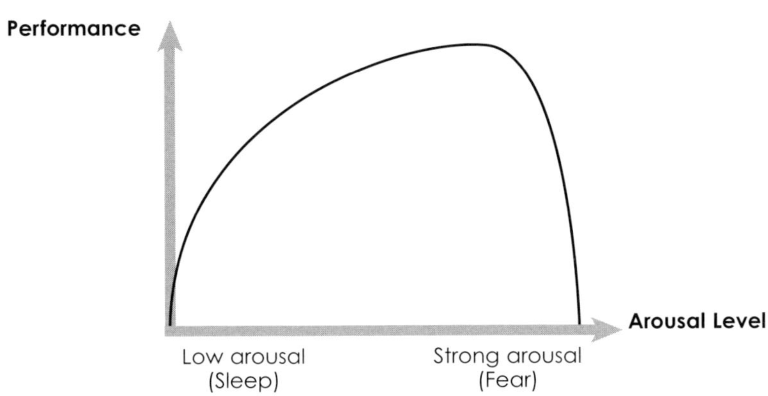

Figure 1.2 The Yerkes-Dodson Law on arousal

which can raise or lower a person's arousal system and thereby alter performance levels. These include, for example, age, sex, health, personality, intelligence and motivation. The amount of stimulation which should be aimed for will depend on the characteristics of the operator(s) and the nature of the task. There are no data providing optimal arousal/stimulation levels; each case needs to be examined individually. It has been found, for example, that 'piped' music increased the performance of operators doing boring, repetitive industrial tasks (Fox, 1983). It is doubtful, however, that similar success would be obtained in a library! It is, therefore, necessary to examine the circumstances in each library, interviewing the operators and assessing the nature of each task. The aim is to achieve a balance, with neither too much nor too little stimulation.

MOTIVATION

Motivation is the term used by psychologists to describe the study of what induces people to act. An individual's level of motivation affects whether incoming information receives attention and whether it will produce a response. In other words, it will affect people's ability to learn new concepts and to respond to different situations.

Motives have been classified by several different researchers. Maslow (1943), for example, classified motives in relation to a hierarchy of needs. These, in ascending order, are: physiological needs; the need for safety/security; social needs; needs of the ego; and the need for self-actualization. He postulated that people are motivated to satisfy these needs in a bottom-up approach – that is, physiological needs are of paramount importance and must be satisfied before those related to safety and security and so on. In other words, hungry people don't think about philosophy. Alderfer (1969, 1972) updated Maslow's concept and introduced the following groups of needs:

- 'Existence' or 'basic' needs – related to the survival of the individual.
- 'Relatedness' or 'contact' needs – related to people's social activities.
- 'Growth' or 'self-actualization' needs – related to people's desire to have influence over their situation.

Managers in the Western world have little need to concern themselves with the basic physiological needs of warmth, food and clothing although they do exercise considerable influence over if, and how, other needs are met, and hence whether people are motivated to carry out their work. For this reason, they need to be sensitive to the factors which affect motivation.

Herzberg *et al.* (1959), while investigating work incentives and the working environment, found that the most important 'satisfiers' or 'motivators' were related to performance, recognition, work content, responsibility, promotion, and pay. Others included organizational policies, relations with subordinates, colleagues and supervisors, and working conditions. Clearly such factors relate to the higher levels of Maslow's hierarchy.

In general terms, people like to feel that they have done a good job, in order to have a sense of achievement. This can only be attained if they have met goals either set by themselves or demanded by the task or organization. Recognition of effort, like other forms of feedback, is extremely important. Sometimes a 'pat on the back' is all that is necessary. In other cases, recognition might be achieved by promotion. Work content involves not only what the job entails, but how rewarding it is, the degree of variety, the level of complexity, how much stress it causes, and the degrees of control and autonomy it provides. All these need to be considered when undertaking job design and are discussed more fully in Chapter 22. The level of responsibility can be a motivator or demotivator, depending on the individual. Responsibility to one person may cause stress and act as a demotivator, but to another may provide an opportunity for self-fulfilment. Consequently, managers must be cautious when using this approach as an incentive or motivator. Most people like to feel that there is some chance, at least, of promotion. Not many people are content with 'dead-end jobs', hence some sort of career structure within organizations is welcomed by most employees. Low pay can drastically reduce motivation and thus the performance of an individual. Unfortunately, low pay is one of the characteristics of much library and information work. Fortunately, in many cases, this demotivating factor may be counter-balanced by the enjoyment many professionals obtain from their work.

Organizational issues, such as how the organization is structured and whether management are caring, accessible and approachable also affect an employee's motivation. Similarly, relations with subordinates, colleagues and supervisors play a part. Poor relations create a bad atmosphere and result in poor motivation. Working conditions – such as the number of hours worked, the amount of holidays, restaurant facilities and the working environment – also affect the degree of motivation. Obviously a safe, spacious, pleasant, noise-free environment is more motivating than one that is not. The last factor mentioned by Herzberg *et al.* (1959) was that of leadership. Research suggests that most employees prefer a democratic rather than an autocratic form of leadership. People, on the whole, like to feel they are involved in an organization and that they can contribute to its overall success.

It can be seen, therefore, that an individual's level of motivation depends on many factors, their relative importance differing for different people. For some, it is sufficient to have good pay and conditions; for others, an interesting job may be more important. Motivating factors also interact and counterbalance each other. For example, librarians may have to work in extremely cramped conditions, be poorly paid and overworked, yet still be fairly motivated because the jobs are rewarding. When assessing whether staff are motivated, managers must therefore consider many factors and determine the overall combined effect. If motivation is low, then ways of increasing it must be sought, such as providing encouraging feedback, improving job design and the work environment or creating more channels for self-enhancement. It must be remembered that increasing the pay of employees can be a useful and welcome motivator in the short term, especially when pay is low, but is rarely satisfactory on its own as a long-term solution to poor motivation.

MENTAL CAPACITY

The mental capacity of individuals varies enormously, but there is one common factor: namely, that if you exceed a person's mental capacity, performance deteriorates rapidly. Brown and Poulton (1961) and Brown (1965) have expanded this fact into a 'mental capacity and performance' model. Essentially, they suggest that performance depends on a trade-off between the demands of a task and of the environment, and the operator's own abilities. They believe that a person has a limited capacity to perform tasks. That capacity may be consumed by the demands of the job, or by operator effort in compensating for an unpleasant environment or by poor equipment, causing performance to drop. Thus, in order to maximize an operator's performance, it is necessary to match the demands of the task and the suitability of the environment to the operator.

PERCEPTION

As we saw earlier, perception is not simply the recognizing of incoming stimuli. Rather, it is a construction based on both incoming data and a variety of internal inputs and processes. People's perception is affected by all these various factors with the result that one person may perceive things differently from another. While this is less obvious (but still true) when considering a written text, it becomes more obvious when we consider social interactions, for example.

Perception is affected by the overall quality of the stimuli presented and a person's general receptiveness at that particular time. The quality of the stimuli is affected by a large number of factors, such as the duration and frequency of their presentation. In the case of written cues, the size of the text, glare, display characteristics, character representation and the format (VDU, book, etc.) also have an effect, as do the duration of an event, its loudness, intonation and accent with verbal stimuli. People's general receptiveness is equally affected by many factors; for example age, motivation, level of arousal, intelligence, attentiveness, health and their familiarity with the stimuli. In addition, environmental factors such as noise level, the presence of vibration, inadequate lighting and poor thermal conditions can affect both the quality of the stimuli and a person's overall receptiveness.

In order to ensure that performance is not degraded, it is necessary to increase the likelihood of incoming stimuli being perceived. Thus, it is important to ensure that the stimuli presented to the person are not distorted in any way and that conditions are such as to enable the person to remain alert and receptive. In other words, managers must consider carefully the design of software, hardware, workstations, jobs and the environment to ensure optimal conditions. All these are discussed in considerable detail in later chapters. Perception can also be increased, to some degree, by increasing stimulus expectancy (Wickens, 1987). For example, online searchers can often read the content of a line of information even though it appears only momentarily on the screen. This is because they expect to receive certain prompts and can recognize them at a glance by the overall shape and pattern of the characters. Stimulus expectancy, in this case, is increased by the familiarity of the material.

To reiterate, then, perception is not just an automatic processing of incoming stimuli; it involves several mental processes and its success depends on many factors. Possibly the only 'certainty' is that people will perceive things differently.

MEMORY

Memory is an abstract and ambiguous word which, 'in its most comprehensive sense, refers to the effects of a person's past on his present' (Hunter, 1964). People are modified by what they do, and these modifications, if retained, will affect their behaviour some time in the future. Consequently, modifications or past experiences that are held in the memory enable people to adjust to new circumstances. Psychologists usually think of memory as having two components: short-term memory (or the better term, working memory) and long-term memory. Working memory is regarded as a temporary and fragile store where we keep our limited 'conscious' information whereas long-term memory is believed to be where we store more permanent facts and experiences. This section concentrates on short-term memory, long-term memory being related very closely to learning which is discussed in depth in section under 'Learning' below. The main characteristics of short-term memory are as follows:

- The length of time information can be held in working memory without repetition or rehearsal is extremely short; approximately 10–20 seconds (Peterson and Peterson, 1959).

- Information may be maintained for longer if rehearsed, but competition from other perceptual and cognitive activities may cause attention to waiver (Klatzky, 1980; Underwood, 1976).
- The number of unrelated items that can be held even when full attention is given is limited to 7 ± 2. A larger number than this exceeds capacity and will result in items being forgotten or transposed (Miller, 1956).
- Even a small number of items will be forgotten within 20 seconds if this interval is filled with some irrelevant, distracting activity (Deese, 1964).
- A technique called 'chunking' may be used to increase the amount of information stored. Chunking involves grouping individual items. It has been found that related 'chunks' of letters or digits can count as single items. For example, words such as 'main exit' may be counted as two items and the area code of a telephone number as one item even though it may contain more than one digit.
- Meaningful information is stored more successfully than meaningless information. Strings of characters, for example, are more readily remembered than strings of numbers (Conrad, 1964).
- Long strings of unrelated digits are more easily remembered if there is a gap after every third or fourth digit. For example a string presented as 333 6195 will be more easily remembered than 3336195 (Wickielgren, 1964).
- Retention is better if related items are grouped separately from unrelated items; for example, telephone numbers are frequently divided, the area code being separated from the rest of the digits.
- Similar items are more likely to be forgotten than dissimilar items. For example, the acoustically similar string EGBPCDT is less likely to be recalled accurately than XFEYWMU (Conrad, 1964).
- A memory load of combined spatial (pictorial/visual properties of the environment) and verbal (linguistic properties) items will be better retained than the same number of either spatial or verbal items exclusively (Kyllonen and Alluisi, 1987).
- Memory for verbal material will be disrupted more by ongoing perceptual, cognitive and other activities that are also verbal and vocal than by simultaneous spatial activities. For example, it is difficult to listen to the radio and hold a telephone conversation – both verbal activities – but relatively easy to knit while listening because knitting is a spatial activity. The converse is true for spatial activities (Baddeley and Hitch, 1974; Baddeley and Lieberman, 1980; Wickens, 1984).
- Working memory can be improved using the technique of visual imagery where a set of items is associated with distinct visual images which are already held in long-term memory (Bower and Reitman, 1972).

- The older the users – particularly over the age of 55 – the less information can be held in working memory at any given time (Deese, 1964).

Some implications for designers and managers

System designers and managers must be aware of the limitations of human working memory and not overload it.

- People should not be expected to memorize more than seven unrelated, transient items at once. Indeed, a good design should not require users to memorize *any* unrelated, transient items.
- To facilitate 'chunking', instructions should always present related information together.
- Accession numbers and other long numbers such as class numbers should never need to be remembered. If such memory *is* called for, the numbers should have gaps after every third or fourth character. This is frequently the case when the Dewey Decimal System is used when the decimal point is placed after the first three digits, but close classification often results in long strings of numbers following the point.
- Where heavy demands are made on working memory, the system or the job needs redesigning.
- Codes such as passwords or logon names present a paradox. In order for them to be memorized they should be meaningful. If they are not memorable, people may be tempted to write them down and leave them near the terminal, which somewhat defeats the objective of security codes. However, it is usually recommended that such codes contain a mixture of upper and lower case characters and non-alphabetic characters so that they are not easily guessed.
- Important messages should be both verbal and spatial (pictorial). For example, handler warnings on floppy disks frequently contain pictures as well as written instructions.

LEARNING

Learning can be defined as 'a process which optimizes performance based on the recognition and use of familiar features in a present situation' (Gibson, 1965). It involves recognizing patterns in a situation, generalizing from one situation to another and making selective responses that optimize performance. The first stage is essentially an exercise in matching selected features in a current task with features experienced in the past. Gibson (1965) suggests two ways this might be done. First, by template matching; features of the current situation are compared and matched with previously learned 'templates'. Second, by property listing; this involves the elements or features of the present situation being compared with stored lists of elements and features. In practice, a combination of the two methods might be used. However, whatever the method used in the recognition stage it is extremely unlikely that a perfect match can be made. Consequently, learning must involve the processes concerned with generalization and selection of appropriate responses, which are linked to decision-making mechanisms (see the following section).

Given the same instruction, some students or trainees learn more quickly than others. In part, the difference can be attributed to age, general aptitude and past experience. However, it is a fact that some people have better learning skills and can assimilate new information more quickly than others with similar age, aptitude and experience characteristics. By examining these learning skills or learner strategies, people can improve their learning capacities. The ability to re-learn is important when people need to operate new systems or adopt different working practices.

There are many cognitive psychology textbooks as well as study guides which detail learner strategy principles, for example: Anderson (1980), Gage and Berliner (1979) and Mace (1968) to name but a few. Kyllonen and Alluisi (1987) also provide a useful overview. The following is a summary of these principles and factors which affect a person's ability to learn.

- *Meaningfulness of information.* Meaningful information is much easier to assimilate than meaningless data. An understanding of the overall system will facilitate the learning of smaller modules. Similarly, the digestion of a manual's overview will help users locate items of interest more quickly.
- *Recognition versus recall.* In a classic psychology experiment, two groups of people are given a list of, say, 50 words to remember in 10 minutes. They return to the laboratory a week later, when one group is asked simply to write down as many of the words as they can. The other group is shown a list of 100 words and asked to mark those which they think were in the original list. The group asked to recognize words will consistently out-perform the group asked to recall words. Well-designed systems take advantage of such findings by, for example, having available options visible on screen rather than demanding users to recall what they can do.
- *Practice (or reinforcement).* The old adage 'practice makes perfect' has been verified in numerous experiments. All show that the more the material is studied, the stronger will be the memory trace for that material (Wickielgren, 1981).

- *Recitation*. Reciting the main points at the end of a training session or a chapter in a textbook aids learning (Anderson, 1980).
- *Spacing of practice sessions*. Evidence suggests that learning is improved if practice or training sessions are spaced over a period of time rather than in one long block. For example, 10 one- or two-hour sessions spaced over a week will produce better results than 10 or 20 hours consecutively (Mace, 1968). Adequate rest breaks are essential. Research has shown that concentration wavers after a short period of time. Taught instruction, for example, should last for no longer than 40 minutes at any one time (Mace, 1968).
- *Mix of topics*. An effective learning strategy involves a good mix of topics – for example, one or two sessions on one topic, two sessions on the next, and so on. Concentration on one topic reduces learning capabilities.
- *Parts-learning*. Often learners try to 'run before they can walk'. It is much more efficient to master one or two skills at a time before going on to the next.
- *Expert performance*. One way of facilitating the learning process is to observe experts and then to analyse and copy their performance. Obviously, this method is not appropriate in all situations but 'sitting by Nellie' techniques have been used successfully for decades.
- *Feedback*. Learning is increased if appropriate feedback is given. The feedback may be negative, indicating that an error was made, or positive in the case of correct response. Positive feedback is to be preferred as it serves to strengthen and reinforce a learner's response. The sooner the feedback is given, the more effective it will be (see also 'Response mechanisms' below).
- *Extraneous distractions*. For effective learning to take place, the learner must commit the material (verbal, written or actions) to long-term memory. Anything which interrupts this process – for example, interactions with staff or customers or the need to answer the telephone – will have an adverse effect.
- *Stress*. The overall anxiety state or mood of a person can affect how much is learned; the more stressed people are, the less they learn. The use of relaxation exercises and positive thinking techniques can greatly enhance performance (Meichenbaum, 1977; Velton, 1968).
- *Motivation*. Whether a person is motivated to learn can affect whether incoming information is attended to. Hence motivated people are likely to be more receptive to new data. One way to motivate staff is to make the task interesting. Malone and Lepper (1983) suggest that instructional programmes, for example, can be made more interesting if they incorporate some of the characteristics of computer arcade games, which have proved to be extremely motivating. These characteristics include the availability of short-term goals, the use of immediate feedback (scores), the adaptability

of the task to the user's skill level, and other characteristics such as exciting graphics and animation. Other factors will include the degree of relevance people see between the task to be learned and their daily work as they perceive it.

- *Transfer of training*. The transfer of training is the term used by psychologists to describe the influence of past learning on present learning. If the influence is helpful, then positive transfer of training is said to have taken place and the rate of learning is accelerated. On the other hand, if past learning interferes with the learning of new material then negative transfer of training occurs and learning suffers. An example of the latter is when the commands of one system do not mean the same as those of another, such as is the case between different online hosts and databases, or when a new library system is installed which utilizes conflicting commands to the previous system. Thus, extra time needs to be allowed if there are elements of negative training or, if possible, negative influences should be eliminated.
- *Active versus passive learning*. Active learning is more effective than passive learning. People learn more far more quickly if they become involved in the learning process – for example, if they write notes or ask questions during a training session. Passive observation (sitting back and expecting all the information simply to be absorbed) can increase the probability of boredom and of inattention (Schneider, 1982, 1984). Thus, where possible, staff should be actively involved during training sessions. Even if 'sitting by Nellie' techniques are used, learners should be encouraged to discuss their observations and ask questions.
- *Training context*. The learning process is enhanced if training occurs on-site rather than elsewhere. One of the reasons for this is that the environment provides memory cues for the material learnt. However, it must be remembered that distractions and interruptions reduce learning and if these are likely to occur, as they would on a library desk during opening hours, for instance, consideration needs to be given to an alternative time or place.
- *Safe learning environments*. A learning environment should be a place where it is safe for the learner to make mistakes. Hence, learning a task like the maintenance of borrower records should not be carried out on 'live' data where a mistake might have serious consequences.
- *Pictorial/graphical information*. As explained earlier pictorial and graphical information is generally learnt more quickly than written or verbal information. For effective learning to take place it is advisable to have a mixture of different types of information presented in a variety of ways – for example, overhead slides, films, tutorial sessions and so on.

Some of these characteristics will seem obvious, but it is surprising how easily they can be overlooked when it comes to designing training sessions. Careful planning can result in training itineraries which not only provide more effective learning but happier and more motivated staff. This is discussed again in Chapter 24.

DECISION-MAKING

Decision-making is formally defined as 'the selection of an alternative response to an inferred environmental situation about which complete certainty is lacking' (Schrenk, 1969). Typically, the types of tasks which involve decision-making are those that demand high-level intellectual skills, including setting objectives, allocating resources, diagnosing environmental conditions, defining and choosing courses of action and interpreting patterns of events.

The need to understand more about people's decision-making processes is therefore important. The role of the decision-maker has become complicated by the vast increase in the amount and variety of information that has become available through the new capabilities for gathering, processing and displaying data. Time allowed for making decisions has also been reduced as both workload and communication capabilities have increased. Instant decisions are often the order of the day. As technology and system complexity have increased so, too, have the penalties for making a wrong decision, both from a financial standpoint and in terms of health and safety of fellow employees. Consequently, it is extremely important to be aware of peoples' limitations as processors of information, in order to make allowances for these in system design. Equally important is the knowledge about how decision performance can be improved.

It is well known that people can assimilate and process information only at relatively low rates and are poor at remembering information and doing computations. What is not often taken into account are the following characteristics and limitations:

- People tend to want too much, rather than too little information to assist them with the result that they then 'cannot see the wood for the trees'. This is particularly true with management information from many automated library housekeeping systems which fail to provide information tailored to a library's requirements.
- Too much weight is given to information collected early in the process; this is called the 'primacy' effect.
- Opinions tend not to be revised on the basis of new data; people are conservative in nature.
- People are not always able to make full use of available information, particularly multi-dimensional information.

- Decision-makers often consider too few courses of action; people generally opt for 'least effort'.
- People are not very good at estimating the probability of some types of outcome. They tend to overestimate low probabilities (e.g. 0.01 for 0.0001) and underestimate high probabilities (e.g. 0.9 for 0.9999), for instance. In addition, they tend to commit the 'gambler's fallacy', whereby an event that has not occurred recently is seen as being more likely to occur. For example, if a tossed fair coin comes up 'heads' four times in a row, people show a tendency to predict that the fifth toss should yield 'tails'. In reality, as long as the coin is fair, 'heads' is as likely on the fifth as on the first toss since the events are independent – the coin doesn't have a memory.

Despite these limitations, people seem to perform surprisingly well at decision-making. It appears that they have an implicit understanding of expressed probabilities and can specify outcome probabilities with reasonable accuracy, particularly in the middle ranges. They can usefully apply probability statements for decisions regarding non-frequent events and can judge the potential value of new information. Furthermore, it has been found that people usually give the greatest weight to the most important factors in decision-making problems. Overall, this means that people can give consistent and reasonably accurate results when undertaking some (most notably, simple) decision tasks. However, performance deteriorates rapidly when they have to evaluate many different alternatives and outcomes. Consequently, managers must recognize the possible need to optimize the performance of the human decision element in a system.

RESPONSE MECHANISMS

It was noted earlier that memory affects the way in which we respond to events in the present. Thus the memory of previous feedback (see also 'Learning' above) will affect our actions in the present.

Feedback during a task is extremely important. Without it, responses can become inaccurate and motivation can be reduced. Feedback can be specific, like the number of 'hits' when searching online, or general, for example a few words of encouragement. Although positive feedback should be encouraged, even negative feedback is better than no feedback at all (Mace, 1968).

VDU users often find that not enough 'help' information has been provided to enable them to make a correct response. Adequate feedback in such cases would not only improve performance but make the task much less stressful and irritating. The total time necessary to respond to a stimulus (reaction time) depends on the length of time necessary to recognize a signal, to make a decision and to make an appropriate movement. It will vary not only between individuals but with how aroused, motivated and attentive a person is at a particular time, and with the type of sensor receiving the stimulus. The average time for reacting to a spoken message, for example, is 150 milliseconds compared with 200 milliseconds for the same visual message. People can respond more quickly than this if both sensors are alerted at the same time. Designers need information about reaction times in order to optimize emergency procedures and system-paced tasks (as opposed to self-paced tasks) so that stress is avoided. Reaction times also have important implications for software design (see Chapter 17).

While a person's response mechanism is affected by all the characteristics discussed so far in this chapter, it is nevertheless important to emphasize the effects of fatigue and boredom. Fatigue is defined as 'a reduction in a muscle group's ability to contract' (Van Cott and Warrick, 1972). Symptoms of the onset of fatigue are not only physiological in nature; they can include tension, stress, tremor, eye strain and, in some cases, pain. Fortunately, much fatigue can be reduced and sometimes can even be eliminated by redesigning the job, modifying the environment or simply changing the number and spacing of rest breaks. These are extremely important in VDU work and, for this reason, are considered in more detail in Parts II, V and VII. Boredom, like fatigue, may adversely affect performance. Although boredom is difficult to measure, it is known that well-designed tasks, software and equipment help to postpone the onset of boredom and reduce its severity and duration. In the case of library and information work, for example, tasks which some may consider boring and repetitive, such as issuing books, may be made more interesting if interspersed with other tasks, like answering enquiries.

In summary, the speed and accuracy of a person's response mechanism depends on many individual and general factors. Improvements can be obtained if people are given adequate feedback to responses, auditory and visual senses are alerted at the same time – especially applicable in emergencies – and fatigue and boredom are kept to a minimum.

HUMANS VERSUS MACHINES

Most of the previous sections in this chapter have focused on human characteristics and, in particular, human limitations. This section concentrates on the characteristics and limitations of the computer and how jobs should be allocated to humans or to computers in order to optimize performance of the total human–computer system.

Many authors have compiled lists comparing human capabilities with machine alternatives (for example, Chapanis, 1960; Murrell, 1971; and

Fitts as referenced in Shackel, 1984). Summarizing these, it appears that people:

- are better at making complex decisions, particularly when unexpected events occur;
- are capable of improvising;
- have a high tolerance for ambiguity, uncertainty and vagueness (although there are large individual differences);
- can rely on a wealth of past experience, not all of which has obvious relevance to the task in hand;
- are able to interpret complicated visual material which involves depth, space and pattern;
- are able to communicate in a variety of ways (touching, seeing, speaking, and even through bursts of emotion);
- are adaptable;
- are able to maintain themselves.

Computers, on the other hand:

- can process vast amounts of data quickly;
- are able to deal with predictable events in a reliable and efficient manner;
- are useful in hazardous environments;
- have infinite patience;
- do not have temperamental moods or social problems;
- can store and access vast quantities of information accurately and quickly;
- can maintain a steady performance – that is, they are not distracted or affected by fatigue or boredom.

In short, computers are eminently suitable for routine tasks that involve the accurate storing and retrieving of vast amounts of information such as the names and addresses of borrowers, catalogues of books in stock, and so on.

When automating a library or information system, however, consideration should be given to the affect on the human operators, the computer being used only to serve and extend their capabilities (Shackel, 1984). As such, it may be unwise for management to attempt to automate *all* possible tasks without due consideration of the effect it will have. For example, some tasks may be undertaken equally well by both people and computers. In these cases, managers should consider not only the cost effectiveness of the task but also the effect on the job. It may not be cost effective to automate a particular combination of tasks if the result is to reduce staff morale and consequent productivity. The correct allocation of tasks to machines and humans is essential to people's general well-being and overall performance. These points will be expanded in Chapter 22 in the further discussion of job design.

PHYSICAL CHARACTERISTICS

People vary not only in their mental characteristics, but also in their size and shape and their ability to lift and to reach (see Figure 1.3). Obviously, these differences must be considered when designing equipment, environments and tasks.

The study of the structural differences in humans is called anthropometry. Tables 1.1 and 1.2 lists the bodily dimensions which are most useful in designing a workstation. It shows the variation in measurement between men and women and among racial groups. It should be noted that data for some Asian groups show a smaller stature than for the average British or American person. Unfortunately, no figures were given for people from the Afro-Caribbean community who, generally speaking, are taller and longer-limbed than their British counterparts. Libraries with staff and users from these racial groups may need to allow for the differences in size when designing workstations. Disabled people and children also have special

Figure 1.3 Variation in size of library staff

Table 1.1 Anthropometric measurements, application and correction factors (adapted from Pheasant, 1986)

Part of the body	*Application*	*Correction factor*
Stature	Basic reference	Shoes. Add ~25 mm for males and ~45 mm for females
Eye height	Standing posture visual height	Shoes. Add ~25 mm for males and ~45 mm for females
Elbow height	Determines work surface heights	Shoes. Add ~25 mm for males and ~45 mm for females
Fingertip height	Lowest acceptable level for finger operated controls	Shoes. Add ~25 mm for males and ~45 mm for females
Sitting eye height (from seat surface)	Sitting posture visual height	Add 10 mm for heavy outdoor clothing, reduce by 40 mm for 'sitting slump' and seat compression
Sitting elbow height (from seat surface)	Height of arm rests; determines work surface height	Add 10 mm for heavy outdoor clothing, reduce by 40 mm for 'sitting slump' and seat compression
Buttock–back of knee length	Defines maximum acceptable seat depth	Add 20 mm for heavy outdoor clothing
Floor to back of knee height	Defines maximum acceptable seat height	Shoes. Add ~25 mm for males and ~45 mm for females
Hip breadth	Width of seat should not be much less than this	Add 10 mm for light clothing; 25 mm for medium clothing; 50 mm for heavy outdoor clothing

requirements which need to be taken into account. The variability within all these populations argues for adjustable workstations, an argument now supported by legislation in the European Union. The nature of such equipment will be covered in detail in Chapters 10 and 14.

The data in Table 1.2 show the static measurements for the 5th and 95th percentile of the population. If a randomly selected group of people were measured and the figures plotted on a graph, the shape (normal distribution) shown in Figure 1.4 would probably appear.

The majority would fall into the middle range – that is, an 'average' person. At the top end of the measurement scale, the number of people registering that height (say greater than 1710 mm) would be very small. Similarly, at the bottom end of the measurement scale, few people would be recorded. Rather than include the very tall and very short people in the figures, which could complicate matters, a point a little way up on the measurement scale, below which lie only 5 per cent of the population and a point near the top of the scale, below which lie 95 per cent of the population are chosen. These are known as the 5th and 95th percentiles respectively.

Functional measurements are those taken when people assume common movement positions, such as reaching out in a horizontal plane, reaching as high or as low as possible, or bending over. Since much library work involves stretching and reaching of some kind, for instance when issuing and discharging books and when cataloguing and so on, consideration of functional measurements is extremely important.

SUMMARY

This chapter has shown that people vary in the way they process information. They not only vary in their levels of attention, arousal, motivation, mental capacity and perception, but also in the way they can remember and learn things. There are also differences in how people

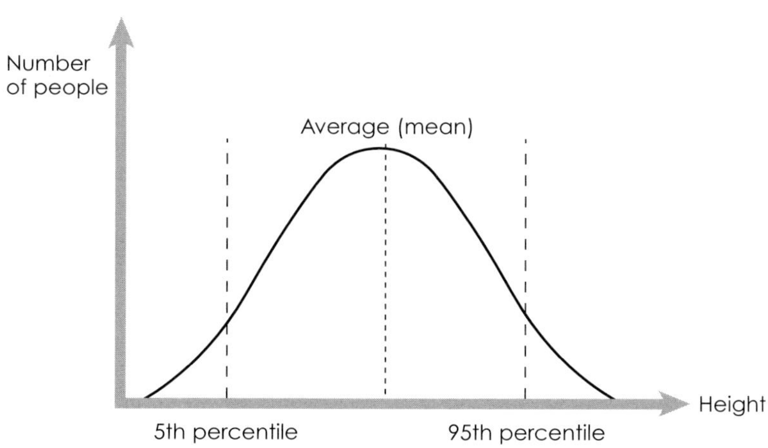

Figure 1.4 Normal distribution

Table 1.2 Anthropometric estimates for various categories of library staff and users (adapted from Bridger, 1995 and Pheasant, 1986)

Part of the body		*British*		*Americans*		*Japanese*		*Indians*	
		5th percentile	*95th percentile*	*5th percentile*	*95th percentile*	*5th percentile*	*95th percentile*	*5th percentile*	*95th percentile*
Stature	Women	1 505	1 710	1 520	1 730	1 450	1 610	1 415	1 615
	Men	1 625	1 855	1 640	1 870	1 560	1 750	1 535	1 745
Eye height	Women	1 405	1 610	1 420	1 630	1 350	1 500	1 320	1 520
	Men	1 515	1 745	1 595	1 825	1 445	1 635	1 440	1 630
Elbow height	Women	930	1 085	945	1 095	895	1 015	875	1 025
	Men	1 005	1 180	1 020	1 190	965	1 105	950	1 110
Fingertip height	Women	560	685	540	660	540	660	525	655
	Men	590	720	565	695	565	695	560	680
Sitting eye height (from seat surface)	Women	685	795	690	810	690	780	635	745
	Men	735	845	740	860	735	835	685	795
Sitting elbow height (from seat surface)	Women	185	280	185	285	215	285	175	265
	Men	195	295	195	295	220	300	180	280
Buttock–back of knee length	Women	435	530	440	540	405	495	405	505
	Men	440	550	445	555	410	510	415	515
Floor to back of knee height	Women	355	445	360	450	325	395	330	420
	Men	395	490	395	495	360	440	370	460
Hip breadth	Women	310	435	310	440	270	340	270	390
	Men	310	405	310	410	280	330	275	365

respond to events and in how they make decisions. People vary, too, in their physical characteristics; their size and shape and their ability to lift and reach. When designing workplaces or selecting software it is essential that these differences are taken into account.

PART

II

HEALTH AND SAFETY CHARACTERISTICS

CHAPTER 2

THE IMPORTANCE OF HEALTH AND SAFETY

WHY HEALTH AND SAFETY MATTERS

As the automation of libraries and information centres has become more commonplace the need to monitor the health and safety aspects of VDUs has become increasingly important. In 1993, the Union of Communication Workers (UCW, 1993) published the results of a large survey of VDU workers based on questionnaires from over 13 300 respondents. They found that 50 per cent of those surveyed reported eye irritation and some respondents also complained that their eyesight had worsened since using VDUs. Over 60 per cent of the respondents also suffered from headaches. Necks and backs were the site of most muscular problems.

In addition, reports of adverse health effects resulting from VDU work are being reported in the library literature (Morris, 1992), and will continue to be so unless librarians become aware of the causal factors and seek to address them. It was noticeable, in some of the libraries visited during the preparation of this book, that there was a lack of awareness of health and safety factors.

There are three good reasons why library managers should be concerned with the health effects of automation on their staff. The first is that employers have a legal obligation to protect their workers from health hazards (see 'Legal aspects' below). The second and third reasons are suggested by Edward Cluff (1985, p. 30):

> [the second] is that we . . . should care about others. [The third] is that it is irretrievably bound up with productivity. So if we should be insensitive enough not to care about others, we should at least care about matters which affect productivity. If staff feel tired or stressed, whether real or imagined, it will always affect their productivity. They may even have days off from work.

Whatever the reason for wishing to reduce adverse health effects, some understanding of the problems which have been reported and the arguments surrounding them will be helpful, if only to allay staff fears.

Research into these factors is continual and new discoveries are constantly being made. In order to obtain a complete and up-to-date picture, therefore, it will be necessary to scan the relevant journals. The following, it is hoped, will provide a basic explanation as well as the known position at the time of going to press.

LEGAL ASPECTS

All employers in the UK have a legal obligation under the Health and Safety at Work, etc. Act 1974 (the HSW Act) to ensure, so far as is reasonably practicable, the health, safety and welfare of their employees at work (HSC, 1992). More recently, the European Union has also given rise to the following associated Regulations which are relevant to all or most workplaces:

- The Management of Health and Safety at Work Regulations 1992 (HSE, 1992a).
- The Workplace (Health, Safety and Welfare) Regulations 1992 (HSE, 1992b).
- The Provision and Use of Work Equipment Regulations 1992 (HSE, 1992c).
- The Health and Safety (Display Screen Equipment) Regulations 1992 (HSE, 1992d).

Although some aspects of all of these Regulations are applicable to libraries, only those relating specifically to automation and the use of computers, as detailed in the Health and Safety (Display Screen Equipment) Regulations 1992, will be discussed here.

The Health and Safety (Display Screen Equipment) Regulations 1992 cover users of any alphanumeric or graphic display screen, regardless of the display process involved. This means that microfiche readers, as well as computers, are covered by the legislation. A 'user' is defined as 'an employee who habitually uses display screen equipment as a significant part of his normal work' (HSE 1992d, p. 4). The Guidance notes issued with the Regulations clarify this a little further. They state that it is appropriate to classify a person as 'a user or operator' if most or all of the following criteria apply:

- the individual depends on the use of display screen equipment to do the job, as alternative means are not readily available for achieving the same results;
- the individual has no discretion as to use or non-use of the display screen equipment;
- the individual needs significant training and/or particular skills in the use of display screen equipment to do the job;

- the individual normally uses display screen equipment for continuous spells of an hour or more at a time;
- the individual uses display screen equipment in this way more or less daily;
- fast transfer of information between the user and screen is an important requirement of the job;
- the performance requirements of the system demand high levels of attention and concentration by the user, for example, where the consequences of error may be critical.

While this clarification is useful, this part of the legislation is open to interpretation. Which library staff member should or should not be regarded a 'user'? For example, should the library assistants working at the issue desk who use computers but also do many other tasks be regarded as 'users'? Library managers need to consider cases individually and be particularly careful in examining the amount of time an employee spends using a computer. If in doubt, employers should always err on the cautious side and class borderline cases as 'users'.

Essentially, these Regulations require employers to:

- Assess and record the identification of hazards, the evaluation of risks and the extent of health risks. 'A **hazard** is something with the potential to cause harm; **risk** expresses the likelihood that the harm from a particular hazard is realised. The **extent of risk** takes into account the number of people who might be exposed to a risk and the consequences for them' (HSE, 1992d, p. 13). The term of assessment should be systematic, appropriate to the likely degree of risk, comprehensive and include information provided by both employer and employee. One of the easiest ways of obtaining information is to use a checklist such as the one provided with the Regulations (HSE, 1992d).
- Comply with the minimum requirements for workstations covering equipment, the environment and the software interface. Most, if not all, new European-made computers now comply with the Regulations. Further details about the requirements may be found in Parts III, V and VI.
- Examine eyes for defects and provide eyesight tests and pay for the cost of special corrective appliances needed for VDU work. It should be noted, however, that employers are only required to pay for a 'basic appliance' prescribed to correct vision defects of the viewing distances specifically for VDU work. They are not liable for designer frames nor for 'normal' corrective appliances. Further details may be found in Chapter 5.
- Provide adequate health and safety training in the use of workstations. This is 'to increase the user's competence to use

workstation equipment, safety and reduces the risk to their or anyone else's health' (HSE, 1992d, p. 25). The training must cover:
- the user's role in the recognition of hazards and risks;
- the causes of risk (for example by poor posture);
- how risks can be minimized (for example, by adopting a comfortable posture, adjusting furniture, arranging workstation components to suit the person and task, stressing the need for proper maintenance and cleaning of equipment, and the importance of breaks);
- communication channels by which employees can report symptoms or problems;
- information on the Regulations;
- the user's contribution to assessments.

Further information on these aspects can be found in Chapters 6, 9, 10, 14 and 24.
- Provide employees with adequate information on all aspects of health and safety relating to their workstations and the Regulations. Essentially, the information should reinforce the training provided by the employer and should include 'reminders of the measures taken to reduce the risks such as the system for reporting problems, the availability of adjustable window covering and furniture, and of how to make use of them' (HSE, 1992d, p. 27).

Compliance with the Regulations is obligatory and is policed by the Health and Safety Executive. Should employers fail in their duties, the Health and Safety Executive can insist on changes or, in extreme cases, can prosecute.

The specific requirements of the Regulations are dealt with in the appropriate chapters throughout the book.

THE ALLEGED HEALTH HAZARDS OF VDU WORK

The following chapters discuss the hazards to health purported to be caused by VDUs. Although not mutually exclusive, for ease of discussion, they have been grouped into the following categories:

- *Musculoskeletal and postural* problems, such as backaches and other disorders caused primarily by incorrect posture or workstation design (Chapter 3).
- *Work-related upper limb disorders*, specific type of musculoskeletal problems orginally known as RSI (Chapter 4).
- *Eye and vision problems*, including tired eyes and difficulty in focusing (Chapter 5).

- *Stress-related problems*, such as high blood pressure or irritability caused by the change in the job or fear of automation (Chapter 6).
- *Reproductive hazards*. Alleged and possible causes of miscarriages and birth defects (Chapter 7).
- *Other alleged health hazards* including cataracts, facial rashes, photosensitive epilepsy, cabling, fire, laser printers (Chapter 8).

CHAPTER 3

MUSCULOSKELETAL AND POSTURAL PROBLEMS

INTRODUCTION

Many surveys have confirmed that VDU workers experience pain and discomfort in various parts of the body (World Health Organisation, 1990). One survey, for example, comparing different companies in Sweden and Australia, found that 91 per cent of users complained of painful neck or shoulders, 45 per cent had backache, 73 per cent experienced headaches and 45 per cent suffered from visual fatigue (Johansson and Shahnavaz, 1995). Although this was based on relatively few users and relied upon self-reporting of pain or discomfort using a questionnaire, it does serve to indicate the fear, substantiated or not, of physical problems arising from VDU work. Other symptoms which have been reported in the literature include: pains in the joints, especially in the knees; pains in the legs and ankles; circulatory problems, producing a loss of sensation in the extremities; varicose veins; and bowel and bladder problems.

Not all the physical problems can be attributed directly to VDU work. Indeed studies investigating musculoskeletal complaints among non-VDU office workers have been contradictory (Carter and Banister, 1994). Some reports have indicated that VDU users experience more discomfort than non-VDU office workers and that the frequency of discomfort increases with the amount of VDU usage. Conversely, other studies have found few differences in the number and type of physical problems in VDU and non-VDU users. Whether VDU users are at a higher risk than non-VDU office workers from musculosketal problems remains unproven. Nevertheless, it is clear that a large percentage of VDU users, estimated to be between 25 and 35 per cent by Carter and Banister (1994), report musculoskeletal discomfort and pain. The discomfort can vary in frequency and be in the form of soreness, aching, stiffness, fatigue, cramps, numbness, tingling or tremors. Many of the problems are caused by incorrect loading of the muscles as a result of poor workstation and job design, and poor posture which causes restricted circulation. Several factors have been associated with musculoskeletal problems in VDU users. These include:

- forward leaning posture;
- static work posture;
- repeated work movement;
- intensity of work;
- psychosocial factors;
- individual aspects.

FORWARD-LEANING POSTURE

The spine adopts a natural curvature of the lower back while standing or lying on one side. Sitting for a long period can cause many problems particularly if the lower back is not supported. If this happens or the operator adopts a forward leaning posture, an abnormal curvature of the spine (called kyphosis) might result and lead to painful lower back problems.

The operator may lean forward for a variety of reasons, such as to view the screen or source document, to compensate for poor vision or for a poorly designed workstation, or to eliminate glare from the screen. Studies using SAMMIE, a computer-aided ergonomics package for modelling workplaces, have shown that postures with the operator leaning slightly back using a reclined backrest with lumbar support minimizes disc loading and are preferable to erect or forward-bending postures (Kayis and Hoang, 1995). A forward-leaning posture causes a strain on the muscles in the neck and shoulders because more effort is needed to support the head. This posture also compresses the abdomen and lungs, reducing the efficiency of the digestive systems and causing shallow breathing which reduces the supply of oxygen to the body.

STATIC POSTURE

A static posture, especially over long periods, can also result in pain in the lower part of the body, particularly in the knees. It may also impair the flow of blood, possibly resulting in a loss of sensation in various parts of the body. Good circulation is necessary to remove waste products from the muscles and thereby to reduce fatigue. Circulation can be improved by minor postural adjustments, such as stretching and changing position, but the operator may not be aware that this is necessary and, when under extreme pressure or concentration, may be less likely to fidget. People with disabilities, too, may find this difficult.

A poorly designed chair can compound the problem of a static posture by causing poor weight distribution on the thighs. This restricts the blood supply and may cause pooling of the blood leading to varicose veins.

The adoption of static work postures are common among VDU users where the hands remain on the keyboard and the eyes fixed to the screen. The best way to avoid problems caused by static postures is to encourage staff to periodically stand up and flex backs and legs. Dainoff (1986) even recommends that chairs should carry a warning: 'Caution: Prolonged sitting may be hazardous to your health.'

REPEATED WORK MOVEMENT

Incorrect positioning of the screen and source document can cause unnecessary, repetitive movement of the head and neck, thus tiring the neck and shoulder muscles, especially if this occurs in a forward-leaning position. Similarly, repeated keying without a break can cause problems in the hands and arms. This can be further exacerbated if the keyboard is at the wrong height (see Chapter 4).

INTENSITY OF WORK

Hünting (1984) studied four groups of bank employees: 53 data entry operators; 109 conversational terminal operators; 78 typists (copy and audio); and 55 traditional office workers. The highest incidence of upperbody pain was in the data-entry group and the lowest in the group of traditional office workers. The noticeable contrast between the two groups was that data-entry operators worked intensely at high speeds and traditional office work involved a great diversity of movement. Hünting concluded that the constrained posture, encouraged by intensive keyboarding (Hünting's subjects used only the numeric key pad, with the other hand in a fixed position, turning over source documents) was a causal factor of the symptoms. MacKay (1985), referring to Hünting's work, suggested that both the nature and the duration of the task are to blame. Bergqvist *et al.* (1995a) also found that data entry operators working more than 20 hours per week at a VDU reported significantly more neck, shoulder, arm and hand discomforts compared to other VDU users.

In addition to forcing the operator to adopt a static position, intensive VDU work may encourage rapid and repetitive movement of the upper body, particularly the hands, which may increase the risk of work-related upper limb disorders (WRULDs) originally known as RSIs (repetitive strain injuries, see Chapter 4).

PSYCHOSOCIAL FACTORS

A number of psychosocial factors appear to influence the incidence of physical problems in VDU users (Bergqvist *et al.* 1995b). VDU users with limited opportunity to take breaks or do varied work are more at risk. VDU users with poor social support are also more likely to have problems. Lim and Carayon (1995) found that having to meet production quotas or standards and lack of control over work undertaken can affect musculoskeletal discomfort.

INDIVIDUAL ASPECTS

Studies reported in Bergqvist *et al.* (1995b) showed a higher prevalence of musculoskeletal problems in women compared to men. The authors attributed this to women having to do a 'double shift' – at work and at home! Bergqvist *et al.* (1995b) also found that VDU users who are stressed at either work or at home are also more likely to report musculoskeletal or postural problems. The use of bifocal or progressive glasses was also found to cause a higher neck muscle tension during data entry work than the use of monofocal lens.

HOW TO MINIMIZE POSTURAL PROBLEMS

In most cases, musculoskeletal or postural discomfort will not lead to any permanent damage. Over time, however, an operator's health and productivity may be affected and eventually persistent exposure may cause a work-related upper limb disorder (WRULD). MacKay (1985) identifies the following causes of postural discomfort in VDU users:

- abduction of the hand (bending or twisting the wrist);
- pronation of the forearm (turning the palm downwards), often close to the anatomical limit;
- raising the shoulders;
- lack of appropriate support for the back;
- prolonged work spells;
- lack of support for the hands and wrists;
- sharp edge of desk cutting into wrist or forearm.

He also makes a number of recommendations to help to prevent postural problems:

- improvements in keyboard design (for the most part, out of the library's control, but for details see 'The keyboard', Chapter 10);
- altered work practices including job redesign (see Chapter 22);
- attention to postural requirement through workplace design (see Chapter 9);
- selection and training to improve or modify technique;
- occupational health care.

One library implemented a series of exercises for their staff during rest break periods and formed aerobic exercise groups (Ralli, 1988). The same library formulated a return to work programme for staff who had been suffering from work-related upper limb disorders, which incorporated a variety of tasks and frequent rest breaks.

However, in the same way that the causes of some physical symptoms cannot be distinguished, other causes compound the issue. Although in many cases the problems will be attributed to poor design of the workplace or working environment, they might also be the result of a more basic dissatisfaction at work, for example, with the type of work being done, the work pressure, the management style or type of supervisors.

In addition, MacKay (1985) argues that confusion can arise in that similar symptoms may be experienced with no bearing at all on causes in the workplace.

New Zealand's Occupational Safety and Health (OSH) (1992) offers the following advice for VDU users:

- know your chair (how to adjust seat height, backrest height and backrest angle)
- remember the basic postural points:
  - let your shoulders relax (shrug them occasionally);
  - let your elbows swing free;
  - keep your wrist straight where possible (both sideways and up and down);
  - pull your chin in to look down (rotate your head about a point just under your ear);
  - keep the hollow in the base of your spine;
  - try leaning back in the chair;
  - don't slouch or slump forward;
  - alter your posture from time to time;
  - every 20 minutes, get up and bend your spine backwards;
- get yourself comfortable by:
  - setting the seat height for your desk (ideally when using a keyboard your elbow should be bent to allow your forearm and upper arm to be at right angles with one another);
  - using a footstool if your feet are not flat on the floor and you can feel pressure on your thighs;

- adjusting the backrest height so that it supports the hollow of your back (your buttocks should fit into the space between the backrest and seat pan);
- using a copyholder to reduce one of the needs to tilt forward.

Other points are discussed in Chapter 14 on workplace design and in the next chapter on work-related upper limb disorders.

CHAPTER 4

WORK-RELATED UPPER LIMB DISORDERS

INTRODUCTION

Work-related upper limb disorders (WRULDs) refer to a number of musculoskeletal disorders that appear to be caused or worsened by tasks undertaken at work. Originally, these disorders were known as repetitive strain injuries (RSIs) and are still often referred to as such by the media. The terminology was changed, and adopted by the Health and Safety Executive, to more accurately reflect the types of medical conditions found. The term 'RSI' was regarded as confusing because not all the disorders it covered were caused by repetitive movements and not all were 'strains' or 'injuries' in the medical sense. The term 'RSI' has also been dropped in other parts of the world. In New Zealand, for example, 'occupational overuse syndrome' is the favoured term. In America the preferred term is cumulative trauma disorder (CTD), or alternatively 'cumulative trauma injury'. Sweden and Japan prefer the expression 'occupational cervico-brachial disorder', while in Australia the latest suggestion is 'reversible fatigue syndrome'. In this book the term WRULD will be used.

RANGE OF CONDITIONS

The range of conditions associated with WRULDs are characterized by discomfort or persistent pain in muscles, tendons and other soft tissues. The range of symptoms experienced by sufferers include (London Hazards Centre, 1988):

- pain
- swelling
- tenderness
- pins and needles
- crepitus ('cracking' feeling when tendons or swelling are pressed tightly)

- ganglion (cyst-like swelling near a tendon, sheath or joint)
- loss of sensation
- muscle weakness
- muscle spasm
- joint restriction/loss of movement

There are two main categories of WRULDs: localized conditions and diffuse conditions. Localized conditions are recognized medically and have well-defined symptoms. They include:

- carpal tunnel syndrome
- tenosynovitis
- tendinitis
- trigger finger
- tennis or golfer's elbow
- bursitis
- writer's cramp
- De Quervain's disease
- Dupuytren's contracture
- peritendinitis
- epicondylitis
- frozen shoulder.

Diffuse conditions are much less understood and have no recognized medical terms. Sufferers may complain of any of the symptoms listed above. However, there is frequently no sign of physical injury or abnormality other than tenderness. Even signs of tissue damage such as swelling, bruising or increased temperature are often not present. Because of the difficulty of diagnosing these conditions some medical authorities have declared that they are 'all in the mind'. Despite this, recent research indicates that the more diffuse symptoms frequently do have a clinical basis but may be aggravated by psychological factors such as stress and anxiety (Ballard, 1993).

To confuse the issue, it is possible that diffuse muscle discomfort and pain may exist alongside localized conditions, with the two seemingly merging. Diffuse conditions can also mimic localized ones (Occupational Safety and Health, 1992).

THE CAUSES OF WRULDS

There are a number of theories about the causes of WRULDs. The most common one is based on the restriction of blood supply to tendons and muscles caused by tension. Muscles and tendons acquire blood through capillaries passing between the muscle fibres. A tense muscle squeezes these vessels and restricts the flow of blood. Muscles can cope with brief periods of tension but if it continues it leads to a build-up of acid waste products (lactic acid). As the acids build up in the muscle, it causes pain and stiffness. Other neighbouring muscles may also tense up in sympathy by a reflex action. An inadequate blood supply to nerves may also cause tingling and numbness.

Muscles and tendons can recover from fatigue given rest and variety of tasks. Long-term problems may occur, however, if muscles and tendons are frequently over-used beyond their normal capacity.

The main causes of muscles and tendons being over-used are:

- awkward static or constrained postures;
- repetitive movements of parts of the body;
- extreme force required by parts of the body (HSE, 1990a).

In some cases only one of these may be responsible, in others a combination of the above could be the cause of the problems. Either way these actions are often imposed on people because of:

- poor ergonomics;
- an inappropriate workload;
- psychosocial factors at work.

WHO SUFFERS FROM WRULDS

While the true prevalence of WRULDs sufferers is unknown, it has been suggested that keyboard operators, cleaners, hairdressers, or machine operators are at greater risk of sustaining damage to the soft tissues of their arms, hands and wrists than those in other occupations (Tindall, 1993).

WOMEN AND WRULDS

Studies of the incidence of WRULDs frequently show an increased incidence amongst women. Figures published by the Health and Safety Executive in 1993, quoted by Ballard, 1993, show a ratio of 1.66:1 for women to men on the incidence of work-related RSI. This result can largely be explained by the fact that far more women than men are employed in unskilled, repetitive tasks, with a low level of autonomy, the exact type of work which is known to carry high WRULD risk factors.

However, some doctors suggest that hormone levels may account for the discrepancy between men and women. It is known, for example, that some 25 per cent of females experience carpal tunnel syndrome at some stage during pregnancy. Despite this, it is unknown how hormonal changes and pregnancy may affect the undertaking of everyday tasks, whether at home or at work and in the studies of WRULDs there is, as yet, little specific evidence to prove the link categorically (*Croner's Health and Safety at Work Bulletin*, 1993).

Another possible reason for the high incidence of WRULDs among women may be that working equipment is often designed to suit a male body, which is generally larger and able to exert greater physical effort. Similarly, some large tools may overstretch a small hand (Ballard, 1993).

INDIVIDUAL RISK FACTORS AND WRULDS

Some personal characteristics have been associated with the development and exacerbation of WRULDs in individuals (*Croner's Health and Safety at Work Bulletin*, 1993). These include:

- rheumatoid arthritis;
- endocrinological disorders, e.g. diabetes mellitus;
- vitamin B6 deficiency;
- wrist size and shape;
- a previous knock or blow to a vulnerable area of the upper limbs.

There is, however, no conclusive evidence to prove a definite link between these factors and WRULDs.

Another popular theory is that of the 'hypersusceptible worker' (London Hazards Centre, 1988). According to this theory some people and certain racial groups are said to be 'hypersensitive' because of their physical and genetic make-up. Such individuals are said to be predisposed to development of WRULDs. Some companies in the USA have introduced pre-employment screening for genetic or physical susceptibility to disease. However, there exists much doubt as to the efficacy of such tests. An assessment of pre-employment screening tests carried out at the Department of Engineering Production at the University of Birmingham appeared to show that such tests were useless. A test performed on the same person gave one result one day and a completely different result the next (London Hazards Centre, 1988).

Hughes (1985) reported an increase of WRULDs among mature people, not trained as typists and starting to use keyboards when their joints were becoming less supple. A situation perhaps not untypical of many library staff.

LIBRARY STAFF

Although data input is rarely intensive in libraries, WRULDs are known to be a problem. One library in Australia, for example, during a retrospective conversion project where targets were set, found that all six staff suffered from WRULDs. At a later stage in the same library, a new group of keyboarders was used, but no targets were set. Operators worked for 30 minutes at a session. No complaints were reported by this group, but the overall performance was three times as high as the first group (Ralli, 1988). This indicates that anxiety to achieve goals may not only cause unnecessary discomfort to the operators, but may also be counter-productive.

One of the authors, involved in the design of issue counters, also found several library assistants at one particular library suffering from WRULDs. Staff were not only having to process more books with an increase in the number of students but having to work with an issue desk originally used for the Browne system; a system where borrower tickets holding book details are stored alphabetically and retrieved normally from trays. The issue and discharge of books required much twisting and handling of heavy books at the wrong height and angle. Five of the six assistants complained of symptoms associated with WRULDs, two of whom had sought medical advice. Fortunately, the library management, with the help of a consultant, identified the problems and sought measures to rectify them. After analysis, the issue counter was ergonomically redesigned and some management practices changed to reduce the risk of WRULDs. In the long term, self-issue systems might be the answer to staff problems at the issue counter (see 'Issue counter', Chapter 15, p. 195).

Shelving heavy books can also cause or aggravate problems. Library staff should be encouraged to use trolleys for the carrying of books wherever possible and have training on the correct manual handling procedures. The provision of step-stools to enable staff to reach the higher shelves without over-stretching is a must.

TREATMENT OF WRULDS

The treatment of WRULDs depends largely on whether the condition is localized or diffuse, or is a mixture of them both. The stage of development is also a factor. When symptoms are mild and occur only at work, a change in work practices or the design of the workplace may be all that is needed to cure them. For more advanced cases, more wide-ranging treatment may be needed which might include:

- relaxation and the use of micropauses when working;

- rest (of the affected part);
- exercise (so long as it does not induce pain);
- splinting (temporary basis to control pain);
- physiotherapy;
- massage;
- steroid injections;
- acupuncture/acupressure;
- surgery (in extreme cases and as a last resort).

WRULDS AND THE LAW

The Health and Safety at Work, etc. Act 1974 (HSW Act) places a general duty upon an employer to 'ensure as far as is reasonably practicable, the health, safety and welfare at work of all his employees' (HSC, 1992); see also Chapter 2. The more recent regulations which were introduced under the HSW Act, the Management of Health and (Safety at Work) Regulations 1992 (HSE, 1992a), and the Health and Safety Display Screen Equipment Regulations, 1992 (HSE, 1992d) require employers to *show* that the correct procedures are being taken to comply with the Act. Under common law, employers also owe their employees a duty of care. Breach of the Regulations and/or common law gives rise to criminal and civil proceedings.

Employees may have a claim for damages for WRULDs if they can prove they have suffered (RSI Association, 1993):

- an injury (which must be properly diagnosed);
- that the injury was attributable to their work;
- that their employer was guilty of negligence.

Negligence occurs where there has been a breach of the duty of care. The breach may be of a statutory duty or of a common law duty. In establishing whether negligence occurred the following issues will arise (Tindall, 1993):

- Does the work process carried out by the plaintiff carry a foreseeable, inherent, specific and not insignificant risk of the plaintiff developing WRULDs?
- Have the employers discharged all their duties to the plaintiff which arise in these circumstances?
- Have the employers set out the work in a reasonable fashion eliminating all unnecessary sources of danger by applying proper ergonomic considerations?

- Have the employers organized working conditions upon the same basis by instigating a rotation system and proper rest periods?
- Have the employers instigated and operated a proper system of warning employees of the risk of injury both pre-employment and post-employment?
- Are the physical conditions from which the plaintiff suffers (which must be properly diagnosed) attributable to their employment?

A victim of WRULDs may also seek compensation under the Industrial Injuries Legislation but only if they can satisfy certain strict criteria – with the result that many sufferers are prevented from claiming benefit. Other benefits which may be available to WRULDs sufferers are Disability Working Allowance and Disability Living Allowance. Details of these may be obtained from social security offices.

MANAGEMENT OF WRULDS

If WRULD cases are suspected, the HSE (1990a) recommend the following steps to rectify problems:

- undertake informal discussions with management/workforce;
- carry out detailed job/task analysis to assess risk factors;
- review medical information/initiate health survey;
- describe the particular problem (specify risky job/task components and identify anatomical sites/nature of disorder);
- select appropriate and practical intervention methods;
- design an intervention strategy (consult/involve workforce and consider retraining needs);
- implement on trial/pilot basis first then on full-scale basis;
- monitor effectiveness.

PREVENTION OF WRULDS

The most effective method of dealing with WRULDs is prevention in the workplace. The most appropriate measures will depend upon the individual work situation, although the following preventative steps will apply in most cases in the library situation (HSE, 1990a):

- *Ergonomic improvements to work stations* – provision of sufficient working space to enable the operator to work in comparative comfort and safety without excessive twisting, turning or arm elevation;
- *Improvement of equipment design* – where practicable old or poorly

designed equipment should be replaced or modified to enable the worker to operate with as high a degree of comfort as possible and to eliminate any physical tension that may be present;

- *Improvement of organizational work factors* – job redesign, rotation of tasks, breaks, time limits placed on VDU work periods;
- *Improvements in managerial aspects* – workers should be warned of the risk of WRULD inherent in certain tasks, early symptom reporting should be encouraged, prompt action should be taken and proper training should be given;
- *Proper risk assessment and planning* – a formal policy on WRULDs should be developed and employees should be made aware of its existence.

CHAPTER 5

EYE AND VISION PROBLEMS

INTRODUCTION

Together with a rise in complaints about physical aches and pains occurring with the use of VDUs, has come an increase in complaints of visual discomfort. The following problems have been reported:

- *Eye discomfort*:
  - soreness, grittiness, burning, itching, dryness, aching;
  - tenderness around the eyes;
  - redness of the eyes;
  - irritation, flickering sensations.
- *Impaired vision*:
  - loss of visual sharpness;
  - difficulty in focusing;
  - double vision;
  - blurred vision;
  - seeing colour fringes;
  - difficulty in looking at one point for a period of time.

These may lead to other symptoms, such as headaches, and muscular aches in the neck, shoulders, back and arms (possibly from adopting a poor posture to obtain legibility of the screen), fatigue, stress or irritability. All these problems may disappear after a time, and have not been proved to cause any permanent ill effects, but may be unpleasant while they last.

The latest research supports a causal connection between VDUs and visual discomfort. Bergqvist *et al*. (1992) analysed eye discomforts and VDU work between 1981 and 1987. A part of the study involved tracking people who had changed from non-VDU to VDU work. The results found that they experienced more eye discomfort using VDUs for their work than previously. The associations between VDU work and eye discomfort are also consistent with other studies, see Bergqvist *et al*. (1995c) and Bergqvist and Knave (1994) which reviews the literature. Several likely factors have been considered to give rise to some visual discomfort and reduce operator efficiency. These include the eyesight of the operator, the technical characteristics of VDUs (contrast, colour, size and shape of characters, flickering), the ergonomic characteristics of the workplace (screen, keyboard, table, chair), and factors specifically related to the task and the way it is being undertaken. The ergonomic characteristics of the workplace and its effects are considered in some detail in Part V. This chapter discusses the other factors which will detract from or enhance visual efficiency.

EYESIGHT OF THE USER

Before looking at the factors affecting vision it is helpful to examine briefly how the human eye works.

HOW THE EYES WORK

Our eyes receive light-waves through the pupil and these fall on the retina, where nerves receive and transmit the message to the brain (Figure 5.1). To assist in sighting and focusing on an object, we have three sets of muscles:

- One set of muscles moves the eyes from side to side and up and down.
- The iris muscles in each eye adjust the size of the pupil according to the varying light intensity. The pupil grows larger in dark conditions to admit more light and smaller if the light is bright to reduce the amount of light entering the eye.
- The ciliary muscles change the shape of the lens to focus the eye. When the eyes focus at close range, the ciliary muscles compress the

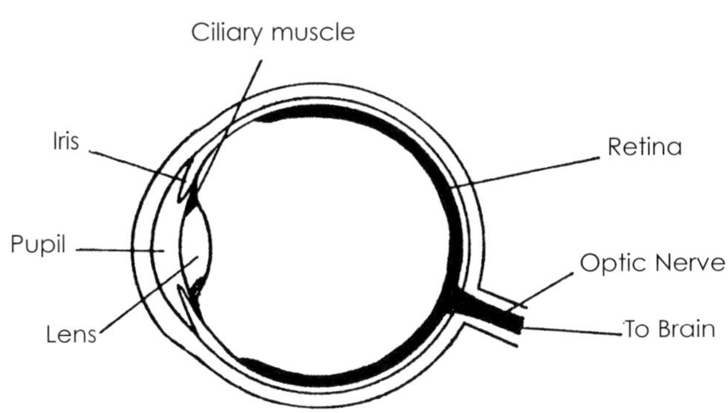

Figure 5.1 Simplified section of the eye

lens. Close-up viewing for long periods therefore is very tiring to these muscles.

So tired eyes can be caused by:

- constant movement of the eyes (up and down or left and right), or too little variety of eye movement and focusing;
- frequently changing brightness, causing the iris muscles to adjust the pupil size;
- close-up viewing for long periods or constant re-focusing.

FACTORS AFFECTING VISION

Various factors affect our eyes:

Age

As people grow older the lens of the eye gradually loses its elasticity, and close-up vision becomes more difficult. Usually this is corrected by glasses, but if it is not, visual discomfort may occur while doing close work.

Lighting

The amount of light available also affects our sensitivity to certain colours. When it is dark, the eyes are more sensitive to blue than when it is light and, conversely, as it becomes lighter, the eyes become more sensitive to red. Green and yellow fall in the middle of the range. This in turn, therefore, affects our perception of colours on a display, green and yellow standing out more clearly under normal conditions than either blue or red.

The level of illumination will also affect our ability to read printed text and, as we get older, we need more light to read. However, it is also true that as age increases, so does sensitivity to glare.

The clarity of VDU and printed text needs to be of high quality under conditions of poor lighting or for older employees.

Different lighting levels are also needed for reading from a VDU screen than for reading written or printed text. Thus no single recommendation can be made, rather the three factors of operator's age, the source document and text on the VDU screen need to be considered together. The best way of doing this is to allow the operator to experiment with different lighting levels.

Glare

Glare can cause considerable visual discomfort especially as people get older and, in some cases, can even impair vision. It may be caused by light reflected from the VDU screen or the work surfaces or coming directly from a light source. Reflected glare, of the kind which occurs in libraries and offices, may cause discomfort but is unlikely to impair vision. Glare from direct sources, such as lights, should be avoided.

Glare is relative – that is, a light source may be bright compared to its surroundings and therefore cause glare. This can be reduced by bringing the overall level of lighting up to a similar level, but care must be taken to ensure that the overall level of lighting does not cause the screen contrast to be diminished.

The relative position of the glare source to the operator is also important; the closer it is to the centre of the visual field, the more performance is impaired. For more details about lighting and glare see Chapter 11.

Pupil size

In order to adjust the amount of light received by the retina, the iris muscles adjust the pupil size. The length of time taken to do this varies and can sometimes be relatively slow. The pupil aperture is also affected, to some degree, by the focusing of the lens, contracting when the lens is focusing on near objects. Thus, good light is needed for close work because the pupil will not compensate in response to the light in the same way that it would for more distant objects.

Retina adaptation

The retina takes a long time (approximately 25–30 minutes) to adapt from light conditions to complete darkness. Adapting to light conditions is, in contrast, much quicker. However, when a bright light falls on the retina, overall sensitivity to light is reduced. Thus, if a VDU faces a bright light source, such as a window, the sensitivity of the retina to the light on the screen is reduced and reading is impaired.

Defective vision

Few people have 'perfect' vision; indeed, sight for most people deteriorates with age. It has been rumoured that constant use of VDU can cause 'eye strain' and deterioration of vision. It is not possible to strain the eye muscles, although tiredness may result from any visually demanding task, such as proofreading, whether on or off screen. Even such intense work is more likely to reveal an existing defectiveness of vision than to induce it. It is therefore recommended that operators seek a sight test before starting any visually demanding task, including VDU work, and obtain corrective glasses if necessary. In the UK, employers are required to provide VDU users, as defined by the Health and Safety (Display Screen

Equipment) Regulations 1992, with an appropriate eye and eyesight test for newly recruited employees (HSE, 1992d). Existing VDU workers should be offered free eyesight tests at regular intervals or when problems arise. Employers are also responsible for providing corrective glasses specifically for VDU work should they be found necessary.

People who wear bifocals may have particular problems because the part of the glasses intended for close reading is at the bottom of the lens. In order to view text on the screen, the operator may have to tip the head backwards in order to see through the lower part of the lens. If this is sustained for long periods or performed frequently, head, neck and shoulder pains may result. It may therefore be advisable to purchase a separate pair of glasses specifically for VDU work. The basic cost of these may be reimbursed by an employer in the UK under the Health and Safety (Display Screen Equipment) Regulations 1992 Act (HSE, 1992d). A number of computerized vision screeners for display screen users have appeared on the market. These tests are a way of identifying individuals with defective vision who should then be given a full sight test. Vision screeners have the advantage of being quicker and cheaper than giving full optometric examinations to all VDU workers. Horberry *et al.* (1995) tested six screening systems and concluded that they should be regarded as 'an efficient option when measuring the vision of display screen users'.

Contact lens wearers may also suffer, this time from dry eyes. People tend to blink less often than normal when operating a VDU, with the result that the eyes may feel dry and uncomfortable. Eye-bathing or 'liquid tears' (which may be purchased at most chemists) may help.

Reading

Our eyes are constantly on the move, taking in information from our surrounding environment. When we read, our eyes move along the text in short, jerky movements called saccades. The eyes jump from point to point, each point being approximately four to eight letters away. At each point, the eyes come to rest and the word, or words, on which the eyes are focused are read. These are rightward reading saccades (leftward on Arabic and other scripts, which read from right to left). Occasionally, the eyes will 'hop' back a stage to check what has just been read (correction saccades) and, at the end of a line, they swing left and downwards, to start at the beginning of the next line (line saccades).

To assist the reading process, it is important that characters are clear and easily distinguished. There must also be a sufficient gap between the lines to avoid confusion. If lines are too close together, it will take longer for the reader to distinguish the shapes (words) which are relevant and reading speed will be reduced.

Character contrast is also important. If contrast is poor, the reading field is narrower and reading speed is reduced. Coloured letters, too, can cause a reduction in reading speed, although if they are well used they can improve the retrieval of specific pieces of information, such as by highlighting part of a record. Legibility is equally important for the printed or handwritten source document as for the VDU screen.

THE VDU AS A CAUSE OF VISUAL DISCOMFORT

Before discussing the technical aspects of VDUs which may cause visual discomfort it is useful to discuss how a VDU works.

HOW A VDU WORKS

Figure 5.2 shows a section through a VDU. The screen is the flattish part of the glass bulb on the right-hand side of the picture. The inside of the glass is coated with phosphors. The electron gun scans across and down the screen very quickly and, on receipt of a signal, usually from the keyboard, fires an electron beam at the screen. This beam activates the phosphor which then starts to glow, causing a visible image to appear on the screen. The phosphor is activated in dots and several dots together are used to form a character (see Figure 5.3).

Several aspects of the VDU may be to blame for visual discomfort.

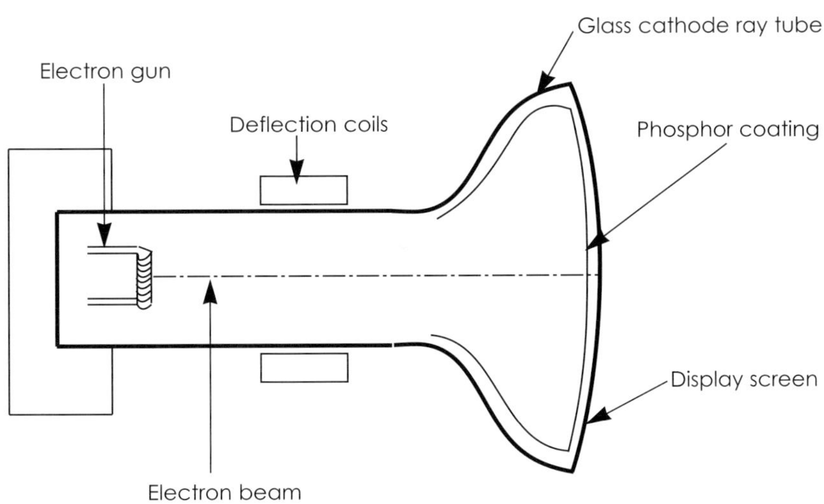

Figure 5.2 Simplified section of VDU screen

TECHNICAL ASPECTS OF THE VDU

Flicker and ghosting

Although the VDU image may appear to be still, it is in fact constantly changing. From the moment the electron gun is fired and the phosphor begins to glow, the glow starts to fade, until it is once more charged by a beam from the gun. The frequency with which the phosphor is restimulated is known as the refresh rate. If the refresh rate is too slow, the screen will appear to flicker.

Most modern VDUs have refresh rates of 60 or 70 Hertz (Hz) or cycles per second and commonly are capable of more than one rate. Generally, the higher the rate, the more stable will appear to be the image. However, in this range (60–70 Hz) it is possible for some flicker to be perceived.

The refresh rate, however, cannot be considered in isolation from the persistence of the phosphor – that is, how long the phosphor stays illuminated after being charged by the beam from the electron gun. A number of different phosphors are currently in use, which all have different decay times (that is the length of time taken for the phosphorous glow to reduce to 10 per cent of its starting level). The slower the rate of decay, the less frequently the screen needs to be refreshed to avoid flicker.

A highly persistent phosphor can, however, bring about the problem of *ghosting* – that is, where the old image persists on the screen for a little while after new text has replaced it or after scrolling (movement) of the text has occurred.

People vary in their awareness of flicker and individuals' perceptions will be affected by their position relative to the screen. Flicker is more easily observed if the VDU is at an angle to the operator and thus seen out of the corner of the eye, than if viewed through the focusing part of the eye. One simple way of assessing if a VDU screen flickers is to hold a straight piece of card or paper under the text. If the characters appear to 'dance', the screen flickers.

Stability of the characters

If the electron beam is well regulated – that is, it travels across and down the screen in an orderly manner – the characters will appear stable; if not, they will appear to move. This movement may be defined in three ways:

- *drift* – that is, where a character gradually moves;
- *jitter* which, as its name suggests, is a jerky change of position of the character;
- *disturbed linearity*, which is where lines in graphs appear to move.

The contrast between the character and the screen background

People vary in their preferences for character brilliance and contrast, some preferring bright characters with high contrasts and others opting for dim characters and low contrasts. Factors such as age, the sharpness of the character, the level of lighting in the work environment and reflections from the screen will affect this. The less clear a character is, the greater the contrast needs to be between it and its background to maintain legibility. The surrounding lighting and any reflections from the screen will similarly affect the amount of contrast needed. It is important that users should have the facilities to change the brightness and contrast on their VDUs. Indeed, the Health and Safety (Display Screen Equipment) Regulations 1992 stipulate that the brightness and contrast of all VDUs (from 1 January 1996) must be easily adjustable by the users (HSE, 1992d). In practice this means that the controls should be located at the front, rather than at the back of the VDU.

The above characteristics point to the need to take care when selecting hardware to ensure a clear and legible screen. Other factors affecting visual discomfort are to be found in the workplace itself.

Reflections on the screen

Reflections on the display screen are a cause of many complaints of visual discomfort. They reduce the contrast of the display, rendering it difficult to read, or even obscure the display altogether. Reflected light may also cause glare resulting in visual discomfort.

Screen colour

Most VDU screens are now colour which allow users to experiment with a wide range of colours. Many operators have a strong personal preference for a particular colour. For the most part, scientists argue that there is no physiological reason why this should be so, although Osaka (1985) concluded that visual fatigue changed as the VDU colour changed. He found that blue and red, which are at opposite extremes of the visual spectrum, cause significant visual fatigue, when compared with green and yellow, which lie in the middle of the visual field. However, Sheedy (1993) concludes that 'the colour of the display is probably not of great significance to visual comfort or performance, so long as the contrast between the letters and background is good'.

Positive versus negative screen images

Screens which have a dark background with coloured characters are referred to as being negative, while positive screens have dark characters on a light background. (There is considerable discrepancy in the literature on this, with the majority of writers opting for the reverse of the above. The authors' justification for going against the stream is that print on paper preceded screens and represents to many people the positive image of dark on light.)

The majority of old DOS-based systems have screens which have a dark background with coloured characters. With Window machines users have the choice of making the screen either positive or negative.

The negative screen requires office lighting that is a little lower than normal to ensure adequate contrast between the characters and the background. As the level of ambient lighting increases, the contrast decreases. Reduced lighting levels may, however, render the source document harder to read. Another problem which may arise from using a negative screen is that, in some tasks, the eyes must move constantly between the light (mainly white) source document and the dark screen. This may result in tiring the iris muscles as they continually change the pupil size to accommodate the varying amounts of light.

In contrast, the use of a positive screen brings no such problems of accommodation and, in addition, is not subject to the reflected light which is common with negative screens. However, it is not entirely free of problems. On a positive screen, the character width appears diminished because of the way in which it is created. The same dots are used as on a negative image, but the characters are formed by the absence, rather than the presence of light. The result is that the light dots surrounding the character tend to impinge on the character strokes, making them appear narrower than they really are. Grandjean (1987) quotes studies by Bauer and Bauer, and Cavonius which concluded that, for strokes to be perceived as equal in width, dark strokes must be 23 per cent wider than bright strokes. Thus the character set on a positive screen must be composed of sufficiently thick strokes to combat the slimming effects of the light background.

Positive screen images also are more susceptible to flicker than negative screens. This may be due to the white phosphor which is used, which is not as persistent as some of those used on negative screens, and to the fact that there is a larger area of activated phosphor. Indeed, it has been reported (Grandjean, 1987) that a refresh rate of 92 Hz (or cycles per second) or 100 Hz is required to ensure a flicker-free positive screen.

Some research has shown that operators perform better with positive screens, feeling more comfortable and operating more efficiently (see below).

The legibility of VDU characters will affect the efficiency of operation and the visual comfort of the operator. Factors which affect legibility are described below.

Screen size

Generally speaking, the larger the screen, the easier it is to read (although other factors also affect legibility). As libraries move towards full text retrieval, screen size becomes more important. High-quality large screens (that is, A4 or A3) are highly recommended for reading large amounts of text on screen.

Character set

In 'How a VDU works' (above), characters were described as being composed of dots. This is something of a generalization as, in fact, there are four methods of creating characters on a VDU screen:

- facsimile generation in which a 'Master' letter is copied on to the screen;
- stroke generation in which the character is created by a series of strokes (as in writing a character);
- lissajous generation in which the shape of the individual character is built up by a series of segments of curves (lissajous curves);
- dot matrix generation in which the shape of the character is made up of individual dots highlighted within a rectangular matrix.

The most common method is dot matrix generation.

Matrix size

The dots are divided into rectangular matrices, typically 5 x 7 or 7 x 9. In Figure 5.3, two matrices, 5 x 7 and 7 x 9, are used to create the letter 'R'. The pattern on the right is clearer than the one on the left, because there

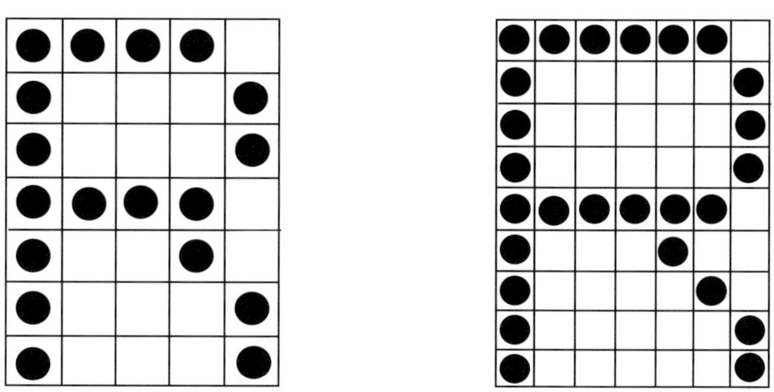

Figure 5.3 Two dot matrices used to create the letter 'R'

is relatively less spacing between the dots. Because the right-hand image is clearer, it can be recognized more quickly. Similarly, on a VDU screen, the clearer the image, the faster the likely reading rate. Thus, the larger the matrix (within limits), the greater likelihood that the characters will be legible, and the more easily will the text be read by the operator. However, very large letters can reduce reading speed.

Character font

Most modern software packages allow users to select the font they require but this too can affect legibility. To some extent the font will be affected by the size and shape of the matrix. The more dots in the matrix, the clearer the image will tend to be and the more obvious will be the ascenders and descenders. Some fonts do not have proper descenders, the tail of the q or the y lying above the line. This makes the text harder to read. While it is important to have ascenders and descenders which are in proportion, serifs only hinder the reading process as they tend to clutter the screen image. Slanting characters are also harder to read than an upright Roman script, and capital letters exclusively are less legible than a mixed case font.

Whatever font is used, it is possible for confusion to occur between the characters. An 'I' for example may be misread as '1' (one) or vice versa. Knave (1983, p. 32) lists the following characters as possible candidates for confusion. The first column, 'mutual', contains those characters which may be mistaken for each other. In the second column, 'one-way', it is more common for the characters on the left to be mistaken for those on the right than vice versa. The entries marked with an asterisk (*) are those characters which together cause more than 50 per cent of the total confusion:

Mutual		*One-way*		
			mistaken for	
O	Q	C	,,	G
T	Y	D	,,	B
S	5	H	,,	M, N
I	L	J, T	,,	I
X	K	K	,,	R
I*	1	2*	,,	Z
		B*	,,	R, S or 8

Character size

The legibility of the information on the screen will be affected by the size of the characters used. Appropriate character size is dependent on viewing distance, among other factors, so that, for example, the text of a viewdata system should be large enough for visitors to the library to read easily at a glance when standing at a distance of, say, two metres. Text intended for close-up reading, however, can be smaller.

The height-to-width ratio also affects legibility. Characters which are tall and thin or short and fat are harder to read than those which are of squarer proportions. Other factors affecting character legibility are: character contrast; the dot matrix used; lighting; and, of course, the operator's eyesight.

For a viewing distance of between 50 and 70 cm the recommended proportions for characters are as follows (Grandjean, 1987):

Height of capital letters	3–4.3 mm
Width of capital letters	75% of the height
Height of lower case letters	70% of the height of capital letters
Width of lower case letters	60% of the height of capital letters
Stroke width	20% of the height
Intercharacter distance	25% of the height
Interline space	4–6 mm or 100–150% of the height of the capital letters

Cursor

The cursor is essential for interactive work to indicate the position at which data are to be entered. Vartabedian, as quoted by Knave (1983), lists five qualities necessary for a cursor. A cursor should:

- be easily found in random location in an alphanumeric display;
- be easily tracked as it moves through the display;
- not interfere with the reading of symbols;
- not impair search time for non-cursor related information;
- not use a graphic technique or shape used elsewhere in the display.

Cursors come in a variety of forms, some of which may be altered with the terminal. The most common for DOS computers are a long underline (_) and a box (□) in inverse video whereas Apple Computers use arrows and vertical lines. After studying a variety of cursors, Vartabedian found that a box cursor flashing at 3 Hz was optimal. With Windows there is a much wider choice. Frequently several different cursors are used in one software package.

Interline spacing

The gap between the lines of text on a screen affects reading performance. It is recommended that a minimum interline distance of $^1/_{30}$ of the line length should be used; thus as the lines increase, so should the interline spacing. Grandjean (1987) concludes that it would be advantageous to use A4-oriented screens enabling shorter lines and hence smaller interline distances to be used. This would accommodate more text per screen.

Character sharpness/resolution

It is not just the size and shape of a character that is important, it is also the clarity which is necessary for legibility. A sharply defined character will be easier to read and will thus give a higher reading performance and reduce visual discomfort. Some VDU characters have a blurred edge which minimizes the clarity of the character. If the blurred edges are wide relative to the width of the stroke which forms the character, then the blurring will be clearly visible. Studies have shown that VDU users read faster when using higher resolution monitors compared to the older style VGA monitors. They also find them less tiring (Sheedy, 1993).

Screen display

This is discussed in more detail in Part VI on the 'Human–Computer Interface', but it is important to know that, even if the VDU screen offers all the correct features for legibility, poor screen display brings its own problems. The Civil and Public Services Association Research Department (1983) reports the use of the Graphix Optical Character Recognition System. The system was installed in the DHSS in 1976 to perform data entry tasks and ran so successfully that bureau work was undertaken, in this particular instance to do cataloguing work for the British Library. The process involved the system displaying all the text which has been read in and highlighting text that has not been recognized. Unfortunately, the British Library data contained a great deal of text that was unrecognizable to the system, including the Cyrillic alphabet. Hence, large proportions of the screen were highlighted at any one time resulting in operators being forced to look at a bright screen for long periods and to correct the brightly lit, unrecognized, text. As a consequence, fatigue set in quickly (after about 20 minutes) and frequent rest breaks were required.

WAYS OF REDUCING VISUAL DISCOMFORT

There are two factors which cause light to be reflected from the VDU screen. The first is the irregular surface of the phosphor and the second is the polished surface of the glass. Various methods of eliminating sources of reflected glare are discussed in Chapter 11, but if it is not possible to remove the sources of glare it may be helpful to consider the use of filter panels, screen coatings or VDU hoods, which have been designed to help reduce the problem.

Filter panels

Manufacturers have created a number of panels to fit over the front of the screen for a variety of reasons: to reduce glare; to change the apparent colour of the screen; to reduce the effects of static; or lead-impregnated screens to provide protection from alleged X-ray emissions. The last claim is misleading. Radiation emission from VDUs is negligible and the minute amount that is emitted occurs from the back rather than the front of the screen. Furthermore, the move towards positive screens has reduced the need for filter panels. Putting a filter over the display will also reduce the brightness of the characters. Grandjean (1987) reports a study which concluded that, when a filter was used, the VDU needed to be operated on a higher brightness setting, which caused the phosphor to age more rapidly thus reducing the lifespan of the cathode ray tube (CRT).

Filters vary in how they are made and the benefits they purport to offer. Some of these are described below. Many are offered by suppliers of computer equipment (see Useful Addresses at end of book), the most common being the micromesh filter. Others, such as etched screens, are created during the manufacturing process.

- *Micromesh filters* are fine fabrics which are placed directly over the screen. Only light which falls perpendicularly on the mesh can penetrate and so reach the screen. Most light in the surrounding environment will fall at an angle to the mesh and thus fail to penetrate it, hence much reflected glare will be eliminated. However, the light emitted by the characters is also reduced by the filter, especially if the screen is viewed from an angle. The contrast between the characters and the background is reduced and, if the micromesh collects dust, the problem is further exacerbated. Cakir *et al.* (1980) conclude that 'Micromesh filters are an effective way of reducing reflections but not without certain penalties, some of which outweigh the advantages to be gained to such an extent that their use is often of questionable value', but this is contradicted by reports cited by Grandjean (1987) which indicate a strong preference by operators for micromesh filters over other forms of anti-reflective device.
- *Neutral density filters* reduce the amount of light which passes through them. Light from various sources in the room is reduced on its way through to the screen and again on its return as reflected light. Light emitted by the characters only passes once through the filter and is therefore reduced less than that of the reflected light, so causing it to stand out more clearly. It is important, however, that the filter should be able to withstand the wear and tear of normal use and cleaning, as a scratched surface reduces the clarity of the characters.

- *Coloured filters* operate in the same way as neutral density filters, but they change the light which passes through them and hence the apparent screen colour. Studies have shown them not to be particularly effective (Grandjean, 1987).
- *Polarization filters* polarize the light passing through them and effectively reduce glare. Unfortunately, they are usually made of an extremely smooth substance called plexiglass which itself reflects light! It is, therefore, necessary for the filter to have a non-reflective coating. This kind of filter can be expensive.
- *Etched screens* have an irregular surface which reflects light in a variety of directions, thus reducing the amount which is transmitted back to the operator. This may be achieved by roughening the display screen glass either by chemical or mechanical means. However, some etched screens also disperse the light emitted from the phosphor and hence blur the characters. This is by no means true of all etched screens, however, and many effectively reduce glare while retaining clarity of characters. Etched glass placed in front of the VDU screen may be less effective because the greater the distance between the light source and the roughened surface, the greater the dispersion of the light. The distance between the characters and the etched glass will cause the clarity of the characters to be diminished. Interestingly, in a study reported by Grandjean (1987), a roughened screen display gave a better performance than either a micromesh filter or a coloured plastic filter.

Screen coatings

A variety of screen coatings are available which reduce reflected glare in a number of ways:

- *Spray-on coatings* deposit a rough film over the display screen, causing reflected light to be dispersed, but also reduce the definition of the characters.
- *Vapour-deposited coatings* work on a similar principle to spray on coatings, but create a finer surface with less detriment to character legibility. They are, however, more expensive.
- *Quarter-wave coatings* are so called because the deposited film has a thickness of a quarter of the wavelength of light. The result is that light passes through the filter but is inhibited on its return. However, the sharpness of the characters is not diminished. The disadvantage of these coatings are, which is the same type as those used on good quality cameras, is that they are easily soiled and particularly susceptible to fingerprints.

Tube shields/hoods

Various shapes of CRT shields and hoods have been designed to reduce the glare, particularly from overhead lighting. These can be very effective, but can also have the drawback of causing the operator to adopt a rigid position.

In conclusion to this discussion on anti-reflective products, it is pertinent to quote Grandjean (1987, p. 86)

> All antireflective techniques have serious drawbacks. If efficiency is weighed against the drawbacks, the quarter-wave coatings and the etching/roughening procedures are preferable, whereas micromesh, polarization and coloured filters cannot be recommended. The adequate positioning of lights and the appropriate positioning of the screen with respect to windows remain the most efficient preventative measures [as well as the cheapest!].

THE TYPE OF WORK PERFORMED

Different aspects of VDU work impose their own visual demands. Some of these are considered here.

THE TASK PERFORMED

The more visually demanding a task is, the more tiring it will be. Some tasks are more visually demanding than others. Online cataloguing, for example, may require the operator to search a full MARC record to locate relevant information. This is visually more intensive than the use of a brief catalogue record. During such a task, it is desirable for operators to be able to rest their eyes from time to time – for example, by looking at a distant view or a picture.

THE SOURCE DOCUMENT

Some tasks, such as word-processing, require much concentration on a source document. If its characters are poorly formed and difficult to distinguish, or if the page layout is poor, reading speed will be diminished as, too, may accuracy. Other factors which affect the readability of the source document are the reflectivity of the paper surface, the colour of the paper and the colour and density of the ink.

The angle of the light on the source document also affects its readability. If the paper is at right angles to the light source, the white background will appear bright in contrast to the dark letters. As the light moves relative to the source document, the contrast becomes less marked and the text, therefore, less legible.

READING FROM A VDU

Much work has been done comparing the performance of people reading text from a VDU screen with the same people or others reading from hard copy (Dillon, 1992). Five possible differences between reading from screens and paper have been identified: speed, accuracy, fatigue, comprehension and preference. Although the use of different methodologies, procedures and stimulus materials in the experiment work make definitive conclusions difficult it seems certain that reading from a VDU screen is slower than reading the equivalent text from hard copy. Accuracy also appears to be adversely affected, particularly if the task is unusually demanding, such as proofreading. Fears of increased visual fatigue and reduced comprehension, however, appear to be groundless. With respect to preference, readers still prefer to read from top quality hard copy rather than from VDU screens, but this may change as technology improves and readers become more experienced with the new medium (Dillon, 1994).

SCROLLING SCREEN

Online searching may necessitate reading information from a scrolling screen. Far from being a cause of visual discomfort, it has been found to present few problems to the operator. Sekey and Tietz (1982), as cited by Hamdy (1987), found little difference in observing eye movements between reading a static display and scrolled text, especially when the scrolled display was rapidly paced. Hamdy also cites Kolers *et al.* (1981) who found that scrolling could actually assist more effective reading, with the operator making fewer and briefer glances at the text. They found an increase in reading speed with no apparent loss of comprehension. As the rate of scrolling gradually increased, performance correspondingly increased to the point where operators performed best at a speed which was 20 per cent faster than the pace at which they felt most comfortable. Dillon *et al.* (1990) also reports that a scrolling mechanism was the most frequently cited improvement suggested by subjects assessing their reading interface. Despite those research results it must be remembered that not all users will feel comfortable with scrolling screens.

SUMMARY

Research has shown that eyes and eyesight are not damaged by using VDUs, nor do they make existing defects worse. However, there is a strong indication that there is a causal correlation between VDU work and visual discomfort leading to a range of symptoms such as impaired performance, red or sore eyes, dizziness and headaches. In a colloquium organized by the French Society of Ophthalmologists (Boyle, 1986), the following factors were cited as the principal causes of visual discomfort affecting a claimed 70 per cent of users:

- frequent eye accommodations – both movements between screen, keyboard and paper (with an estimated 30 000 such movements being needed during a working day) and the constant adjustment between positive and negative contrasts;
- screen image flicker (refresh rate);
- dazzle effects caused by poor lighting;
- screen reflections;
- poor character readability.

Not surprisingly, the group recommended that ophthalmic examinations should be made on the employees before they start to work on VDUs and that these should be repeated regularly. This recommendation has since become law in many European countries as a result of the European directive, No. 90/270/EEC of 29 May 1990, on minimum safety and health requirements for work with display screen equipment.

Further steps include:

- The use of good quality high resolution colour monitors which have a high refresh rate and the facility for making the screen positive.
- The use of appropriate adjustable furniture, set in the correct position to enable operators to adopt good working postures.
- The placement of reference material as close to the screen as possible to avoid frequent large eye and head movements, and at the same distance from the eyes as the screen to avoid having to change focus when looking from one to the other.
- Eliminating bright sources of light and reflections from the eye of the operator.
- Giving workers frequent rest breaks: the Health and Safety (Display Screen Equipment) Regulations 1992 recommend breaks of 5–10 minutes after 50–60 minutes continuous screen and/or keyboard work preferably away from the workstation (HSE, 1992d).

For further information on how to adjust the work environment, see Part IV; and for information on job design, see Chapter 22.

CHAPTER 6

STRESS

INTRODUCTION

Stress is commonly understood to be a work-related health hazard. Some employers now routinely examine their staff for early warnings of impending heart disease or strokes which are known to be partly caused by stress. For the most part, though, little is done in the workplace to reduce the cause of stress on people's lives. Indeed, in the US, the National Association of Working Women, which has undertaken considerable research into occupational health, found that the following beliefs about stress existed (Apex, 1985):

- People with the greatest responsibility, who make a lot of important decisions, have most stress.
- People bring stress with them from home into work and, if they are under stress, it is because of family or personal problems.
- Certain people are most susceptible to stress; this is not due to the job, but to the individual.

A common misconception seems to be that stress only affects important, highly paid jobs and that it is, in a sense, the price the executive pays for the kudos of this career. Sadly, stress does not confine itself to a set of well-paid individuals, but often also affects those who are at the bottom of the ladder.

Sometimes, routine work brought about by automation can produce stress, or it may serve to reduce it. Our concern here is to identify the areas where automation may create stress for some individuals.

WHAT IS STRESS?

Stress is our bodies' biochemical response to a threatening stimulus. Different stimuli will cause us to react differently; we may blush, jump or hide the reaction under a confident mask, but the biochemical response of our bodies is the same – an increase in the heartbeat and breathing rates, increased secretion of stomach acid and the release of certain hormones. An increased flow of adrenalin and a rise in cholesterol and blood sugar cause a surge of energy to the muscles, while the stomach becomes inactive. Emotionally, we may experience frustration, anger, excitement or anxiety.

Some stress is good. Without thrills or excitement, life would be dull and unproductive. Indeed, after a short period of stress, our bodies return to equilibrium. It is stress which is prolonged and continuous that poses a threat to health, as our bodies remain poised for 'fight or flight' and the resultant imbalance can wear our bodies down.

The experience of stress through work is 'associated with exposure to particular conditions of work, both physical and psychosocial, and workers' realization that they are having difficulty in coping with important aspects of their work situation' (Cox, 1993).

STRESS AND HEALTH

As a result of prolonged stress, a number of health problems may occur. Coleman (1978) cites no fewer that 194 symptoms and diseases which may be stress-induced. The following are the more common problems:

- *Short-term stress symptoms*:
  - indigestion, loss of appetite, nausea, vomiting
  - irritability, restlessness, behavioural changes, boredom
  - nervousness, anxiety
  - chest pains
  - insomnia
  - headaches, migraines
  - backache
  - toothache (from grinding one's teeth!)
  - susceptibility to colds and 'flu
  - menstrual irregularities
  - sexual difficulties
  - fatigue, memory blanks
  - depression, low self-esteem
  - accidents
- *Long-term stress symptoms*:
  - ulcers, bowel disorders, spastic colon
  - vaginal and bladder infections
  - high blood pressure
  - coronary heart disease
  - hypertension

- chronic mental health problems, nervous breakdown
- decreased resistance to chronic disease.

The various symptoms are not inseparable, and may even create more stress for the person, thereby exacerbating the situation. They may also cause the sufferer to increase consumption of tobacco, alcohol and drugs in order to relieve the symptoms. This, in turn, can compound the health problems.

THE CAUSES OF STRESS

During the last decade libraries have experienced not only major financial restraints but massive technological changes. This has led to increased pressures on staff which in turn has caused more stress.

Work-related stress caused by the introduction of new technology can be divided into four main groups: environmental factors; job design faults; employee–employer relations and the nature of VDU work.

ENVIRONMENTAL FACTORS

Unpleasant environmental conditions caused by smoking, crowding, excessive noise, inappropriate provision of heating, lighting, ventilation and inadequate furniture and equipment can add to job stress. Recommendations for improving the environment and workstation layout are made in Parts IV and V.

JOB DESIGN FAULTS

Poor job design, resulting in too much or too little work, time pressures, having to make too many decisions and the over- or under-utilization of skills, can cause stress. Kompier and Levi (1994) quote Rosch who identified the following further job design factors as causes of occupational stress:

- inadequate time to complete the job to one's satisfaction;
- lack of clear job description or chain of command;
- absence of recognition or record for good job performance;
- lots of responsibilities but with little authority or decision-making capacity or the converse;
- lack of control over what is to be done.

The HSE (1995) list among their causes of stress:

- pressures that are piled on top of each other or prolonged;
- feeling trapped or being unable to control demands;
- confusion from conflicting demands;
- high degrees of uncertainty;
- inflexible or over-demanding work schedules;
- lack of clear and achievable work objectives;
- prolonged periods of repetitive or monotonous work.

Stressors specifically applicable to libraries include:

- having to deal with theft, mutilation and destruction of library materials;
- the constant need for speed in response to reference enquiries;
- the high demand for innovative and creative problem solving;
- the reliance on other organizations to supply information/ documents;
- the need to keep abreast of technological advances and services (document delivery services, Internet, CD-ROMs, on-line public access catalogues (OPACs), virtual library concepts) and the training implications.

Holmes (1995) also found that one of the main stressors for academic library assistants was having to do lots of repetitive unstimulating jobs which they did not perceive as commensurate with their experience, qualifications or skills. Other potential stressors at work found by Holmes were: work overload, lack of adequate emergency back-up procedures, insufficient pay and lack of promotion opportunity.

Research investigating stress among middle managers in public libraries has shown that the main sources of stress include: feeling undervalued in the library organization; feeling the library service is undervalued; increasing competitiveness among staff; bureaucratic inertia; poor management; low pay; lack of communication; and having to deal with changes both inside the library and inside local government (Hodges, 1990).

Further details about job design may also be found in Part VII.

EMPLOYEE–EMPLOYER–PATRON RELATIONSHIPS

The introduction of an automated library system changes the traditional function in the library and the roles which each employee is expected to play in the system.

> Fundamental organizational change evokes feelings of apprehension at all levels . . . Uncertainty surrounding the need for one's own job in the new system or its possible restructuring with consequent changes in

> pay, promotion and training, etc., inevitably lead to anxiety. (MacKay and Cox, 1984)

Bunge (1991) reports on the changes brought about by the introduction of CD-ROMs and the increased stress that this has caused reference librarians. CD-ROMs are immensely popular, but library clients often demand and expect help from the reference librarian with getting into a database, selecting terms and printing out results even though user guides and online help systems are available. Coupled with this are the extra non-reference demands such as clearing paper jams, sorting out computer failures, distributing CD-ROMs and producing training guides. Often extra staff have not been forthcoming to compensate for these extra demands. Bunge says that staff often feel inadequate in their knowledge of how to use CD-ROM sources or how to teach their use to others. He recommends that managers try to reduce stress by emphasizing the positive aspects of the CD-ROM service and addressing the frustrations by giving librarians time and privacy for learning sources, reminding them that patrons are only human and campaigning for more standardization between products.

Caputo (1991) discusses stress and burnout in the library service. She points out that dealing with patrons can be extremely stressful. Social interaction can be very tiring. Librarians have to remain calm, caring, helpful and patient even when faced with computer failure, having to do several jobs at once, lengthening queues and with frustrated, angry patrons. They also have to be seen to be upholding rules without being tyrannical.

Social isolation

Library work does not always require intensive use of VDUs, and yet it can bring with it a sense of isolation. For example, although the Browne issue system is cumbersome and time-consuming, some library assistants we interviewed remarked that it had given them the opportunity to talk to the user while searching for the book card. Automated systems, they felt, encouraged the assistant to 'process' the reader more rapidly and reduce personal interaction.

In some libraries VDU users can be totally isolated. In one library, staff were expected to work with a VDU in a small office on their own all day every day. Not surprisingly, staff turnover was high with many opting to return to manual work.

Failure to solve grievances

This may occur because management does not respond or there is a failure to understand the underlying causes and thus treat only the symptoms. Other problems may arise because the solutions may be seen as being costly to implement: finances for suitable furniture, for example, may be difficult to justify in a library that has only just won a hard-fought battle for a computer system.

Fear of adverse health effects

Unless the fear of potential health risks is handled sensitively and knowledgeably by management, it can increase out of proportion.

Threat of job losses

Automation is sometimes introduced partly to cut the staff budget, which may imply that existing staff are to be axed. Even if this is not the case in any individual library, the plan to automate may create anxiety.

Lack of supervisory support

Yang and Carayon (1995) found that VDU workers without adequate supervisory support experienced higher stress levels.

The nature of VDU work

Professor Frank Zingrone of York University, Toronto, has propounded the theory that the nature of the VDU and of the tasks performed on it cause conflict between the right and left hemispheres of the brain (Else, 1985). The left side of the brain controls language and logic – skills which are frequently called into play during VDU work. The right side of the brain controls visual spatial capacity, imagination and intuition and thus the viewing functions at the VDU. This conflict, Zingrone suggests, causes stress, but this has been questioned by other researchers. He suggests no easy solutions, but recommends only short spells (20 minutes) on the VDU at any time.

THE NEED TO REDUCE STRESS

Unlike postural effects, stress effects do not go away when the person has stopped working; rather, they may have an insidious influence on the rest of the person's life – for example, increasing the time needed to wind down at the end of the working day. As this will obviously have a direct effect on the individual's private life, work-related stress can have a bearing on all employees' quality of life.

Similarly, not all stress is work related. Holmes and Rahe (1967) formulated a table of stress points which are related to various lifestyle changes (see Table 6.1). The higher the number of points, the greater the

Table 6.1 Social readjustment rating scale (adapted from Holmes and Rahe, 1967)

Death of spouse	100
Divorce	73
Marital separation	65
Jail term	63
Death of close family member	63
Personal injury or illness	53
Marriage	50
Fired at work	47
Marital reconciliation	45
Retirement	45
Change in health of family member	44
Pregnancy	40
Sex difficulties	39
Gain of new family member	39
Business readjustment	39
Change in financial state	38
Death of a close friend	37
Change to different line of work	36
Change in number of arguments with spouse	35
Large mortgage or loan	31
Foreclosure of mortgage or loan	30
Change in responsibilities at work	29
Son or daughter leaving home	29
Trouble with in-laws	29
Outstanding personal achievement	28
Spouse begins or stops work	26
Begin or end school or college	26
Change in living conditions	25
Revision of personal habits	24
Trouble with boss	23
Change in working hours or conditions	20
Change in residence	20
Change in school or college	20
Change in recreation	19
Change in church activities	19
Change in social activities	18
Moderate mortgage or loan	17
Change in sleeping habits	16
Change in number of family get-togethers	15
Change in eating habits	15
Holiday	13
Christmas	12
Minor violation of the law	11

level of stress, and the points are cumulative, so that if there is more than one stress-giving situation in a person's life, and points related to each should be added together to give the total stress level. More than 100 points at any time is likely adversely to affect the health of a person. Thus it is important to see that work-related stress will add to the pressures experienced elsewhere. It should be noted that this table relates only to stress caused by changes in people's lives; other stressful factors are excluded, and that some of the changes, such as 'change in church activities' may not feature so highly now.

The adverse effects of stress will impinge upon the running of the library. Unnecessary tensions may be created, staff–user relationships may deteriorate, inaccuracies may develop in work, and so on. More serious effects could include employee absenteeism, which in turn would increase the load on fellow workers. Consequently, library managers should be aware of the need to reduce work-induced stress.

STRESS AND THE LAW

The EU Framework Directive on Health and Safety requires the employer to recognize stress factors and to correct them. It states that the employer has 'a duty to ensure the safety and health of workers in every aspect related to the work . . . following general principles of prevention: avoiding risks, evaluating the risks which cannot be avoided, combating the risks at source, adapting the work to the individual, especially as regards the design of work places, the choice of work equipment and the choice of working and production methods, with a view, in particular, to alleviating monotonous work and work at a predetermined work-rate and to reducing their effect on health' (EU, 1989, pp. 2,3).

In addition to this general framework many EU member states have more detailed legislation aimed at improving the level of safety in the workplace and the mental and physical well-being of employees. Kompier and Levi (1994) describe the Working Environment Act in the Netherlands and the Swedish Work Environment Act (1991). The psychological aspects of legislation in the Netherlands, for example, recognizes the need for organizing work according to personal skill, working at an employee's own pace, being informed about the purpose and results of work, keeping in contact with colleagues, avoiding monotonous and repetitive jobs and taking into account differences between workers' personal characteristics.

In the UK, the HSE (1995) has produced formal guidelines on good practice for managing stress. There are no specific detailed standards or requirements which employers must meet but the guidelines recognize that 'stress should be treated as any other health hazard'. As such employers must take steps to reduce stress under the Health and Safety at Work, etc. (HSW) Act 1974. This places a general duty on employers to ensure, so far as is reasonably practicable, the health, safety and welfare at work of all their employees. The Management of Health and Safety at Work Regulations 1992 take this one step further and require detailed risk assessment to be carried out (HSE, 1992a). Hence, although the guidelines do not have the legal force of Regulations or Approved Code of Practice, the contents are likely to be relied upon by experts in civil proceedings as exemplifying best practice and what should be done under the HSW Act.

HOW TO MONITOR STRESS LEVELS

Library managers are responsible for identifying and monitoring health and safety risks (including stress) in the workplace. Under current legislation employees must be involved in the process. At the very least they must be informed about how, why, when it will take place and what will happen to the results. Not involving staff may increase stress and induce distrust and resistance.

The most common way of identifying hazards and monitoring health and safety risks is with the use of checklists, questionnaires and analyses of administrative data. Example checklists can be found in the Health and Safety (Display Screen Equipment) Regulations 1992 (HSE, 1992d) and in Kompier and Levi (1994). Questionnaires focusing on identifying stress risks and stress-related health problems can also be found in Kompier and Levi (1994). With regard to administrative data the most important indicator of stress is absenteeism. Other useful data to consider is turnover of staff, productivity rates, number of accidents, amount of overtime and the backlog of untaken leave. By comparing data over a period of time managers can identify trends.

HOW TO REDUCE WORK-RELATED STRESS

There are four basic strategies to preventing or coping with stress (Kompier and Levi, 1994):

1. Removing the person from the stress-producing situation.
2. Obtaining more resources to meet pressures.
3. Modifying the work and/or environment to reduce the number of stressors.
4. Strengthening the person's resistance to stress.

The first three strategies are concerned with the working situation. The last is person-oriented.

WORKING SITUATION

Although staff respond differently to the process of automation there are a few points which may help to avoid, or reduce, stressful situations:

- Optimize the environment and workstation layout (see Parts IV and V).
- Ensure good job design (see Chapter 22) but, in particular, that the job is 'do-able' and, where possible, is matched to the employee's ability.
- Communicate with all staff, telling them what is happening and explaining their role within the new system; also listen sympathetically to their comments, complaints, and fears.
- Involve employees in decision-making.
- Where possible, sort out their grievances. If a solution is not possible, for financial or other reasons, tell them why.
- Monitor employees' health, changes in work situation and their interrelationship.
- Increase awareness, inform, train and educate staff.
- Provide adequate supervisory support.

PERSON-ORIENTED MEASURES

Not all stress is a problem. As Cakir *et al.* (1980) affirm, the ideal is not the absence of stress, but that the amount of stress should be optimized. Not all people will have the same reaction to stress, since several factors will be involved.

- *An individual's personality.* A person's ability to cope is affected by their emotional stability and their degree of introversion or extroversion. People with a so-called 'Type A' personality are more likely to suffer from stress. Typical traits of a Type A are: competitiveness, haste, impatience, obsessive concern with doing more and more in less and less time, aggressiveness, restlessness, alertness, lack of cooperation and hard-driving effort.
- *An individual's ability to cope with situations.* Coping generally means employing one of two strategies: *altering* the situation or changing the *perception* of the situation (without it being changed). There is no ideal coping behaviour, it depends on the situation, circumstances and the targets required (Kompier and Levi, 1994).
- *The age of the person.* Young people starting work in a library may be

forced to accept a job that is not commensurate with their qualifications. They may be given work, for example, that is too simple and unstimulating with little or no responsibility and control. Elderly workers may be at risk too, particularly if they are expected to do physically demanding tasks such as shelving for long periods or standing at the issue desk for several hours a day. Smith *et al.* (1984), however, found that younger and more highly qualified reference librarians were more at risk than other reference staff from stress and burnout.

- *Motivation.* Some people are driven by the prospect of increased responsibility or a greater financial reward or other factors (see 'Motivation', in Chapter 1).
- *The relevance of the stress to the work being undertaken.* As a reference librarian, the challenge of a difficult enquiry may be considered an enjoyable part of the job and thus enhance job satisfaction. The need to learn how to operate an automated system or to cope with computer failure may be seen as outside the person's remit and thus become a negative source of stress.
- *The employee's attitude to the previous system.* Some library assistants interviewed remarked that the laborious manual issue systems coupled with long queues of readers created more stress for them than learning and operating the computerized issue system.
- *The employee's attitude to the new system.* If the previous system caused problems for them, they may welcome the computerized system. Alternatively, they may prefer to stay with a system they know and resent the change.
- *Other stress effects in the employee's life.*

Strengthening a person's resistance to stress can be achieved in a number of ways:

- *Physical exercise.* Physical exercise can help to burn off the pent-up energy produced as normal reaction in stress situations.
- *Relaxation/meditation methods.* There are a number of relaxation and meditation methods which can help people to unwind and reduce their tendency to react to stress.
- *Medication.* In severe cases of stress, prescribed medication may be needed. Some drugs act on the brain which affect mental and/or emotional functions. Others block the streams of nervous impulses from the brain to various organs to reduce reactivity to stressful stimuli. The use of drugs should only be used with caution and under the guidance of a doctor. After all, they will not take the stress problem away, it may still be there once the person has stopped taking the drugs.
- *Co-counselling.* Co-counselling is where two people are brought

together to share each other's experiences and demands which might cause stress and to offer each other advice. To obtain maximum benefit both employees should have received a short training course in co-counselling before practising the techniques.

- *Counselling.* Counselling by a doctor or therapist can help some people to improve their ability to cope with stressful situations. It focuses on changing a person's perception of the demands placed on them, their perceived capabilities and what the consequences might be if demands are not met. It may also include aspects of time management.

There are also a number of other alternative medicine approaches, such as acupuncture, herbal medicine and aromatherapy, which might also be useful.

SUMMARY

Stress is becoming a serious threat as libraries become more automated and pressurized to provide instant services with reduced staffing levels. Managers, therefore, need to recognize the causes of stress and to be active in stress prevention. They need to involve staff and adopt a systematic approach, that is planning, monitoring, risk assessment, choosing a strategy, implementing improvements and evaluation (Kompier and Levi, 1994). The most common stress prevention strategies involve reducing excessive job demands, increasing worker control and providing more social support.

CHAPTER 7

REPRODUCTIVE HAZARDS

INTRODUCTION

The press has been keen to report stories of 'clusters' of birth abnormalities among female VDU workers and some cases of sterility among men. The finger of accusation has been pointed at X-rays and at extremely low frequency radiation (ELFs) emitted by VDUs. In defence of the VDU have come equally numerous articles reassuring the user that all is well. The following is a summary of the discussions to date.

REPORTS OF ADVERSE PREGNANCY OUTCOMES

The massive expansion in the use of VDUs over the last twenty years, has been accompanied by fears that these machines cause reproductive problems. To a large extent this has been fuelled by media reports about 'clusters' of VDU operators where the occurrence of adverse pregnancy outcomes have been unusually high. One of the early reports, for example, highlighted a bank in Grimsby, England, where the pregnancies of five women resulted in three miscarriages, one stillbirth and one deformed baby. All women were said to be low level users of VDUs. Official investigations of these clusters, however, have rejected the findings as insignificant because of the small sample size and short study periods. In addition, other variables, such as the environmental conditions, the stress induced by the work, the social class of the women and their general level of health, were often not taken into account. It has also been argued that many clusters are to be expected on the basis of chance alone. Bergqvist (1986) and Ong *et al*. (1990) give more details about the clusters.

Abnormal pregnancy outcome is an emotive issue, however, so it is unwise to leave the matter as 'unproven'. In order to confirm or deny the allegations, a number of studies have been, and are being, undertaken. Table 7.1 summarizes the main epidemiologic studies undertaken to date and presents the type of study, the study period, the reproductive effects examined and the results obtained.

The adverse outcomes of pregnancy most commonly investigated include spontaneous abortion, stillbirth, perinatal death, premature delivery, low birth weight, and congenital abnormalities. It is well known that many factors are capable of increasing the risk of reproductive problems. For example, spontaneous abortion, which occurs in as many as one in seven pregnancies, has been linked to race, age, general health, reproductive history and suspected environmental factors such as smoking, alcohol consumption, low quality prenatal care, low socio-economic status, stress, heavy lifting and radiation exposure (Ong *et al.*, 1990). The incidence of stillbirth, perinatal death, premature delivery, and low birth weight is much less but the risk of each is known to increase with lack of education and with poverty (National Radiological Protection Board, 1994). Cigarette smoking also increases the risk of women having babies with low birth weight. Thus any comparison between women working with VDUs and those who do not must take these factors into consideration.

Most of the studies investigating the possible link between VDU exposure and adverse pregnancy outcomes have involved surveying women by postal questionnaire or by interview to obtain details about their activities in their first three months of pregnancy and the average number of hours each week they spent using a VDU. A few studies have also examined 'passive' exposure to a VDU and other variables such as stress, lifestyle and electromagnetic energy emissions. The main differences between the studies are the way the samples were obtained, the size of the samples, the length of time that had elapsed between the survey and the pregnancies, response rate and the reproductive effect examined (see Table 7.1).

Of the 14 studies examined, in only three was any association with VDU use and adverse pregnancy outcomes found, and these studies have weaknesses in their design. Lindbohm *et al.* (1992) identified 368 women who had experienced a spontaneous abortion and 1 098 controls (women who had live births) who were employed in three Finnish companies with extensive VDU use. The study showed a hint of an association between spontaneous abortion and VDU use of 20 hours or longer each week, though not significant, and an increase in spontaneous abortion among workers using VDUs with high magnetic fields which was significant. The results of this study could be largely discounted, however, because first, the response rate was very low (only 52 per cent for cases and 36 per cent for controls) and second, the women were asked to recall the extent of VDU usage as long as 10 years before the survey. Furthermore, the electromagnetic fields of the VDUs were not measured directly. Lindbohm *et al.* relied on estimates based on the women's reports of VDU use and thus are subject to recall bias. The other two studies that reported a significantly increased risk in women who used VDUs (McDonald *et al.* (1988) and Goldhaber *et al.* (1988)) were set up for other purposes.

Table 7.1 Research into adverse pregnancy and VDU use

Author	*Year*	*Data collection*	*Locality*	*Type of study*	*Reproductive effect examined*	*Results*
Bjerkedal and Egenaes	1986	1967–84	Norway	Secular trend study of pregnancy outcome in 1 820 women employed during a period of increasing VDU usage	Birth defects Still births Perinatal deaths Neonatal deaths Multiple births Prematurity	No indications that the introduction of VDUs into the working environment has led to adverse outcomes among female employees
Bramwell and Davidson	1993		England	A two-stage questionnaire sent one year apart to female civil servants aged 16–35 years. 3 711 valid responses obtained in part 1, 2 349 in part 2. VDU use and associated ergonomic and stress variables measured	Birth defects Low birth weight Spontaneous abortion	No significant association for any of the variables examined

Table 7.1 Research into adverse pregnancy and VDU use (cont.)

Author	*Year*	*Data collection*	*Locality*	*Type of study*	*Reproductive effect examined*	*Results*
Brandt and Nielsen	1990	1982–85	Denmark	A study of commercial and clerical employees; 661 cases (women who had babies with birth defects) and 2 252 controls. Postal questionnaires were used to determine VDU exposure, lifestyle, stress, and solvent exposure	Birth defects	No association of birth defects with VDU exposure or lifestyle, stress and solvent use. However, there was a correlation of the amount of VDU use and job stress
Bryant and Love	1989	1984–85	Canada	334 women were interviewed after experiencing a miscarriage. Control group size was 648, half prenatal, half postnatal	Spontaneous abortion	No associations found. Possible bias in sample, only women admitted to hospital included
Ericson and Kallén	1986a	1976–77 1980–81	Sweden	Groups drawn from Swedish censuses of 1975 and 1980, same socioeconomic strata, but with low, medium, and high occupational exposure to VDUs examined. A total of 10 025 pregnancies were investigated	Birth defects Stillbirths Neonatal deaths Low birth weight Spontaneous abortions	No associations found

Table 7.1 Research into adverse pregnancy and VDU use (cont.)

Author	*Year*	*Data collection*	*Locality*	*Type of study*	*Reproductive effect examined*	*Results*
Ericson and Kallén	1986b	1980–82	Sweden	531 cases and 1 032 controls were selected from a birth registry by job classification. Cases had abortions, birth defects, perinatal deaths, and low birth weight	Spontaneous abortions Birth defects	No associations found
Goldhaber *et al.*	1988	1981–82	California	1 583 pregnant women attending hospital were asked to recall exposure in their pregnancy an average of 2.5 years after their exposure	Spontaneous abortion Birth defects	Significant increase in spontaneous abortion in women using VDUs for >20 hrs/week in first three months of pregnancy. No significant difference in birth defects reported
Kurppa *et al.*	1985	1976–82	Finland	Interviewed 1 475 women who delivered children with birth defects and similar number of women who had normal births	Birth defects	No associations found

Table 7.1 Research into adverse pregnancy and VDU use (cont.)

Author	*Year*	*Data collection*	*Locality*	*Type of study*	*Reproductive effect examined*	*Results*
Lindbohm *et al.*	1992	1975–85	Finland	191 cases who experienced spontaneous abortions were compared to 394 controls who had live births. Particularly interested if magnetic fields caused spontaneous abortion	Spontaneous abortion	No statistical association between VDU work and spontaneous abortion. However, study showed a significant increase in spontaneous abortions among workers using VDUs with high magnetic fields (>0.9 MT in the ELF range)
McDonald *et al.*	1988	1982–84	Canada	Data obtained from 11 hospitals on 56 012 current and 48 608 past pregnancies. Substantial use of VDUs in 17 632 cases	Birth defects Spontaneous abortion	Study does not support the suggestion that work with a VDU in pregnancy increases the risk of congenital defects or spontaneous abortions

Table 7.1 Research into adverse pregnancy and VDU use (cont.)

Author	*Year*	*Data collection*	*Locality*	*Type of study*	*Reproductive effect examined*	*Results*
McDonald *et al.*	1988	1982–84	Canada	VDU usage was recorded in 4712 current and 2164 previous pregnancies in women in full-time employment at conception	Birth defects Renal defects Spontaneous abortion Stillbirth Prematurity Low birth weight	Increased risk in birth defects and renal defects among women who used VDUs more than 15 hours per week, but the authors put this down to the possibility of recall bias. Interviews conducted at a time of public concern about VDUs
Nielson and Brandt	1990	1982–85	Denmark	Postal questionnaires sent to 6312 women randomly selected from a clerical union membership list and 426 employers. Ergonomic factors, lifestyle and job stress were considered in addition to VDU usage	Spontaneous abortion Low birth weight Preterm birth Stillbirth Perinatal death Infant death	No associations found

Table 7.1 Research into adverse pregnancy and VDU use (cont.)

Author	*Year*	*Data collection*	*Locality*	*Type of study*	*Reproductive effect examined*	*Results*
Nurminen and Kruppa	1988	1972–82	Finland	1 475 mothers who experienced birth defects were divided into office workers (239) and non-office workers (805)	Threatened abortion Bleeding Length of gestation Birth weight Placental weight Spontaneous abortion	No associations found
Roman *et al.*	1992	1987–89	UK	Interviewed 150 women who had spontaneous abortions and 297 controls	Spontaneous abortion	No associations found
Schnorr *et al.*	1991	1983–86	USA	2 430 women were interviewed; 882 pregnancies met the criteria for inclusion in the study. Compared VDU users to non-users. Measured electromagnetic energy emissions from VDUs and usage	Spontaneous abortion	No excess risk among women who used VDUs in the first three months of pregnancy

Table 7.1 Research into adverse pregnancy and VDU use (cont.)

Author	*Year*	*Data collection*	*Locality*	*Type of study*	*Reproductive effect examined*	*Results*
Tikkanen *et al.*	1990	1982–84	Finland	Examined 500 cases (all registered cardiovascular malformations reported during 1982–84) and 1 055 controls	Cardiovascular malformations	No significant increase in the occurrence of cardiovascular malformations in offspring of women who used VDUs during pregnancy
Westerholm and Ericson	1987	1980–82	Sweden	4 117 full-term pregnancies of women clerks working for the national social security agency were investigated. Questionnaire used to obtain information on population obtained from census data	Birth defects Spontaneous abortions Stillbirths Perinatal deaths	No associations found

Table 7.1 Research into adverse pregnancy and VDU use (concluded)

Author	*Year*	*Data collection*	*Locality*	*Type of study*	*Reproductive effect examined*	*Results*
Windham *et al.*	1990	1986–87	California	Examined 628 cases who were county residents >20 years of age who had a spontaneous abortion before 20 weeks gestation. 1308 controls were obtained from the same source who had normal births. Data collection was via telephone interview	Spontaneous abortion Low birth weight Intrauterine growth retardation	The results were inconclusive. Call for further studies to be undertaken

McDonald *et al.*'s study was designed to investigate the general occupational associations of adverse outcomes of pregnancy. Although a slight increased risk was found for women who used VDUs for more than 15 hours a week the investigators put this down to possibility of recall bias and the fact that the interviews were conducted at a time of public concern about VDUs. Again the study had a low response rate and different methods of interviewing were used for the cases and the controls; women with live births were personally interviewed in hospital while women who had experienced spontaneous abortions were interviewed by telephone at home. Goldhaber *et al.*'s study was originally designed to examine the reproductive effects of pesticides. They identified cases of birth defects and spontaneous abortion from hospital records and then sent the women questionnaires. The researchers reported an increased risk of spontaneous abortion associated with VDU exposure for more than 20 hours per week. Unfortunately, this study also suffered from low response rate and recall bias; women had to recall VDU usage some two to four years previously.

Although the remaining studies which failed to find a link between VDU usage and adverse pregnancy outcomes were not perfect, they had the fewer shortcomings. Some of the more recent studies are discussed below. The studies by Bryant and Love (1989) and by Roman *et al.* (1992) were similar in design. Both were hospital-based case-control studies with high response rates and the interviews were conducted soon after the pregnancy so that the details of the VDU usage of the interviewees would be fresh in their minds. The cases and controls in Nielsen and Brandt's study (1990) were identified in a similar manner from hospital records but restricted to females having certain occupations in the clerical and commercial field. Although the response rate was not high, the researchers did make considerable effort to validate their results and reduce possible bias between the cases and the controls. Schnorr *et al.* (1991) studied telephone operators in the USA specifically to investigate the relation of spontaneous abortion and VDU use at work. Unlike the other studies, VDU use was obtained from company records and the electromagnetic energy emissions from different types of VDUs was measured directly. The down side of the study was that it relied on the self-reporting of spontaneous abortions some years after the event. Another approach was used by Bramwell and Davidson (1993). They questioned female civil servants in two stages, one year apart. In stage one the women were asked about VDU usage, health aspects, ergonomic factors, job stress, job satisfaction and previous pregnancies. Stage two was used to identify pregnancy outcomes of those women who had conceived or had babies since the first stage. The main advantage of this study was that it was less subject to recall bias.

Overall, the well-designed studies indicate that there is little evidence for a link between VDU usage and adverse pregnancy outcomes. Nevertheless, studies are continuing and possible links are not discounted entirely.

Although the evidence points to no association between VDU usage and adverse pregnancy outcomes, if such an effect exists there must be a mechanism. What factors could explain a link should one to be found to exist?

ALLEGED CAUSES OF REPRODUCTIVE HAZARDS

It has been proposed that the following factors may be hazardous to pregnant women:

- *Radiation and other emissions*:
  - X-ray radiation;
  - ultraviolet radiation (combined with an excess of vitamin A);
  - microwave and infrared radiation;
  - low frequency electromagnetic fields;
  - electrostatic fields in combination with light air ions;
  - ultrasound;
  - polychlorinated biphenyls (PCBs).
- *Other factors*:
  - stress and worry;
  - sedentary work, postural problems.

RADIATION AND OTHER EMISSIONS

Radiation is a beam of energy. Electromagnetic radiation is electrical energy in two parts, electric and magnetic, travelling in the form of waves. These waves differ in frequency (the number of times they go up and down every second). They also vary in wavelength (the distance from the peak of one wave to the peak of the next). Frequency and wavelength are related; the shorter the wavelength, the higher the frequency. Frequency is usually measured in Hz or cycles per second (see Figure 7.1).

Thus radiation in its various forms stretches along the spectrum. At one end are long waves with low frequencies and at the other, short waves with high frequencies. The higher the frequency, the more powerful the radiation (see Figure 7.2).

In the low-frequency end of the spectrum, the two parts of the wave – the electric field and the magnetic field – begin to separate. Different sections of the radiation spectrum have been highlighted from time to time as being a possible cause of birth abnormalities. The various types of radiation and their alleged effects are discussed below. The first to be accused was X-ray radiation.

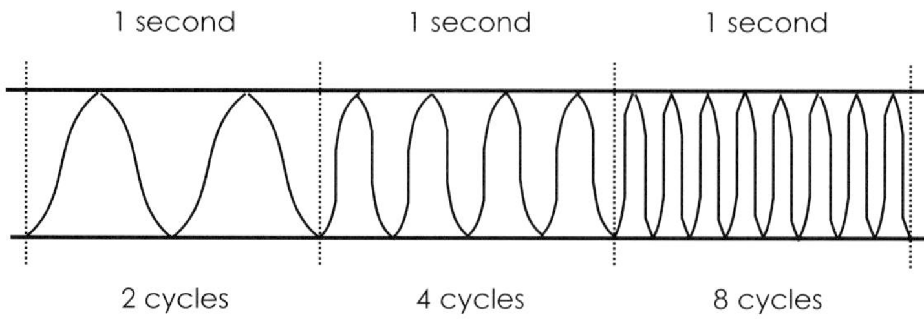

Figure 7.1 *Radiation patterns, indicating the inverse relationship between frequency and wavelength. Adapted from Apex, 1985*

X-rays

- *Where do the X-rays come from?* No gamma or hard X-rays are produced by VDUs. Soft X-rays are generated inside the cathode ray tube as the electrons from the gun (Chapter 5) are slowed down by the screen phosphor. The glass of many tubes has a high lead content, which blocks the passage of X-rays, and, in addition, the number of X-rays produced may also be reduced by correct circuit design.
- *What harm could they cause?* The effects of X-rays are cumulative and irreversible, so it is the total level of absorption rather than the level of emission which needs to be considered. If a person absorbs more than a certain amount, X-rays can cause cancer, genetic damage

<table>
<tr><td colspan="7">NON-IONIZING RADIATION</td><td colspan="3">IONIZING RADIATION</td></tr>
<tr><td>extremely low frequencies</td><td>very low frequencies</td><td>radio frequencies</td><td>micro-waves</td><td>infra red</td><td>visible light</td><td>ultra violet</td><td>soft X-rays</td><td>hard X-rays</td><td>gamma rays</td></tr>
<tr><td>e.g. high voltage lines, food mixers</td><td>e.g. navigation equipment</td><td>e.g. radio, CB, TV</td><td>e.g. micro-wave ovens, radar</td><td>e.g. sun, heat-lamp</td><td>e.g. sun, candle light</td><td>e.g. sun, strip-light</td><td>e.g. TV, medical X-ray</td><td colspan="2">e.g. nuclear fall-out</td></tr>
<tr><td colspan="8">may be emitted by VDUs</td><td colspan="2"></td></tr>
<tr><td colspan="4">possible reproductive hazards</td><td colspan="3"></td><td colspan="3">reproductive hazards</td></tr>
<tr><td colspan="3"></td><td colspan="2">cataracts</td><td colspan="5"></td></tr>
<tr><td colspan="6"></td><td>skin cancer</td><td colspan="3">many cancers</td></tr>
<tr><td colspan="4">possible general effects on health</td><td colspan="3"></td><td colspan="3">severe general health effects</td></tr>
<tr><td colspan="8"></td><td colspan="2">death</td></tr>
</table>

Figure 7.2 *Types of radiation and alleged hazards. Source: London Hazards Centre, 1993*

and other problems. They are also known to affect cell growth, and could therefore cause abnormal cell reproduction and harm embryos.

- *What are the safety standards?* In order to allow for the cumulative effect of X-rays, the measurements include an allowance for quantity and time (milliroentgens per hour, or mR/h). Distance from the source of an X-ray also makes a difference – for example, 100 mR/h at one foot would be reduced to 25 mR/h at two feet (Apex, 1985). Based on an assumption of a five-day, 40-hour week, official regulations for occupational safety in Europe and the United States set a maximum exposure of 2.5 mR/h at a distance of 40 cm from the screen (Grandjean, 1987).
- *Do VDUs emit X-rays?* X-rays were originally a cause for concern because some computer equipment from the 1950s was found to emit this type of radiation. Measurements of large numbers of more modern CRT equipment in use, however, have indicated that such emissions no longer occur (Doerfel *et al.*, 1986), although recent cases of radiation leakages have been recorded under test conditions (Apex, 1985). Bergqvist (1986) argues that the test methods themselves often include creating a faulty condition in order to examine the CRT when faulty as well as when operating correctly. He argues that X-rays emitted under these conditions are not a fair assessment of the VDU in a normal operating environment.

Grandjean (1987, p.195) cites the conclusions of the US National Academy of Sciences following a number of studies:

> Taken collectively, these studies have examined a wide variety of models and hundreds of terminals. Measurements of emissions from older and newer VDT [VDU] models have not differed significantly. Measurements have been made under normal operating conditions and under conditions designed to maximise potential emissions . . . These studies have concluded that the levels of all types of electromagnetic radiation emitted are below existing occupational and environmental health and safety standard limits of exposure . . . The level of X-ray radiation emitted by VDTs is far less than the ambient background level of ionizing radiation from natural sources to which the general population is exposed.

The last part was borne out when one of the authors' VDUs was given a routine check. It was a concrete pillar in the office, *not* the VDU, that caused the meter to react! From the studies undertaken so far, it is likely that VDUs do not emit X-rays and, if X-rays are emitted, these fall within the safety standards. The report from the National Radiological Protection Board (1994) puts this succinctly: 'The exposure arising from VDU x-ray emission is many thousands of times below normal levels of ionising radiation.'

- *How would I know if my VDU was leaking X-rays?* It is possible to arrange for equipment to be tested periodically for unwanted emissions; a health and safety officer can advise. But perhaps you need look no further than the screen itself. VDU screens are prone to static but this would be destroyed by X-rays if they were present. Dust on the screen is a good indicator of static so, if it is not present prior to cleaning, it may be wise to have your machine tested.

Ultraviolet radiation

Large amounts of ultraviolet radiation are emitted by the sun, most of which is absorbed into the atmosphere. What remains is beneficial to life on Earth, helping growth and producing vitamin D in the body. Excessive doses can, however, lead to sunburn and may cause cancer.

It was suggested by Mawson in 1985 that a combination of ultraviolet radiation from VDUs and an excess of Vitamin A, could be the cause of a large number of adverse effects from VDUs, including adverse pregnancy outcome (Bergqvist, 1986).

The amount of ultraviolet radiation emitted by VDUs is, however, negligible compared with that from fluorescent lights, and sunshine both outdoors and through glass. Therefore, if ill effects were to be caused by a combination of ultraviolet radiation and vitamin A, it is likely that a source of ultraviolet radiation other than VDUs is to blame.

Microwave and infrared radiation

These non-ionizing rays can cause heat to be generated in the substances absorbed by them, hence the effectiveness of microwave ovens, and infrared heating appliances such as electric blankets.

Concern exists about the heating effects of these rays on the body. This has been caused by reports of sterility and sexual problems among men who have been exposed to microwaves. The testicles are particularly sensitive to heat, having inadequate heat regulation themselves as a result of poor blood flow in and around the organs. The levels of microwave and/or infrared radiation emitted from VDUs, however, are not high enough to cause such effects. Measurements of infrared radiation in the environment of a working VDU are less than 1 per cent of outdoor and indoor levels and microwave levels are one thousandth to one billionth the level of the strictest safety regulations (Grandjean, 1987).

Low frequency electromagnetic fields

The radio frequency band is divided into bands ranging from ultra high frequency (UHF) to extremely low frequency (ELF). The high-frequency range can produce heating effects similar to the microwave and infrared frequencies, but only in excessive quantities. As yet, little is known about the effects of very low and ELF radiation.

Low frequency radiation is emitted from the flyback transformer in a VDU. This controls the electron beam, sending it back to the top of the screen, approximately 100 times per second. The result of the activity is the emission of ELF radiation. Manufacturers, however, usually shield these emissions, resulting in only minute quantities being present. Indeed, the level of radio frequency radiation in the vicinity of a VDU is comparable to the ambient levels generated by radio transmitters in urban areas. These levels are some 2 000 times lower than those permitted by the strictest safety regulations (Grandjean, 1987).

The concern about low-frequency pulsed magnetic fields was generated by a group of Spanish researchers who reported that chick embryos had been deformed by exposure to such fields. Faults were found with the researchers' methodology, however, and similar experiments have not produced the same results (Mild, 1986). Brent *et al.* (1993) review the literature in this field and conclude that 'there does not appear to be a measurable risk of reproductive failure and birth defects from electromagnetic field exposures in humans'. Nevertheless, the debate persists.

Electrostatic fields and light air ions

An electrostatic field is caused when ions flow from a negatively charged body to a positively charged body or vice versa. In the case of the VDU and its operator, the VDU is positively charged – from between zero to 10 000 volts (Jeffreys, 1984) – and the human body is also charged, this time negatively (up to over 1 000 volts in some cases). The electrostatic fields which may occur in front of VDU screens may vary according to different makes and the distance of the operator from the screen (the strength of the field diminishing as the distance increases). A majority of measurements have shown electrostatic fields around VDUs in the order of 0 to 20 kilovolts per metre (kV/m), with a figure of 2.4 kV/m at normal working distance. The combination of an operator and a VDU can generate fields of between 0 and 15 kV/m (Grandjean, 1987).

Apart from an occasional mild electric shock, the presence of static electricity at these levels is harmless. Concern has been expressed, however, about the effect of static on ions in the air, causing a decrease and/or an imbalance. This, in turn, has been suggested as causing stress reactions which may interfere with pregnancy.

What impact does exposure to excessive positive ions, ion deprivation and increased deposition of airborne particles have? Jeffreys (1984) refers to two studies: one, in 1972, examined stress and irritability among air traffic controllers and pointed to the problems of an imbalance in atmospheric ions; the second, in 1976, anticipated a number of medical symptoms which would be experienced by one-third of the population if negative ions were withdrawn from the atmosphere. These included: irritability and anxiety, headaches, neck/back pain, asthma, apathy, depression, nausea, fatigue, allergies and anorexia. These symptoms were found to occur more frequently among women rather than men, and among older rather than younger, people. However, Bergqvist (1986) argues that there are no facts to support the assumption that static causes an imbalance in ion concentrations or that any such change can have adverse effects.

Ultrasound

The claim that ultrasound emission from VDUs could cause adverse pregnancy outcomes is not supported by any facts according to Bergqvist (1986).

Polychlorinated biphenyls (PCBs)

PCBs are used as an insulating fluid in some VDUs, and the chemicals have been linked to the occurrence of abnormal foetuses in sea birds and sea mammals. It has been suggested by one study that VDUs leak such chemicals, but this has been contradicted by other studies (Bergqvist, 1986).

OTHER FACTORS

It should, perhaps, be noted that many of the arguments listed above are kept alive by the media. This may be due to the newsworthiness of a possible link between VDUs and adverse reproductive outcomes; more likely, but less newsworthy, causes are stress and worry, and sedentary work.

Stress and worry

Stress may occur in an automated workplace for a number of reasons (see 'The causes of stress', Chapter 6), in addition to which may be added the fear of adverse pregnancy outcomes. The possible link between stress and spontaneous abortion was first mooted some time ago and apparently remains unsolved. To compound the issue, as stress increases so, too, does smoking and drinking, each with known harmful effects on foetuses.

Sedentary work

MacKay (1987) suggests the possibility that sedentary work may adversely affect the pregnancy outcome, although this has (to the authors' knowledge) not yet been confirmed or repudiated. It is possible, however, that extended periods of poor posture might reduce the blood flow to the foetus. Also, prolonged sitting is known to cause back pain and regardless of whether this contributes to pregnancy problems, comfortable well-designed workstations should be provided for all employees, whether pregnant or not.

MINIMIZING RISK TO PREGNANT (VDU) WORKERS

The recent epidemiological evidence has not demonstrated an association between VDU usage and adverse pregnancy outcomes. Surveys show that radiation levels emitted from VDUs are well below recommended maximum continuous exposure levels (see Table 7.2) and are thus unlikely to be the cause of any association should it exist. The problem of stress and poor posture is difficult to assess. It is possible both may have an effect on an unborn child, but the nature and implications of this linked to VDU use are, as yet, ill-defined. Until they are resolved, managers should take all precautions to minimize any possible risk to pregnant VDU workers.

By law, employers have no special duty for their employees during pregnancy (given the current state of knowledge), nor do they have any special duty to protect the foetus. If, under the Congenital Disabilities (Civil Liability) Act 1976, they were to have special duty, the employers would possibly be liable to a child born defective because of their negligence.

Thus, although, at present, this is not a problem because no proof exists that employees are at risk, if this changes employers could become liable (Suter, 1985). However, lack of proof is no justification for lack of sympathy. The case of the dismissed library assistant, Hazel Johnstone, hit the headlines in both the regional newspapers and nationally in the library press. The Highland Regional Council dismissed Mrs Johnstone for refusing to operate a VDU while she was pregnant. An industrial tribunal was critical of the unsympathetic handling of the case by the Council and ruled it as unfair dismissal. It was considered unfair, because the journal literature contained (and still does) conflicting reports on the alleged health risks. Suter (1985) makes three points concerning this case. The first is that there were other staff available to take over Hazel Johnstone's duties during her pregnancy; the second, that she operated the VDU for only two hours per day, thus making it relatively easy to find alternative duties; and third, that, as with all tribunal rulings, no precedent can be taken from this case. A second case, however, in which Lesley Bradley was awarded £8 000 compensation by an industrial tribunal (VDU Award, 1986), might indicate that, despite the absence of precedents, tribunals look sympathetically on such cases.

It is suggested, therefore, that employers should, where possible, be sympathetic to employees' fears, although it is recognized that not all

Table 7.2 Permitted radiation levels

	Regulations (units)	*Maximum levels reaching VDU operators*
X-rays	2.5 $MSvh^{-1}$	0.1 $MSvh^{-1}$
Ultraviolet	10 Wm^{-2}	0.124 Wm^{-2}
Visible	100 Wm^{-2} sr^{-1}	3 Wm^{-2} sr^{-1}
Infra-red	100 Wm^{-2}	4 Wm^{-2}
Microwaves and radio frequencies	60 Vm^{-1}	0.5 Vm^{-1}
VLF electric fields	20 Vm^{-1}	10 Vm^{-1}
VLF magnetic fields	38 Am^{-1}	0.17 Am^{-1}
ELF electric fields	20 Vm^{-1}	10 Vm^{-1}
ELF magnetic fields	38 Am^{-1}	0.6 Am^{-1}
Static electric fields	?	300 Vm^{-1}

Note: The Regulations quoted are either European, WHO or American; the lowest permitted amount for continuous exposure has been recorded. The quoted maximum levels of emissions reaching VDU operators have been obtained from numerous studies, cited in Blackwell and Chang (1988) and Kavet and Tell (1991).

situations lend themselves easily to finding alternative work for VDU operators. Nevertheless, it is desirable that a pregnant employee should, if they wish, be moved to suitable alternative employment during the course of their pregnancy, with no loss of pay or seniority. Indeed, laws to this effect have been passed in Canada and Ireland, and unions, such as the Council of Civil Service Unions, are arguing for it to become law in the UK. However, the employer should not take all the responsibility. Braidwood (1988, p. 35) cites Ross Edwards, an electronic engineer studying the biological effects of electromagnetic radiation, who argues that there is little point in 'coming off a VDU at work unless you are also prepared to review your home environment, because so much there – the radio, the TV and the microwave, for instance – could also be having an effect'.

CHAPTER

8

OTHER ALLEGED HEALTH HAZARDS

CATARACTS

INTRODUCTION

Another risk to health allegedly caused by VDUs is cataracts. Cataracts are opacities which occur in the lens of the eye: they can cloud the vision and, in some cases, even cause blindness.

Cataracts may occur spontaneously, especially in people above the age of 60, although it is said that approximately 25 per cent of people have lens opacities which do not affect vision (Grandjean, 1987). Thus the disease is fairly common and by no means always a threat to sight. Indeed, cataracts which impair vision are rare, affecting only approximately 7 per cent of the population.

It is considered likely that 20–30 per cent of VDU operators aged between 20 and 50 and 40–50 per cent of those over 50 will in the normal course of events develop cataracts but, of these, only a small proportion will suffer visual defects.

THE CAUSES OF CATARACTS

The causes of most cataracts, especially in the younger age groups, are not well known. They may be present at birth or may form in early childhood, linked perhaps with the occurrence of an intra-uterine infection or genetic metabolic deformities. Later in life, the causes may be due to trauma, metabolic and degenerative disorders or exposure to an agent such as high steroid intake, previous eye surgery or radiation (Hamdy, 1987).

Electromagnetic radiation at very high microwave frequencies ($>1\ 000\ Wm^{-2}$) has also been found to cause cataracts in animals and possibly humans (National Radiological Protection Board, 1994). Undoubtedly, the damage is caused by the over-heating of the eye lens. These types of cataracts, which form as a result of radiation, can usually be distinguished from spontaneously occurring opacities. Whether high exposure at much lower intensities can produce these types of cataracts has been the subject of controversy.

The possible link between cataracts and the use of VDUs was first raised by Zaret in 1984. He described 10 cases of VDU operators who developed cataracts and implied that they were caused by microwave emission. Little credence is now attached to this investigation, however. Four of the cases had cataracts or related eye problems before they started work with VDUs, or had a previous illness or had been exposed to X-ray treatment which could account for the disease. The other six had minor lens opacities, not cataracts, which did not affect their vision and could have occurred naturally.

All other studies have failed to substantiate Zaret's claim. The cataracts that Zaret examined appeared to have been caused by radiation, but as the radiation emissions from VDUs are well below those of the surrounding environment, it would seem unlikely that the VDU would be the cause in all but the most extreme of cases. Grandjean (1987, p. 201) puts it more strongly: 'There is therefore not the slightest reason to suspect that VDTs [VDUs] have a cataractogenic effect.' This is further supported by the National Radiological Protection Board (1994) which states that:

> There is no evidence that work with VDUs results in a predisposition to the formation of cataract . . . Minor opacities of the eye lens that do not affect visual acuity are to be expected among a large fraction of the population whether or not they use VDUs.

FACIAL RASHES

INTRODUCTION

During the early 1980s there were various anecdotal reports in the literature of people becoming 'sunburned' at work or experiencing a facial rash while working with VDUs. The rashes usually disappeared a few hours after finishing work. It was generally believed, at that time, that facial rashes were caused by electrostatic fields produced by VDUs.

POSSIBLE CAUSES OF FACIAL RASHES

As stated in Chapter 7, an electrostatic field may surround a VDU and its operator. This can cause a strong attraction of negative ions in the atmosphere towards the VDU screen and, at the same time, a propelling of positive ions towards the operator's face, and then out of the immediate vicinity. Initially, therefore, there is an increase in the positive ion concentration inhaled by the operator and also in minute airborne particles which hit the operator's face at a faster than normal pace. The transfer of ions can be further accelerated if surrounding objects are also highly charged, for example synthetic carpeting or if the relative humidity is low, that is below 20 per cent.

Several epidemiological studies have investigated the suggestion that electrostatic fields from VDUs cause skin reactions. The general consensus now is that facial rashes do not appear to be caused by the electromagnetic fields from VDUs (National Radiological Protection Board, 1994). Despite this, there does appear anecdotal evidence that in conditions of low humidity the associated electromagnetic fields may aggravate existing skin conditions and that psycho-sociological factors concerned with stress at work may add to the problems.

STATIC ELECTRIC SHOCKS

The presence of a field of static electricity around the VDU and the operator can result in mild electric shocks being received when the operator comes into contact with an earthing device such as the computer or a metal filing cabinet. While not dangerous, such an experience can be unpleasant and unnerving. The methods of reducing static listed below, if implemented, will reduce or eliminate the occurrence of static electric shocks:

- The problem may be in the type of carpeting, so a change of carpet to a low static one can help, although it is very costly. A cheaper alternative may be to use a rubberized anti-static mat. These can be purchased in a range of shapes and sizes and are designed to fit under the VDU and keyboard or under the chair and the desk. Spraying the carpet with a static-reducing chemical or even with fabric conditioner which has static-reducing properties may also help.
- Low air humidity can also encourage a build up of static. This can be rectified in various ways (see 'Relative humidity', Chapter 13, p. 166).
- Anti-static sprays for use on the display screen can be used, but this can become expensive over a long period.
- Earthed screens which are placed in front of the VDU screen can reduce static and glare.
- It may be helpful to rearrange the workstation so that the operator is not directly facing the VDU (Rycroft and Calnan, 1984); in some instances, improved workstation design may occur by positioning the VDU screen slightly to one side (see Chapter 14).
- Another cause of static may be the operator. Clothes, particularly

underwear, are often made from synthetic fabrics which can encourage a build-up of static. A change in the type of clothes or underwear worn may, therefore, cause less static. It is left to the reader's discretion to decide how to phrase this delicately to staff!

PHOTOSENSITIVE EPILEPSY

Epilepsy is an illness which, thankfully, is experienced by only a small proportion (0.5 per cent) of the population (Newmark and Penry, 1979). Of these, only between 1 and 3 per cent are subject to photosensitive epilepsy – that which is triggered by a flickering light or a particular kind of pattern.

The use of VDUs at work, with their pattern of lines, is a possible trigger for such an attack. An estimated 50 per cent of photosensitive epileptics, however, experience their first seizure while watching television (Cakir *et al.*, 1980). This event is likely to have occurred in the 10–14 year age group which is typical for the onset of the illness. A person is therefore likely to be aware of their illness prior to starting work. Anyone who is unsure should seek medical advice, and known photosensitive epileptics should avoid VDU work. The Employers Medical Advisory Service (see Useful WWW Sites and Addresses) can offer help to photosensitive epileptics.

CABLING

Loose cabling can be hazardous and cause accidents. It should be hidden to avoid tangling with the feet of the operator or of passing staff and users.

FIRE

Any electrical equipment could short-circuit and start a fire and a computer is no exception. Fire extinguishers (carbon dioxide or powder variety – not foam) or fire blankets should be near at hand and staff should know how to use them. Large computer installations may be situated in a single, easily sealed room, in which case a fire prevention system, such as halon gas, can be used.

LASER PRINTERS

Laser printers have also been rumoured to be a health hazard, reports in the literature claiming that the ozone released can cause breathing problems and blindness but, as yet, there is no evidence to support this.

RECOMMENDATIONS

Much space has been given to the discussion of possible health hazards arising from working with VDUs because there are many underlying fears and uncertainties concerning them. If the possible hazards and the solutions are more clearly understood, efforts to reassure staff may be more effective. It is hoped that some of the preceding discussion will offer a basis from which further research may be undertaken. In this field, as in many others, discoveries are continually being made and it is necessary to keep abreast of developments.

It must be remembered that many positive effects will also result from automation. Jobs may be enhanced, reducing stress and some health risks may even be eliminated! The three-part issue system, for example, often used carbonless paper which has been reported to cause headaches, eye, skin and respiratory irritation (Hedge, 1987). It is thus important to recognize that automation is not necessarily the bringer of health problems but that it, like many other aspects of modern life, has its own problems which need to be recognized in order to be overcome.

Many of the complaints, especially the postural problems, have arisen from intensive VDU work. At the present time, much library use of VDUs is not intensive, with the obvious exceptions of data input and word processing. Other uses tend to be interactive and offer frequent breaks. Nevertheless, unless steps are taken to prevent them, health problems do still occur and will do with increasing pressures on staff. It is equally important, however, that in reacting to the problems, matters are not made worse by encouraging a change which is inappropriate or by causing unnecessary alarm and concern (Scalet, 1987). For example, some have argued that pregnant operators should automatically be transferred from VDU work. This may mean moving an employee working at the issue counter to photocopying or shelving jobs which require a lot of standing and possibly lifting of heavy items; both tasks pose a greater known risk to the unborn child than the VDU. Fears may also be aroused rather than laid to rest if action is taken to shield the operator or VDU. The logical conclusion here will be that there is something which needs shielding.

In conclusion, a variety of recommendations may be made to help reduce the real or imagined health problems:

- Recognize that managers have a legal obligation to take 'reasonable care' for the safety of their employees and other obligations as prescribed in the Health and Safety (Display Screen Equipment) Regulations (1992) (HSE, 1992d).
- Listen to staff complaints and fears sympathetically and do what is required to boost morale or reduce fear, but remember that complaints about fear of health problems may mask other, deeper anxieties.
- Develop good workstation design and ensure that staff know how to rearrange the furniture to suit themselves.
- Ensure good job design.

PART

III

WORKPLACE DESIGN: WORKSTATION CONSIDERATIONS

CHAPTER 9

BASIC CONSIDERATIONS

INTRODUCTION

Library staff are one of the most expensive and complex components of a working library system, and it is therefore important that they are given an environment which maximizes their ability to work. This means, in effect, ensuring that the workplace is designed to maximize their comfort, safety and well-being. As a result, productivity is also likely to increase. Studies reported by Dressel and Francis (1987), for example, show that 'ergonomically designed' furniture can increase productivity by approximately 20 per cent.

Throughout Part III, the terms 'workplace' and 'workstation' will be used. 'Workplace' refers to the complete working environment including the heating, lighting and decoration, while 'workstation' is used to describe the more limited concept of the furniture and other components which together make a unit for one or more operators to perform a particular task.

In addition to increasing productivity, good workplace design can reduce, or even eliminate, many of the health complaints listed in Part II. Halpern and Davis (1993) found that the redesign of workstations to fit each user's anthropometric dimensions resulted in a significant improvement in worker health, an increase in the perceived efficiency and enhanced usability of the office furniture and equipment. They concluded (p. 821) that the 'use of office equipment properly adjusted to fit the user through ergonomic and anthropometric principles may enhance productivity and performance, reduce awkward postures, and the associated cumulative trauma injuries'.

The importance of ergonomics has been filtering through the library literature but, sadly, has been slow to penetrate many libraries. The probable reasons for this are: lack of money, lack of expertise to apply ergonomics, and the plethora of old library buildings – over a quarter of libraries in the UK were built before 1940 (Morris and Dennison, 1995). Certainly, it is true that the initial outlay on a new or upgraded computer system is difficult for many libraries to obtain and, faced with the recurrent costs of maintaining the system, the idea of extra expenditure on compatible furniture and adapting old buildings is often rejected. There is also, it seems, a belief among some managers that people are infinitely adaptable and able to adjust to work with a new system; with other managers there may simply be a lack of awareness of what to do to provide an ergonomically sound environment.

Whatever the reasons, many examples of poor workstation design can be found throughout the library and information world. Morris (1992, p. 214) performed a small survey of serial librarians and found that:

> less than half of the group had adequate chairs or tables, fewer than 17 per cent were provided with footrests or document holders, less than half were provided with a screen cleaning kit which was used regularly, over half complained about the presence of trailing cables and other trip hazards, noise and their office temperature. Few of the employers provided eye tests or information on health and safety aspects and only 9 per cent trained their employees on how to adjust equipment. Only a quarter of employers assessed the health and safety aspects of all workstations and less than 40 per cent ensured that workers using VDTs had adequate breaks or allowed users to participate in the planning, design and implementation of work tasks.

Fortunately, the Display Screen Equipment Regulation (HSE, 1992d) and similar laws in other countries, have brought ergonomics to the fore. Managers are now being forced to consider workplace design. The following chapters outline the factors which need to be considered. This chapter discusses some of the basic considerations which need to be examined such as users of the workstation, the tasks to be undertaken, and posture.

USERS OF THE WORKSTATION

Ideally, the workstation should be designed to suit the person who works there. This is relatively straightforward with a single-user workstation but, where several people share a common work area, it is more difficult. People vary in their physical characteristics (see Chapter 1) and in their opinion about what is comfortable. Slovak and Trevers (1988) found, for example, that people such as designers, programmers and scientists who used the VDU for 10–50 per cent of their working day could tolerate ergonomic and postural problems more easily than could secretaries and accounts personnel who spent 80–100 per cent of their time performing VDU work. They suggested two possible reasons for this. In part, the increased tolerance may have been due to the relatively small amount of time spent at the VDU, but it may also have been affected by the attitude of the staff towards their job. It was suggested that those with more creative work, such as designers and programmers, had higher job satisfaction and used the computer in a creative rather than routine fashion. The happier people were with their working life, the more likely they were to tolerate poor workstation design and its attendant problems. The more dissatisfied they were overall, the less happy they would be with inadequately designed workstations. Because people vary in size, shape and attitude, shared workstations can create particular problems. A shared workstation should therefore be adjustable, especially if a range of tasks is to be performed.

THE TASK(S) TO BE PERFORMED

Libraries vary in size from 'one-man bands' to large academic institutions and thus the tasks performed are equally varied, ranging from one-man operation, where the employee is expected to be 'jack of all trades', to a large organization where job specialization is the norm. The effect of certain tasks on workstation design is considered further in Part V.

SITTING POSTURE

Incorrect posture, as outlined in Chapter 3, is one of the factors which will induce backache, headaches, pains in the neck and shoulders, digestive and circulatory problems. Sitting has several advantages (Grandjean, 1980):

- it takes the weight off the legs;
- it helps to avoid unnatural body postures;
- it reduces energy consumption;
- it places fewer demands on the blood system.

Nevertheless, it brings with it a set of problems:

- it causes a slackening of the abdominal muscles ('sedentary tummy');
- it leads to curvature of the spine which in turn is bad for the digestive organs and lungs.

The most severe problem, however, is for the back. The spine is formed by separate bones (vertebrae) which are linked together. Between each vertebra and the next is a kind of cushion or disc. These discs can gradually become worn away or flattened, or the viscous fluid, which causes the cushioning effect, may be squeezed out. This, in turn, can cause the tissues and nerves to be trapped or strained and can lead to various pelvic disorders, and lumbago. Bad seating can accelerate deterioration of the discs by increasing the pressure on them. The following relative pressures have been recorded (Grandjean, 1980):

Posture	*Disc pressure (%)*
Standing erect	100
Lying flat on the back	24
Sitting, trunk erect	140
Sitting, bent forwards	190

which indicate that the pressure on the discs is greater when sitting than when standing.

The ideal position for the back is one which allows the spine to adopt a gentle 'S' shape, which is the norm when we are standing. What happens when we sit slumped is that the small of the back curves outwards rather than inwards thus increasing the pressure on the discs. However, in order to sit up straight, muscular effort is involved, which can be tiring over long periods. Thus the ideal posture is one which enables the back to retain its 'S' shape, but does not require muscular tension to retain it.

The pressure on the discs is also increased if weights are being lifted. Cataloguing, acquisitions and issue desk work require frequent lifting of piles of books. This can compound the pressure on discs if an incorrect posture is adopted. Yates and Karwowski (1987) found that men could lift greater weights when standing than when sitting and, whilst seated, could lift a greater weight in a forward position than when twisting the trunk through 90° to lift something from the side. Therefore care needs to be exercised when designing a workstation to ensure that a combination of simultaneous lifting and twisting is avoided.

It is not just lifting or twisting which affects the pressure on the discs, however. The sitting posture itself can have an effect, and there is much debate in the literature on what is deemed to be the ideal posture. There are currently three main schools of thought:

- those who favour the 'traditional' upright posture, with hips and knees at right angles;
- those who argue for a forward-tilting chair, extending the angle between the spine and the thigh;
- those who prefer a backward-leaning posture.

UPRIGHT POSTURE

The traditional view of the VDU worker's posture is of an erect back, horizontal thighs, and knees at right angles. The elbows should be bent at

90° and the forearms should run parallel to the thighs, a short distance above the work surface. However, studies have shown (Tougas and Nordin, 1987) that only 10–20 per cent sit in the recommended upright position.

FORWARD-TILTING POSTURE

Naqvi (1994) argues that the seat should not be horizontal, but should slope forward 15°. This means that the top part of the body can lean forward to perform tasks, such as keyboarding or writing, while, at the same time, the downward slope of the thighs retains the natural curve of the spine and prevents cramping of the internal organs. Unfortunately, if too steep, the downwards slope of the seat can cause the body to slide, enforcing the operator to hold still by pushing with their feet. If the chair is on castors, this pushing action can cause the chair to move away from the desk. Despite this, there does seem to be support for office chairs with an optional forward tilt (about 5°) controllable by the user (Pheasant, 1991).

An alternative is the kneeling (or 'Balans') chair, which has a forward-tilting seat and a kneeler on which to rest the knees and take the weight of the body (see Figure 9.1). Anecdotal evidence, however, indicates a split of opinion between those who find the kneeling chair more comfortable than the traditional office chair and those who find that, after a period, their leg muscles tire and their bodies begin to slump. Pheasant (1991) states that there may be a minority of people with low back problems who find knee-support chairs extremely beneficial but suggests that the disadvantages outweigh the advantages for most people. He concluded that the main disadvantages of the kneeling chair were: the absence of a back rest which can cause muscle fatigue and slumping; the fixed position of the lower limbs which restricts movement; and the difficulty of standing-up and sitting down from and to a kneeling position. In addition, Brunswic, as reported by Pheasant (1991, p. 227), found that:

- When seat angle and knee angle were independently varied, forward-tilted seat position did not result in a lumbar posture which was significantly different from that obtained with a horizontal seat and the knees at right angles.
- The lumbar posture of subjects using a kneeling chair in writing and typing tasks was not significantly better than when they used a conventional office chair.

He concluded that 'the kneeling chair does not offer any measurable advantage over well-designed and correctly adjusted chairs of the conventional type'.

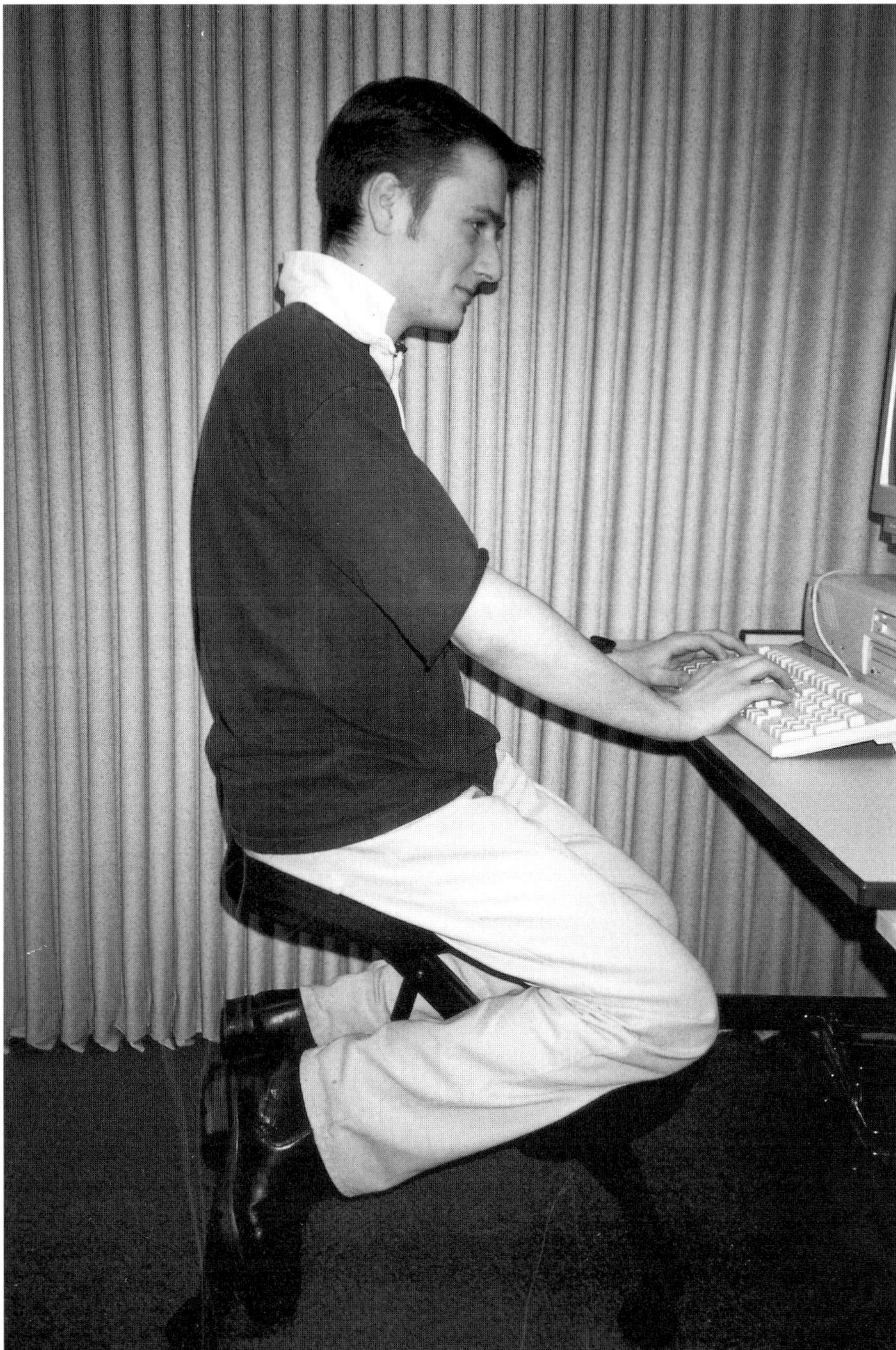

Figure 9.1 The kneeling chair

BACKWARD-LEANING POSTURE

Research has shown that a backward-leaning posture not only reduces the disc pressure, but is one in which the back muscles are more relaxed (Kayis and Hoang, 1995). An angle of between 110° to 130° between the seat and the back has been found to be both comfortable and relaxing. Certainly, the backward-leaning posture is one which is adopted by several operators, even though the chair is not designed to support them.

It has also been found that a lumbar pad – that is, a support in the small of the back – can help to reduce disc pressure. When this pad is used in combination with a backward-sloping angle of between 110° and 120°, disc pressure is lower than when standing (Grandjean, 1987). However, Mandal (1986) argues that the backward-sloping position required greater bending of the neck which could cause pain in the neck over time.

THE IDEAL POSTURE?

The traditional upright posture has been selected as the basis for the workstation designs which follow, although some reference is made to the backward-leaning posture. This is not intended as a recommendation of either posture but is to form a working base until the experts reach a conclusion. This position cannot, however, be assumed to be 'ideal' (Cakir *et al.*, 1980, p. 153):

> no matter how well the ergonomic criteria for VDT [VDU] workplace design are satisfied in practice this, in itself, is no guarantee against postural discomfort. If the job itself required the individual to maintain a static position or permits only a few essential movements, some degree of postural discomfort is still to be expected.

In other words, to design a workplace which forces an individual to maintain a so-called optimal posture *cannot* be optimal.

Fleischer *et al.* (1987), in a study looking simply at the weight distribution of 16 seated subjects during an experiment, concluded that, even in this limited study where operators performed identical tasks, the sitting behaviour of each of the operators was different. Given a range of ages and tasks, it is even more likely that operators will exhibit a range of preferred postures: leaning back, crossed legs, one foot tucked under the body, bent forward with feet crossed behind and so on. Thus, the optimal workplace is affected by two factors: the need to be able to adjust the workstation to suit all the staff who will use it and the need for those operators to express their individual preferences in sitting behaviour and still have freedom of movement.

CHAPTER 10

WORKSTATION COMPONENTS

In order to attain a comfortable and efficient workstation, detailed consideration needs to be given to the components which make up the workstation and to how these may be assembled. The following items together form a workstation for VDU work: chair(s); desk(s); VDU; keyboard; mouse and/or touchscreen; printer; and document holder and, in some cases, a barcode reader. For details of copy or document holders see Chapter 14; the other items are described below.

CHAIRS

The chairs should be able to support the person's body in a position which minimizes disc pressure and muscular tension. As explained earlier, for the purposes of this book, the upright posture with the knee and hip joints at right angles has been chosen. This position may be attained using the conventional typist's chair, although high-backrest and tilting chairs are now available. Investment in new office furniture may be required, as few libraries already have such chairs. Indeed, in some, even traditional typist's chairs are not available.

Grandjean (1987) performed studies with office workers using the traditional typist's chair and a high-backrest chair. The latter was favoured by 75 per cent of the group studied and 89 per cent of the group favoured a high-backrest chair which tilted backwards.

RECOMMENDED FEATURES FOR CHAIRS

There are a number of recommended features for chairs. These include the following:

- The chair should be adjustable in height between 380 and 540 mm to cater for different operators.
- Chairs must be able to support forward, upright and backward-leaning positions. This means that the backrest must be adjustable

in angle. It must also be possible to lock the chair at the required angle to ensure support in the forward-leaning and upright positions.

- The backrest should be adjustable with a minimum height of 480–500 mm (between the surface of the chair and the top of the backrest) and between 320 and 360 mm wide. Ideally, it should have a lumbar pad at a height of 100–200 mm above the lowest point of the seat surface, and should be concave towards the top of the backrest to accommodate the outward curve of the spine by the thorax. The whole backrest should be concave in the vertical plane, and should also be adjustable forwards and backwards to ensure that equal support is given to all sizes of users.
- A swivel seat is preferred to prevent the operator twisting unnecessarily. It may be helpful if the swivel can be locked in position for stability when needed.
- All controls should be easy to operate from the sitting position. Ideally a gas or hydraulic mechanism should be incorporated to make it easy to adjust the seat height. However, it has been reported (Ralli, 1988) that in libraries where gas-lift chairs are used operators do not always adjust them, so some training is required.
- The seat should measure 400–450 mm across and 380–430 mm from front to back. If it is too small, it may cause discomfort and restricted circulation. The seat should be slightly hollow and raised a little towards the front to prevent the operator sliding forward. The front should curve away from the legs (a waterfall edge) to prevent cutting in on the thighs and restricting circulation.
- The back and seat should be upholstered to approximately 20 mm thickness and covered with a non-slip, permeable fabric.
- Chairs should have a stable (five-point) base. It should not be assumed, however, that five points will necessarily guarantee stability, especially when sitting on the edge or leaning backwards. Grandjean (1980), in a study of 261 male and 117 female office workers, found that they spent 15 per cent of the time sitting on the edge of their chairs and 42 per cent of the time leaning backwards. Stability, therefore, needs to be ensured, and this can only really be done by testing the furniture prior to purchase.
- Casters help the operator to move the chair easily.
- Armrests can help to reduce the pressure on vertebral discs but, if poorly designed or too long, can impair operator performance. It must be possible to move the arms comfortably to perform the required task, to be able to move the chair as far under the desk as required and to get in and out of the chair easily.
- Footrests should be provided for short operators to take the weight of the lower leg and to reduce pressure on the underside of the thighs. Footrests should be adjustable in height and angle and have a minimum size of 450 mm x 350 mm (BS 7179: Part 5: 1990).

PTO

Bridger (1995) recommends that they should be at least 50 mm high and tilt towards the user by 15°.

- High chairs or stools may be needed on the issue desk to bring the operator to the same level as the standing customer. With high stools, a footrest is essential both for stability and to prevent restricted circulation on the thighs. If no footrest is provided, the operator will find their own, which may in turn result in poor sitting posture.

The minimum requirements set down by the Display Screen Equipment Regulations (HSE, 1992d) are that chairs used for VDU work must be stable and allow operators freedom of movement, be adjustable in height and that the seat backs must be adjustable in height and tilt. Footrests must also be made available to operators if needed to enable them to adopt a comfortable posture.

Unfortunately, developments in backrests for operators' and typists' chairs have not been echoed in many high stools. Frequently, those which are advertised are meanly upholstered in plastic or are bare wood with a narrow seat and backrest, neither of which are adjustable. An exception is the draughtsman's chair offered by some furniture suppliers. These are, in essence, tall typist's chairs with a built-in footrest.

Many librarians argue that the cost of new chairs is prohibitive. However, it seems unfortunate that, given an expenditure of several thousands of pounds in automated equipment, more could not be found to provide for operator comfort. It even may be cost-effective in terms of increased productivity to purchase appropriate chairs. Paznik (1986, p. 20) reports a study undertaken by T. J. Springer, the president of an ergonomics consulting firm: 'In the study of 120 people, it took 15 per cent less time to accomplish the same amount of work in an adjustable setting as it did on standard desks.'

DESKS

A desk is a key aspect of the working environment, providing both storage and a surface on which to work. In choosing a desk, several factors need to be considered:

- *Is there adequate thigh clearance?* Desks with drawers immediately below the working surface often do not provide adequate thigh clearance. The recommended thickness for the work surface is 25 mm. Much thicker and thigh clearance may be impaired, much thinner and the desk may not be strong enough to support the weight of an operator leaning to write and a VDU.
- *Is the desk surface of a matt finish rather than shiny or white, to cut down reflected glare?*

Figure 10.1 Cluttered library workstations

- *Is the desk surface free of sharp edges and sufficiently large to accommodate all the items necessary for a task* and *a VDU?* In libraries, in particular, workstations tend to become extremely cluttered (see Figure 10.1) so larger rather than smaller desks are more sensible.
- *Is the desk the correct height for using the VDU keyboard?* According to researchers, the 'home row' of keys (ASDF, etc., on the QWERTY keyboard) should be level with the elbow. Too high a desk will cause operators to hunch their shoulders, putting strain on the neck and shoulder muscles, or put the elbows out to the sides, causing stress on the muscles in the upper back. A desk that is too low will not permit operators to get their legs underneath the table and force them to sit in awkward positions (see Figure 10.2).
- *Can the operator sit near enough?* The visual distance is an important consideration. If the desk is too far away for the operator to see comfortably, they will be forced to bend down to bring the source document or keyboard into focus.
- *Does it provide adequate storage for the task to be performed and the operator(s) who will use it?*
- *Is the storage space appropriate (for example, does it incorporate hanging files, shallow drawers, deep drawers) for the task to be performed, and is it easily accessible from a sitting position?* Some

Figure 10.2 The effect of working at low desks: lack of space and knee room at workstation

tasks – for example, online searching – require easy access to heavy manuals. It should be possible to reach these without twisting, bending or stretching.

- *Is there room for operators to move their feet?* Some workstations have spars or trailing wires which restrict movement. Others have boxes, waste bins or shelves under the desk which reduce the amount of space. Lack of movement can cause cramped muscles and aching joints and can restrict the posture of the upper body.

Adjustable desks are also available, but they are more expensive. The use of such desks ensures that the individual preferences of operators may be met. In practice, however, it is unlikely that they will be adjusted by each operator, particularly as some types can be difficult to adjust. The main purpose of having adjustable desks would be to ensure that a workstation could be designed to suit a single operator exactly, rather than anticipating that people will meticulously readjust the furniture every time a change of shift occurs.

The minimum requirements for desks as set down by the Display Screen Equipment Regulations (HSE, 1992d) are:

- The work desk or work surface shall have a sufficiently large, low-reflectance surface and allow a flexible arrangement of the screen, keyboard, documents and related equipment.
- The document holder shall be stable and adjustable and shall be positioned so as to minimize the need for uncomfortable head and eye movements.
- There shall be adequate space for operators or users to find a comfortable position.

RECOMMENDED DIMENSIONS FOR WORKSTATION CHAIR AND DESK

No chair or desk can be considered in isolation as it is possible to have a well designed chair, but one which is too low for the desk and vice versa. Thus it is the relative dimensions of both the chair and the desk which need to be considered (see Figure 10.3).

It is recommended that:

- Fixed height desks for supporting keyboards should be between 660 and 730 mm, with 720 mm preferred (BS 7179: Part 5: 1990).
- Adjustable height desks should have a range between 660 and 770 mm and be easy to adjust without undue force (BS 7179: Part 5: 1990). If adjustable tables are in use then the following rule of thumb may be helpful:
  - for *writing* (and general office work) the desk top should be about 50 mm above the user's elbow height;
  - for *typing*, the home row of the keyboard (i.e. middle row of letters) should be approximately level with the user's elbow (Pheasant, 1991).
- The size of the worksurface should have a minimum of 1200 x 600 mm with 1200 x 800 mm preferred (BS 7179: Part 5: 1990).
- The width clearance must be at least 600 mm and the depth clearance at least 400 mm at the knees and 1 000 mm at the feet (Dul and Weerdmeester, 1993).
- The distance between the floor and the seat surface should be 380–540 mm, although experiments cited in Grandjean (1987) indicate that a range of 320–550 mm is preferred by some operators. The chair should also meet the other recommendations as listed on p. 112.

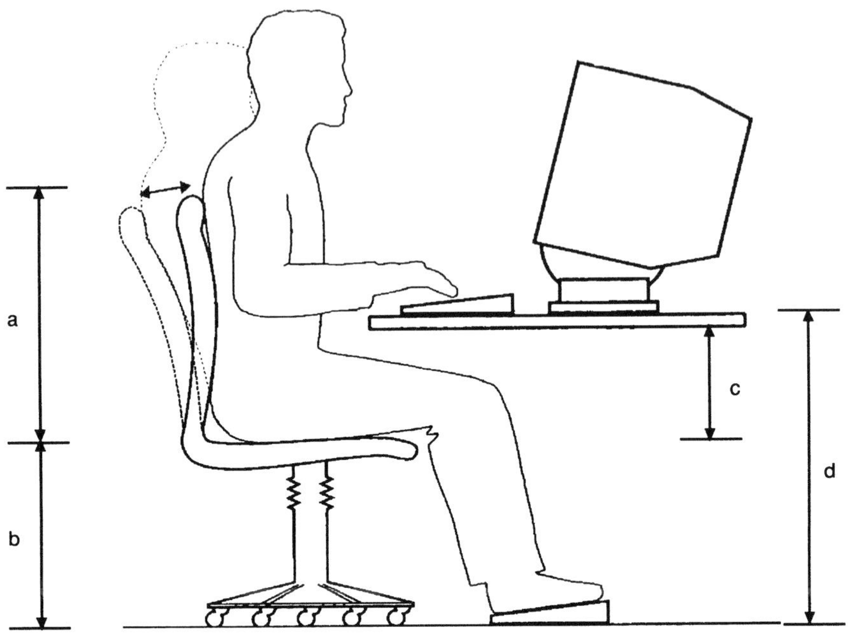

a = 480 - 500 mm
b = 380 - 540 mm
c = > 170 mm (+ 35 mm for heavy outdoor clothing)
d = Keyboarding 610 - 640 mm; writing 690 - 710 mm

Figure 10.3 Desk and chair dimensions

- The footrest, if used, should have a minimum size of 450 x 350 mm (BS 7179: Part 5: 1990).
- The distance between the surface of the seat and the underside of the desk should be at least 170 mm. It may have to be greater for men, or to cater for heavy outdoor clothing worn by library users in winter (add 35 mm for the latter). This is to ensure adequate thigh clearance.

The above recommendations apply mostly to European and North American men and women. In Table 1.2 (p. 24), however, the anthropometric measurements could be seen to vary with different groups of people, such as Asian and Far Eastern races. In libraries employing large numbers of such staff, it may be desirable to use the figures in Table 1.2 and to adjust some of the furniture to suit the relatively small stature of these people. However, as a rule, it is best to cater for the 95th percentile and to assist smaller operators by the use of adjustable chairs and footrests to bring them to the required height.

Public library users offer the widest range of body sizes to be catered for, ranging from children to old people. The needs of people with disabilities should also be considered as desks intended for standing adults will be too high for wheelchair users and may prove tiring for users with other kinds of disability.

Space also needs to be provided for users' personal belongings. Problems can occur when OPAC terminals or microfiche catalogues are housed in individual study carrels (1 000 mm wide), as insufficient space is available for personal belongings such as the ring-binders or lever-arch files typically used by students. It is preferable to place two terminals on a flat desk or double carrel thus offering greater flexibility.

The use of chairs with castors can aid access to public terminals, but it is appreciated that this may not always be desirable, either from the point of view of cost, or because of misuse by playful library users.

THE VDU SCREEN

The VDU screen, and features such as its shape and colour can affect operator comfort and well-being. Many features of the VDU screen and its display – flicker, character stability, character contrast, reflections, screen colour, positive versus negative images and the character set have already been discussed in Chapter 5. This section will deal with aspects of the box which houses the cathode ray tube.

There are a number of standards for the casework design and finish of a VDU, issued by various national and international bodies. Some of these are included in the following list:

- The characters on the screen should be well defined and clearly formed, of adequate size and with adequate spacing between the characters and lines.
- The image on the screen should be stable, with no flickering or other forms of instability.
- Major controls, such as on/off, brightness and contrast should be clearly labelled and within easy reach of the seated operator.
- The screen should be free of reflective glare and reflections liable to cause discomfort to the operator or user.
- It must be possible to tilt the VDU screen to adjust its angle to avoid glare and reflections and allow operators to adopt comfortable working positions. A range of between $-5°$ to $+20°$ from the vertical is suggested. It is also desirable that the operator should easily be able to adjust the angle by simply pushing or twisting the VDU casing.
- It should be possible to swivel the screen easily from one side to another. If older VDUs are being used which do not have this facility

Figure 10.4 Use of VDU stand to alter height of screen

then mounting the screen on a turntable which allows the screen to be swivelled and, often, tilted to the required angle might be an alternative.

- It may be desirable to alter the height of the screen, and a minimum range of adjustment of 110 mm is recommended (see Figure 10.4).
- The outer surface of the VDU (and keyboard) casing should be a light colour and matt or semi-matt to minimize reflected glare.
- The colours for indicator lights should be:
  red – for danger
  yellow – for caution
  green – for safe operation.

Thus the red indicator light used on most disk drives to show that the disk is being accessed, contravenes this standard (although it could be argued that it is a warning that the disk is being accessed and any interference could cause damage of the disk and/or the data). Red lights which are used to show that the shift lock or caps lock keys are in operation are also inappropriate. Green is preferred.

As a minimum the Display Screen Equipment Regulations (HSE, 1992d) require the first six points listed above to be observed.

THE KEYBOARD

The Display Screen Equipment Regulatons (HSE, 1992d) state that the design of the keyboard should 'allow workers to locate and activate the keys quickly, accurately and without discomfort' and list the following recommendations:

- The keyboard shall be tiltable and separate from the screen so as to allow the operator or user to find a comfortable working position avoiding fatigue in the arms or hands.
- The space in front of the keyboard shall be sufficient to provide support for the hands and arms of the operator or user.
- The keyboard shall have a matt surface to avoid reflective glare.
- The arrangement of the keyboard and the characteristics of the keys shall be such as to facilitate the use of the keyboard.
- The symbols on the keys shall be adequately contrasted and legible from the working position.

The keyboard should be lightweight and easy to move around on the work surface, but with grips underneath to prevent unwanted movement during keying. Factors, such as keyboard height, slope, layout and geometry, key design and the use of special purpose keys, and audible feedback can also affect operator performance levels and comfort (see below).

KEYBOARD HEIGHT

The height of the keyboard is a debatable issue. BS 7179: Part 5: 1990 recommends that, if fixed, the keyboarding desk height should be between 660 and 730 mm with 720 mm preferred. The thickness of the keyboard, however, will determine the height of the 'home' row of keys (ASDFGHJKL), which is the important issue. The recommended keyboard thickness is 30 mm (Pheasant, 1991), but many keyboards, particularly older models, are thicker. Grandjean (1987) cites a study that indicates an acceptable range of home row to floor heights of between 730 and 850 mm. Thus it may be possible, even with a thicker keyboard, to adapt the workstation to provide an acceptable keyboard height. However, with thicker keyboards it may be necessary to provide wrist support.

KEYBOARD SLOPE

The slope of the keyboard should be adjustable between 0 and 25° (BS 7179: Part 4: 1990). In practice, the slope of the keyboard can often only be adjusted by 'legs' underneath the back of the keyboard. Usually these enable just two options: with or without the legs – and in some cases, the use of the legs can obstruct access to the disk drives if the keyboard is placed directly in front of the drives. If the slope is not adjustable then it should be fixed at an angle of 10–15° to the horizontal.

KEYBOARD LAYOUT

In Chapter 4 it was reported that repetitive work could cause work-related upper limb disorders (WRULDs). The layout of the alphanumeric keys may be to blame, according to Professor Ferguson of Sydney University (Hobday, 1985). There are various layouts available which are discussed below. The most common and, according to Professor Ferguson the cause of WRULDs, is the QWERTY keyboard.

QWERTY

The QWERTY keyboard (so-called because the top row of keys spells 'QWERTY') is also known as the Sholes layout after its inventors, the Sholes brothers (see Figure 10.5). The QWERTY layout was patented in 1878 and was adopted as international standard (after modification) in 1966.

Originally, keyboards were alphabetical but it was found that keys began to jam. The Sholes brothers therefore designed a keyboard which separated those letters which frequently co-occur in the English language, such as 't' and 'h' and 'q' and 'u'. Their aim was reliability in keyboard operation, not speed. Nowadays, however, typewriters are not prone to jamming and, of course, neither are computers (at least not in that way!). It is therefore technically possible for the keyboard layout to be changed. This may be desirable as there are a number of problems with the QWERTY layout (Gilad and Pollatschek, 1986):

- Of all motions, 48 per cent are one-handed, that is, it is necessary to move one hand to touch two consecutive keys. It is possible to type

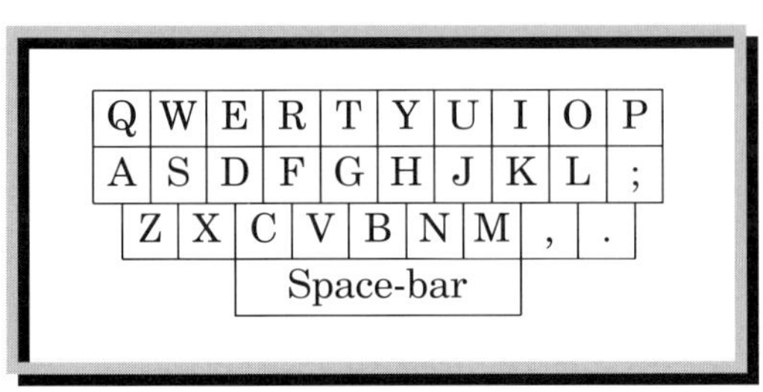

Figure 10.5 The Sholes (QWERTY) keyboard

more quickly if keys are pressed by alternate hands in the same way that it is possible to beat a drum faster with two hands than with one. It has been estimated that one-handed motions take 75 per cent more time than two-handed actions.

- It is necessary to move to a row other than the 'home' row for 68 per cent of key strikes; 70 per cent of strokes should be on the 'home' row.
- The hardest motions in typewriting are one-handed motions to one or two rows above the 'home' row. Some writers are of the opinion that the number of these motions forced by the QWERTY layout is excessive.
- Of all strokes made, 56 per cent are performed by the left hand. For most people, it is desirable to have the majority of strokes undertaken by the right hand.
- The QWERTY keyboard induces awkward, one-handed fingering and inequitable loading on each finger. Consecutive strokes, ideally, should be with different fingers and, preferably, with different hands.

Because of the problems of the QWERTY keyboard, several alternatives have been designed, each with its own group of supporters.

Dvorak layout

In 1932, August Dvorak, Professor of Statistics at the University of Washington, developed the layout shown in Figure 10.6.

The aim of this design was to improve the efficiency of the typist. All the vowels are placed on the 'home' keys of the left hand and the most-used consonants on the right hand set of 'home' keys. Thus more than two-thirds of the key strokes are on the home row (Stobie, 1986). The remaining keys are laid out according to their frequency of use with the more common on the top row and the less common on the bottom. The loading on the little fingers has also been reduced.

Dvorak himself maintained that typists trained on this keyboard were faster and more accurate than those on QWERTY but despite its manifest advantages, the Dvorak keyboard never really caught on.

?	,	.	P	Y	F	G	C	R	L
A	O	E	U	I	D	H	T	N	S
'	Q	J	K	X	B	M	W	V	Z

Space-bar

Figure 10.6 The Dvorak keyboard

Alphabetical layout

It has been suggested that novices might type faster using an alphabetical arrangement. Norman and Fisher (1982) quote other writers who fail to support this view, finding that novice typists typed as rapidly on the QWERTY keyboard as on those which are arranged alphabetically. In Norman and Fisher's own study the operators, on average, achieved a typing speed of 13 wpm on a QWERTY keyboard compared with 7.8 wpm on two alphabetical keyboards and 7.1 wpm on a randomized keyboard layout.

In all these studies, although non-typists were used, the bias in favour of QWERTY may have been due to the slight acquaintance that most adults have with the standard keyboard layout. Children with no experience of the traditional layout may be better subjects for such an experiment.

An alphabetical arrangement may seem logical to those unfamiliar with the QWERTY keyboard. Certainly, it could be argued that it would place less mental strain on the operator as the characters would logically follow one another. Norman and Fisher (1982) used several keyboard layouts (see Figure 10.7) in an effort to assess the effectiveness of alphabetical keyboards.

They concluded that alphabetical keyboards are not always as logical as they might seem. For example, they argue, suppose that one is typing the word 'flower' on the alphabetical-1 keyboard. The 'f' is located on the top row near the middle of the keyboard. Where should one look for the 'l' ? Logically, it should follow to the right of the 'f' but, in this case, one must look to the left to find it. Thus the logical arrangement of an alphabetical keyboard cannot be assumed.

Kathleen Gardner also devised an alphabetical keyboard (Figure 10.8) which was claimed to be so easy to use that 8 year-olds could reach speeds of 90 wpm within a few months (Barclay, 1986). It was also considered to be ideal for the 'hard-pressed executive [who] could learn to use this keyboard in half an hour'. The keyboard was designed to be easy to learn because of the logical positioning of the keys relative to one another. The design also took into account the problem of muscular control of the little fingers, requiring them to press only one key each, A and Z. The Gardner keyboard was patented in Britain and the USA but does not appear to be commercially available.

It can be concluded that, although there are many advantages of altered keyboard layouts (they can be easier for novices to learn and may offer improved productivity through increased typing speeds) there is one overriding disadvantage, which is that QWERTY is the de facto standard. Most typists are already trained on the QWERTY layout and, as Norman and Fisher (1982) showed, even non-typists' acquaintance with the QWERTY layout makes them reasonably proficient. Norman and Fisher

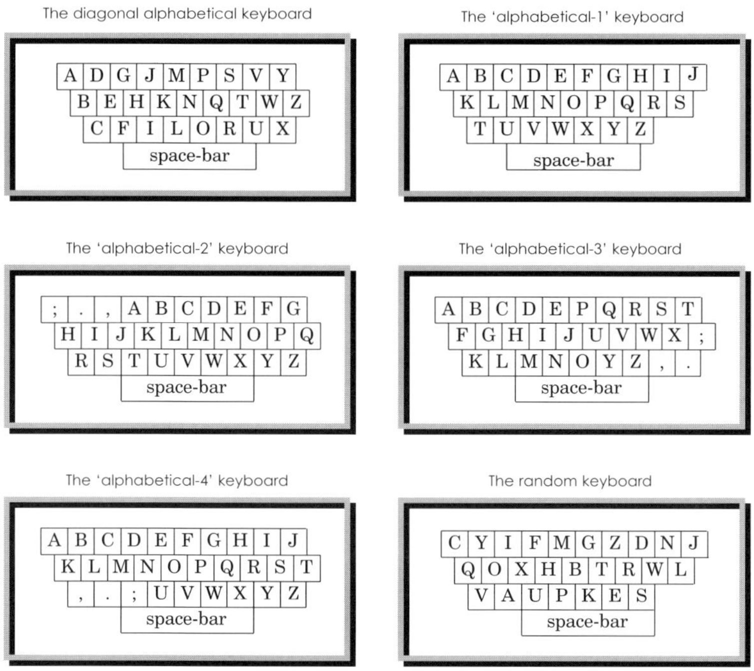

Figure 10.7 Keyboard layouts used by Norman and Fisher (1982)

also concluded that, 'for the expert typist, the layout of keys makes surprisingly little difference'. Similarly, while it is desirable to encourage children to use computers in libraries, there is no evidence to suggest that they are deterred by the QWERTY layout. It is therefore questionable whether they should be encouraged to use a special keyboard type only to have to 'graduate' to another layout on transfer to use of the library's main stock.

For libraries wishing to experiment with altered keyboard layout there is a variety of tools available which customize a standard keyboard which are described by Wallach (1996).

KEYBOARD GEOMETRY

Some writers argue that it is not the layout of the alphanumeric keys which causes WRULDs, but the shape of the keyboard. Indeed, some researchers believe that altering keyboard layout without amending its shape, or geometry, could increase the risk of WRULDs. Most traditional keyboard layouts set the alphanumeric keys in parallel rows, causing the

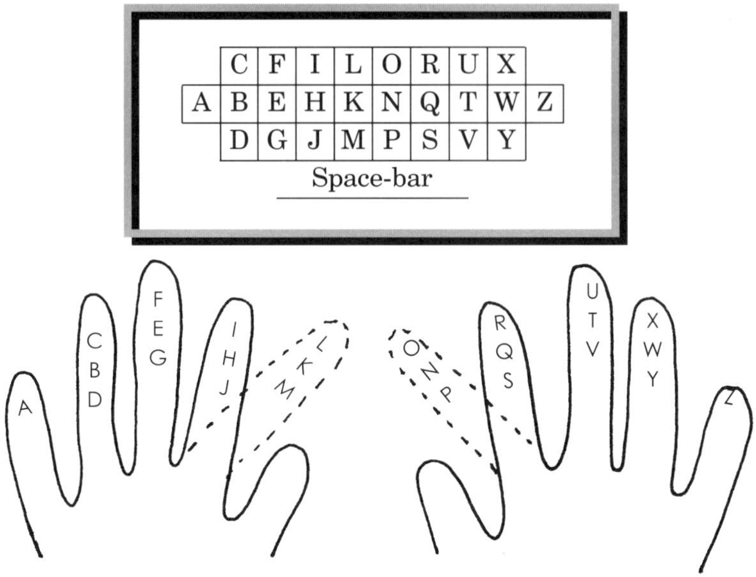

Figure 10.8 Kathleen Gardner's keyboard layout. Adapted from Barclay, 1986

operator's wrist to be twisted during typing. The new lighter keyboards (with the possibility of revised key layouts) encourage faster typing. This, in turn, may cause an extra strain to be placed on the wrists and lower arm. To counteract this problem, ergonomically designed keyboards have been developed which allow the arms to be held in a more natural position.

Split and tilt keyboards

Several so-called ergonomic keyboards involve physically separating the keys into at least two sections and incorporating a tilt mechanism. The first 'split and tilt' keyboard was proposed by Klockenberg in 1926. Since then several designs have been devised and various keyboards marketed. Possibly the most famous of these is the Maltron keyboard. The Maltron keyboard was devised by Lilian Malt and Stephen Hobday in 1977 (see Figure 10.9).

The keys are set into a sculptured curving surface and are divided into five main groups. The main alphanumeric keys are divided into two blocks, one for each hand. Each thumb has eight keys to control, set again in a block for each hand. A separate numeric keypad is situated in the centre of the keyboard. Under each palm, the keyboard casing is raised, offering support for the hands when they are not being used for typing.

The vowels have been split between the right and left hands of the home row and the more commonly used letters are under the first and

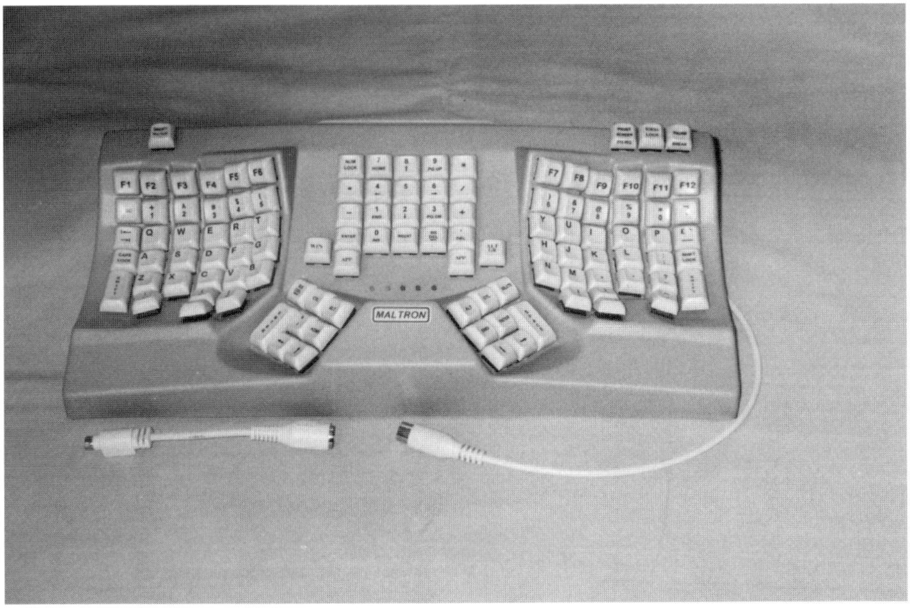

Figure 10.9 The Maltron Keyboard. Photo courtesy of P.C.D. Maltron Ltd

second fingers. Frequently occurring letter combinations, such as *th*, *ion*, and so on can be typed without the fingers leaving the home row.

Other 'split and tilt' designs include the Comfort Keyboard System, FlexPro Keyboard and the Kinesis Ergonomic Keyboard (see Figure 10.10 and Wallach, 1996 for more information on them and other similar products).

Research so far has been encouraging. Cakir (1995) assessed the 'split and tilt' Apple Adjustable Keyboard, which has since been discontinued, and after trials concluded that such designs could improve postural comfort and reduce fatigue.

Chording keyboards

Another style of keyboard design relies on the idea that two or more keys should be pressed to generate a character rather than just one. This is called chording. Kroemer (1995) describes a ternary chord keyboard 'TCK' which has only eight keys. Each key has three states: middle which is 'off' and forward and back which create two different 'on' positions. Each character is generated by a 'chord', that is by activating two or more keys simultaneously. Unfortunately, no evaluation data is reported. There are several chording keyboards commercially available including the Bat Personal Keyboard, which can be used with one hand, and Accukey, which purports that learning time is about two or three hours for getting started

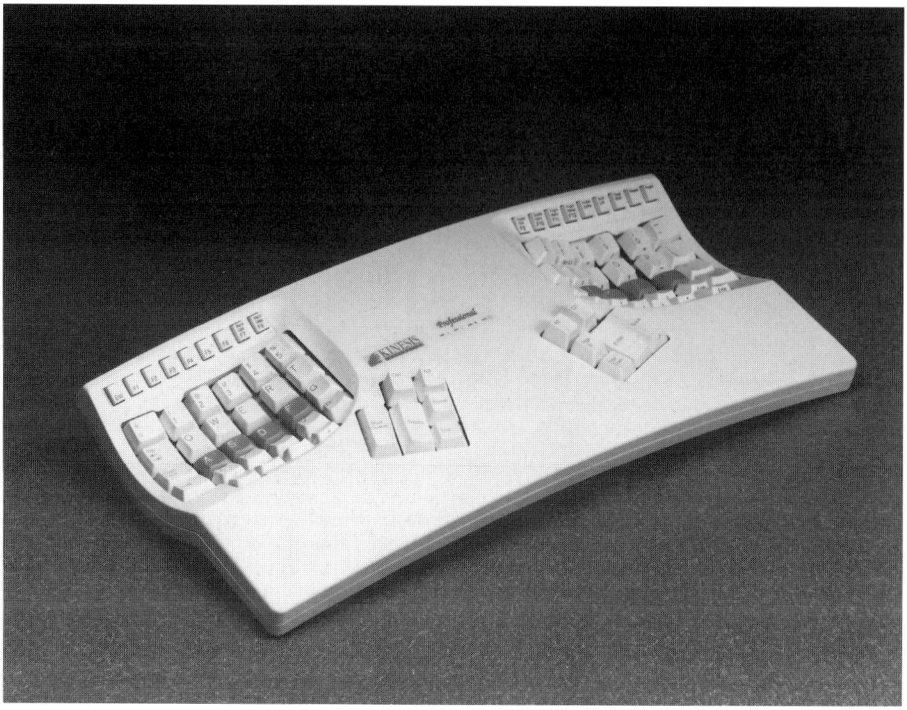

Figure 10.10 Kinesis Ergonomic Keyboard. Photo courtesy of Kinesis Corporation

and about two weeks to get used to it (Figure 10.11 and Wallach, 1996).

Other designs dispense with keys altogether. One example of this is DataHand where each finger has its own 'pod' which is used to operate a number of switches inside the 'pod'. The idea is that your hands never have to move to use the keyboard. Despite appearances, DataHand still retains the QWERTY format (see Figure 10.12 and Wallach, 1996). Even more radical is what is called the 'dataglove' which is equipped with sensors for spatial displacement of each digit (Kroemer, 1995).

It is obvious from the literature that much research is going on in the field of keyboard design. However, there is room for more detailed product testing, particularly of newer designs. What is known to date is that split keyboards appear to be more comfortable to use and preferred by operators and that there is little benefit to be gained from altering keyboard layout (Shackel, 1989).

THE KEYS

Various factors in key design affect the ease of use of the keyboard.

Figure 10.11 *Bat Personal Keyboard. Photo courtesy of Infogrip Inc.*

Key separation

It is recommended that the keys should have a spacing of 18–20 mm between centres (BS 7179: Part 4: 1990). If keys are too close, keyboarding errors may occur and, if too far apart, the hand is forced to travel too far between keystrokes reducing speed and, possibly, inducing fatigue.

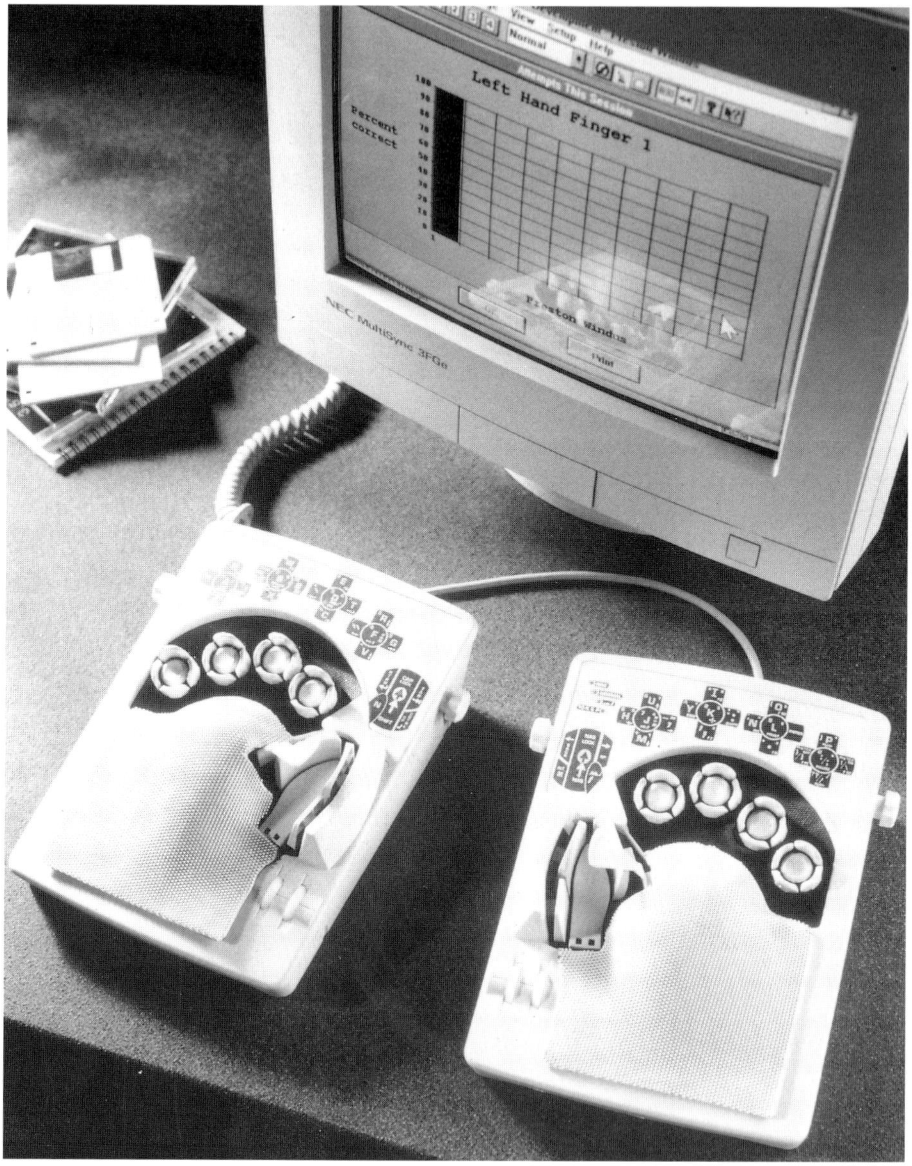

Figure 10.12 DataHand. Photo courtesy of DataHand Systems Inc.

Keytop area

The top of the key should measure a minimum of 12 mm x 12 mm if square (maximum 15 mm) or 113 mm^2 (BS 7179: Part 4: 1990). Anyone who is familiar with using the small keys of many of today's pocket calculators will appreciate that a small key size is not synonymous with speed and accuracy.

The top of the keys should be concave as an aid to accuracy. A concave shape reduces the risk of the operator's finger slipping to an adjacent key in error and confirms to the operator in a tactile manner that the finger has located the centre of a key. The minimum radius of the concave area should be approximately 10 mm (Ferris, 1987).

If keys travel only a short distance before coming to a halt, the keyboard can give the impression of being 'hard' and have the effect of jarring the operator's arm. Conversely, if the keyboard is too 'soft' and the key travels a long way before activating the signal, miskeying may occur. Operators may then think that the keystroke has been accepted and yet have failed to depress the key sufficiently. The optimum distance for the key to travel is between 1.5 and 6 mm, with between 2 and 4 mm being preferred (BS 7179: Part 4: 1990). The amount of force required to depress the keys will also affect the 'hardness' or 'softness' of the keyboard.

Audio feedback (usually a click) or tactile feedback (a sensation through the key) may be helpful and aid keyboard accuracy. Grandjean (1987) suggests that the best feedback quality is achieved when the point of acceptance and pressure is located about halfway down the key displacement. If audio feedback (key click) is used in addition to tactile feedback, it is desirable that it is adjustable and can be turned off (if only for the benefit of other people in the room).

The names of the keys should be easily legible under the lighting conditions in which the VDU keyboard is being used, a minimum height of 2.6 mm being recommended for the letters and a combination of dark letters on a light background preferred (Ferris, 1987). The legends should be resistant to wear and abrasion and should be easy to understand. The word 'ENTER' or 'RETURN' is often more easily understood by novices than the frequently used symbol ↵, and similarly 'CAPS LOCK' is more meaningful than 'TTY CAPS'.

The surface of the keys should be of low reflectivity, as should the legends, to help minimize reflected glare. They should also be dirt-resistant. Keys should be neutral (beige or grey) in colour rather than black or white (or red, yellow, blue or green) as this is more restful on the eyes. For the same reason, there should be a low contrast between the keys and the keyboard. Some contrast in colour, however, is useful between different groups of keys – for example, numeric and function keys.

SPECIAL PURPOSE KEYS

In automated library and information work, several keys are needed in addition to the alphanumeric keys. Retrieval from an online catalogue system, for example, may be by ISBN, thus requiring the use of a numeric keypad for ease of input. Function keys, too, play an important role as they help to reduce the memorizing and input of commands. They are used, for example, on some online public access catalogues (OPACs). The position of these other keys needs therefore to be assessed in order to maximize efficient use of the systems. Each type is discussed below.

Numeric keypad

Most VDU keyboards have numeric keypads to the right of the alphabetical keys with the majority adhering to the following layout:

7 8 9
4 5 6
1 2 3

If, however, a telephone with a numeric keypad is also frequently used, the operator may begin to experience problems, because telephone keypads are laid out in the following format:

1 2 3
4 5 6
7 8 9

The latter arrangement is usually preferred by novices if the keyboard can be reconfigured but either style is supported by BS 7179: Part 4: 1990.

The ESCAPE key

The ESCAPE key can be invaluable as an escape mechanism for the experienced operator, but may have disastrous consequences in some systems if pressed inadvertently. It should be positioned so that it cannot be accidentally pressed. Some systems require the operator to press CONTROL and SHIFT simultaneously with the ESCAPE key, but while this prevents inadvertent use of the key, the simultaneous use of three keys can cause problems for some users.

Frequently used keys

Keys, such as ENTER, which are in constant use, need to be positioned so that they are easily accessible. It is helpful if they are larger than the other keys to assist accuracy.

Cursor control keys

These should ideally be situated in a logical position relative to one another (see Figure 10.13) to minimize the mental loading and to increase speed and accuracy in their use. On some keyboards, the numeric keypad doubles as a set of cursor keys and this usually offers a logical arrangement.

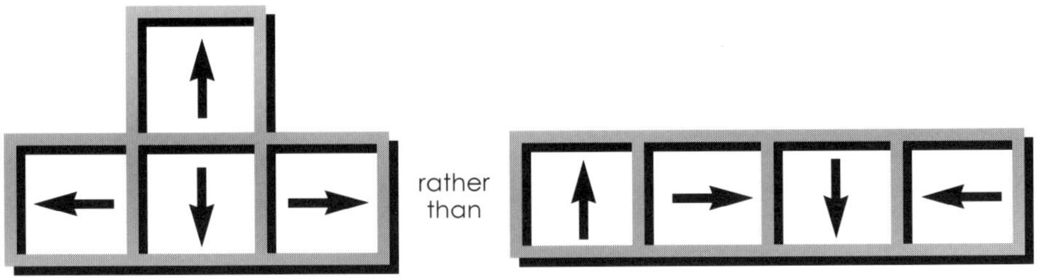

Figure 10.13 Preferred cursor control layout

Function keys

Some keys have functions whose use is specific to particular programs. A set of these should be clearly distinguished by colour and position. Usually they are situated either in a row along the top of the keyboard which correlates well with a prompt line at the foot of the screen, or in a two-column block at the left of the keyboard. The former is often used with a cardboard memory strip which can be placed above the keys, while the latter permits the use of a cardboard template designed to fix round the keys. These are supplied with much commercial software.

FOREARM RESTS

Holding the arms in a raised position for typing for any length of time is tiring and this can be compounded by delays in computer response which force the operator's hands to hover unproductively over the keys. It is desirable, therefore, that operators have something on which to rest their forearms, specially if there are pauses in the keyboarding, such as in a conversational task (see 'The tasks performed', Chapter 14, p. 177).

One form of support for the wrists is the surface of the desk. If the keyboard is positioned a little way back from the edge of the desk, it is possible to use the vacated space as a wrist support. However, the thickness and slope of the keyboard should not force operators to bend their wrists sharply upward nor allow the edge of the desk to cut into their forearms. An alternative is to use a specially designed wrist support which attaches to the keyboard (see Figure 10.14). Other supports have been devised, including forearm and elbow supports. Bendix and Jessen (1986) found, however, that supporting the elbows on armrests mounted on the chair brings about a constrained working posture.

Differences of opinion exist as to what form of forearm support, if any, is appropriate. Grandjean (1987) argues that disc pressure is reduced if the weight of the arms is carried by a support and that less strain is placed

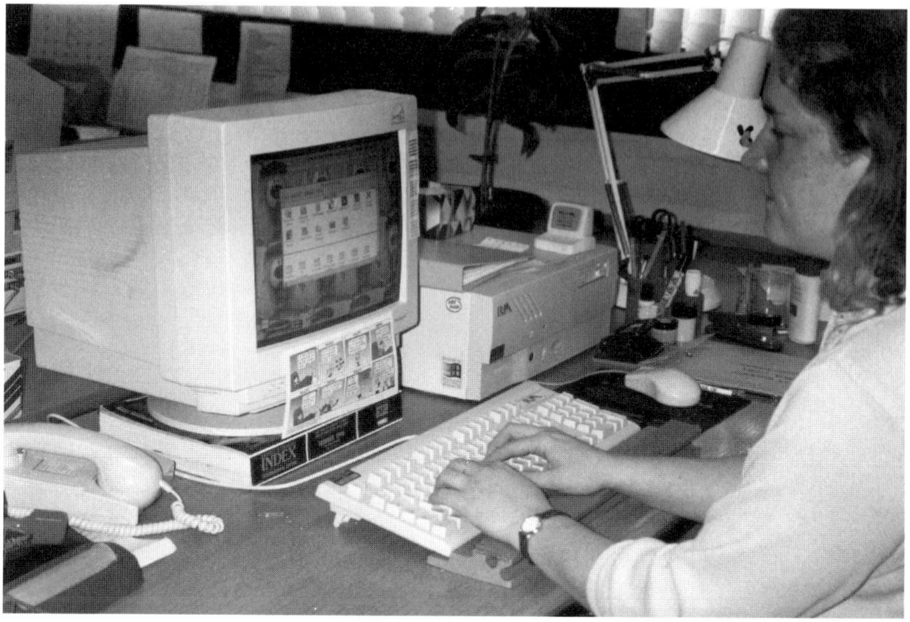

Figure 10.14 Wrist support for keyboarding

on the shoulder muscles if a support is used. In contrast, Bendix and Jessen (1986) found that the loading on the trapezius muscles (those which pull the shoulders back) was significantly greater with wrist support than without and was more marked the higher the support was positioned.

In the absence of any recommendation it is best to allow operators to adjust their own keyboard and forearm support. It is suggested, however (Ferris, 1987), that a palm rest should be used with keyboards with home key heights of more than 30 mm.

TOUCH SCREENS

Touch screens are used for some library functions – in particular some OPACs. As their name implies, touch screens enable a user to select an item by pointing to it. The simplicity of its operation makes it invaluable as a novice user interface; because touch screens offer options for the operator to select from, no commands need to be remembered and keyboard skills are not required.

There are disadvantages with touch screen technology, however, not least that the screens require frequent cleaning! Another problem is that touch screens can only work by offering choices to the user, thus restricting the flexibility available with a command-driven system. In one of the early touch screen OPACs, the user was obliged to select up to 14 different screens before the chosen record was displaced. Thankfully, system designs have since improved. Another aspect to consider is whether the system is activated when the finger first touches the screen or when it is removed. Many older people use their finger to keep their place when reading, this can be an obvious disadvantage if the screen is activated when the finger first touches it.

Use of an OPAC implies that eventually the user will wish to input some data such as author, title or keywords. To do this, a keyboard is required. It is possible to circumvent this problem by displaying a 'keyboard' on the touch screen. The vertical orientation of the screen and its relatively slow speed of response to touch, however, make its use difficult, and for this reason, total dependence on a touch screen is perhaps, inadvisable.

MICE

A mouse is, at its simplest, 'a hand-held device that translates hand movement across a surface into equivalent (scaled) cursor motion on a computer screen' (Whitford, 1986, p. 31). It enables rapid cursor movement, unimpeded by the speed of the key-repeat system or the vertical- and horizontal-only movements of the traditional cursor. It should be of a size and shape which can be held between the thumb and index finger and should be symmetrical for use by right- and left-handed users. The front of the mouse should be lower than the back to accommodate the natural angle of the hand and be rounded rather than angular. Any control buttons should be parallel to the fingers to ensure accuracy. The lead should be on the terminal side of the mouse so that it does not become entangled with the hand and the mouse should be easy to move and precise, both in terms of cursor movement and resistance to skidding across the desk surface.

Prolonged use of the mouse, however, has been found to lead to a variety of musculoskeletal symptoms in the neck, forearm, wrist and hand. These have been termed 'mouse-syndrome' or 'mouse–arm syndrome' by some patients and occupational health care personnel (Hagberg, 1995). Free and shareware software is available which aims to change the way the mouse works to avoid or modify operations that are painful, such as mouse dragging (see Wallach, 1996 for details). Carroll (1996) also lists a large number of alternative ergonomic pointing devices.

PRINTERS

The main characteristics which affect the person at the workplace are noise, print quality, print speed and chemical emissions.

NOISE

One of the most obvious features of a printer as far as its operation is concerned is the noise it makes when printing. In this respect, printers can be divided into two main types: impact printers, such as daisy wheel and dot matrix printers; and non-impact printers, such as laser, ink-jet and thermal printers. Impact printers, as their name implies, create printed output by 'punching' the shape of the letter through a ribbon onto the paper. This action, needless to say, is noisy. Non-impact printers are quieter, but not silent. Novak (1987) warns users 'never [to] underestimate the noise level of a laser printer'.

The situation of a printer will determine the maximum acceptable noise level. If it is located in the reading room – for use with the OPAC, for example – a quiet printer is a necessity (see 'Recommendations' to reduce the effects of noise, Chapter 12, p. 162).

PRINT QUALITY

The print quality is also important if the presentation of information is to be of a high standard. Nowadays, typesetting via a desk-top publishing system, using a laser printer is the norm for the business world. If the library is to compete, a high-quality printer is a necessary component.

Clarity of print is also important to ensure accuracy. Important information in poor-quality print, or produced by a thermal printer whose output fades in time, may be misread or ignored.

PRINT SPEED

Print speed can also be a problem. Too slow a printer can be frustrating for the operator, leading to feelings of underachievement, as well as being inefficient.

CHEMICAL EMISSIONS

Some laser printers emit ozone into the atmosphere. It is therefore important that rooms are adequately ventilated where printers are being used extensively.

BARCODE READERS

The main factors to look for in barcode readers are accuracy in reading and ease of manipulation.

ACCURACY

Frustration can occur if the barcode reader frequently misreads. This may be due to the poor quality of the printed code or the type of code and the way in which the light pen reads. If a barcoding system is to be used it is highly desirable that the accuracy of the reader and the barcodes be tested.

EASE OF USE

Some light pens read successfully in contact with the barcode while others require to be held a short distance away. It is easier for a novice to use the former type and these should be chosen where possible. Similarly, bidirectional codes (which read in either direction) are easier to operate than those which read in only one direction.

The wire connecting the barcode reader to the computer or terminal also affects ease of use. In one library, the barcode reader was connected to the terminal with a stiff wire which made it difficult to hold the pen at the correct angle; coiled wires with flexible coatings are to be preferred.

Laser readers are increasingly used by libraries to read barcodes and, while more expensive, are much easier to use.

SUMMARY

This chapter has discussed the components which make up a workstation. Recommendations have been given to guide purchasers. It is important to realize, however, that buying equipment which meets the guidelines will not in itself lead to an ideal workstation. The way items are positioned relative to one another (see Chapter 14) is extremely important, as is the need to train operators to adjust equipment to enable them to adopt a comfortable working posture.

PART

IV

WORKPLACE DESIGN: ENVIRONMENTAL CONSIDERATIONS

CHAPTER 11

LIGHTING

INTRODUCTION

Under the Health and Safety at Work, etc. Act 1974 (HSC, 1992) all employers in the UK have a general duty not only to ensure the health, safety and welfare at work of their employees, but also a duty not to put visitors or passers by at risk. As part of these duties, employers must provide lighting which is appropriate to the workplace and its activities, as deficient lighting can cause accidents or ill health. A number of other Regulations, designed to comply with EC Directives, also specifically require suitable and sufficient lighting to be provided at work (HSE, 1992b, 1992c and 1992d). Similar regulations also apply in other developed countries.

In any building, whether a library or a factory, lighting plays a critically important role. Good lighting enables people to see hazards and work more clearly and easily, and provides a more congenial workplace. Good lighting will, therefore, help prevent accidents; reduce the likelihood of visual and postural fatigue and discomfort; and may even enhance task performance.

Several factors need to be considered when discussing lighting. These include recommended lighting levels, the use of natural daylight and artificial light, light sources, lighting colour, glare, environmental factors such as the reflective properties of floors, walls and ceilings, emergency lighting and general recommendations for improving the visual environment.

UNITS

In order to understand some of the problems concerned with lighting, it is necessary to be aware of the concepts and terms which are used.

The term ‘illuminance’, or ‘illumination’, is used to describe the amount of light energy which falls on a surface. It is measured, using a photometer, in units called lux. One lux is the illumination obtained from one standard candle at a distance of 1 metre (Pheasant, 1991). In practical terms, the measurements may be understood as follows:

50–500 lux	–	artificial light at night
300–500 lux	–	recommended office lighting level (BS, 1990)
500–700 lux	–	recommended office lighting level (Grandjean, 1987)
100 000 lux	–	outside in the midday sun on a bright summer's day

Luminance is a measure of light energy which is reflected back from a surface. Luminance is measured in candela per square metre (cd/m^2). The following examples illustrate the use of this measure in an office with an illuminance of 300 lux (Grandjean, 1987):

65 watt fluorescent light	–	10 000 cd/m^2
White paper on table	–	70–80 cd/m^2
Table surface	–	40–60 cd/m^2
VDU screen background	–	5–15 cd/m^2

RECOMMENDED ILLUMINATION LEVELS

The amount of light necessary to do a task well depends on the nature of the task, the environment in which the task is being carried out and the eyesight of the person doing the task (Galer, 1987). The minimum illumination needed to read and write is 10 lux, but we need about 30 times this to carry out tasks easily and efficiently (Edholm, 1967). In general, the finer the degree of visual discrimination needed for a task, the higher the illumination needed. This holds true up to illumination levels of about 1 000 lux, after which, studies have shown that people become more dissatisfied with lighting conditions (Grandjean, 1988).

Too high a level of lighting can cause reflected glare and create dark shadows and marked contrasts between those surfaces which are illuminated and those which are not. On the other hand, if office lighting is too low, although the VDU itself may be readable because of its own luminance, the source documents may be hard to read and visual performance, speed and accuracy may decrease.

There are considerable national differences in recommended lighting (illumination) levels. The American Illuminating Engineering Society (IES) prescribe significantly higher levels of lighting than European standards. The recommendations given in Table 11.1 summarize the ranges proposed by the Commission Internationale de l'Eclairage (CIE, 1984), which are essentially the same as the current International Standard (ISO 8995), and those recommended by BS 8206 (BS, 1985). These levels do not, however, take into account the use of a VDU.

Table 11.1 Recommended lighting levels for different work areas

Task/area	*CIE/ISO 8995 Lux*	*BS 8206 Lux*
Storage areas with no continuous work (e.g. stacks)	100–200	150
Routine work (e.g. library workroom, library counter)	500–1 000	500
Demanding work (e.g. studying tables of figures, manual literature searching)	1 000–2 000	750
Fine work (e.g. examining and valuing rare books)	2 000–7 500	1 000

As mentioned in Chapter 5, the eyes must adapt to changing lighting levels. If the source document is well lit, it will appear very bright in contrast to the screen, and thus the eyes will need to adapt continually between one and the other. It has been suggested that office illumination should be reduced to a level compatible with the VDU (approximately 200 lux), which is lower than the recommended levels of office lighting. However, in the same way as office tasks vary in visual intensity, so, too, do VDU tasks (for an explanation of the different tasks, see Chapter 14). The recommended lighting levels are given in Table 11.2.

The above is one attempt to compromise the need for reduced lighting for the VDU with the need for adequate lighting for the source document. Another is to supplement the lighting with desk lamps to illuminate the source document. These can be very effective, but too great a contrast between the darkness of the screen and the brightness of the source

Table 11.2 Recommended lighting levels for different kinds of VDU work (source: Grandjean, 1987, p. 40)

Working conditions	*Illumination levels (lux)*
Conversational task with well printed source documents	300
Conversational task with reduced readability of source documents	400–500
Data entry tasks	500–700

document needs to be avoided. It should be remembered that as we get older we need more light and we become more susceptible to glare. The age of the staff needs, therefore, to be taken into account when adjusting lighting levels.

NATURAL DAYLIGHT

There is little doubt that most people prefer to work in natural daylight rather than artificial light. Natural daylight in a room makes people 'feel good', gives them aesthetic pleasure and, if the window is low enough, enables them to relax their eyes by giving them a distant point of fixation. Windowless rooms are often regarded as isolated, claustrophobic and depressing (Pheasant, 1991).

One of the features of the British climate, in particular, is the varying level of sunshine and daylight. The weak wintry light gives way in summer to bright sunlight, covered intermittently (and sometimes continuously!) by cloud. The variation in natural daylight, which occurs throughout the day as well as over the year, is desirable for human well-being; a uniform unchanging environment is dull and monotonous (Galer, 1987). However, it does mean that natural light is not a reliable source of illumination and needs to be supplemented by a more stable source of artificial light.

Windows, providing natural daylight, can also give rise to problems of solar heat gain in summer, heat loss in winter and excessive glare. Solar heat gain can be reduced by opening windows and by using fans, but this may introduce unwanted noise. The use of double and triple glazing can help reduce heat loss but can also be very hot in summer. Glare from windows is a common problem. Sunlight can be effectively shielded using curtains or blinds. Windows may be coated with a thin film which reflects much of the incoming light allowing only a proportion to pass through the glass, this offers the advantage of allowing staff and users still to see out of the windows. Its main disadvantage is its expense. Net curtains (in a muted shade rather than brilliant white nylon) can be used to good effect and can be left drawn all the time. Lateral or venetian blinds are common in offices, but allow small slits of light to enter, giving a stripy effect. Vertical fabric blinds tend to give a more even, muted shade and are to be preferred. Free-standing screens, or even bookshelves can also be used to cut out glare from windows.

The natural daylighting inside a building is measured by the daylight factor, which is the ratio of the illuminance due to daylight (excluding direct sunlight) at a certain point inside to the illuminance outside. The amount of daylight reaching a point inside a building is affected by the amount of daylight passing through the windows and the amount of daylight reflected from surfaces both inside and outside the building.

Making sure window frames, walls and ceilings are of a light matt colour will, therefore, help to increase the daylight factor and keep glare to a minimum.

The daylight factor immediately adjacent to a window is approximately 20 per cent. This falls to 10 per cent at a distance inside the room equal to the height of the window and to 2 per cent at twice this distance (Boyce, 1981). Given that the exterior illuminance resulting from an overcast sky is about 5 000 lux you would have to sit very close to a window to obtain 500 lux (a 10 per cent daylight factor) which is necessary for routine library work. Roof glazing can provide extra light but it gives rise to glare, shadows and can create thermal problems. The CIBS lighting guide for libraries (CIBS, 1982) recommends a minimum daylight factor of 1 per cent which, it says, represents a reasonable window area but will not provide an adequate level of lighting. It is obvious, therefore, that daylight must be supplemented with artificial lighting.

ARTIFICIAL LIGHT

It is necessary to provide a supplement or an alternative to natural light. This may be in the form of general room lighting or local lighting directed to a particular task. Usually, it is necessary to provide a good level of general artificial lighting, based on the needs of staff when there is no daylight, and additional local or task lighting. Somerville *et al.* (1986) report their use of an automatic dimmer system which reduced output from the artificial lighting when the light shone into the library. They argue that it maintained constant levels of illumination and saved money. As libraries vary, however, so too do the lighting requirements and librarians would be advised to consider the specific needs of individual rooms and desks.

ROOM LIGHTING

A workplace may be lit by directional lighting, indirect lighting or a combination of the two. The choice depends on a number of factors including ceiling height, electrical supply and space restraints. The objective with any installation, however, must be to reduce high luminance reflections from the screen and keyboard and to provide adequate levels for moving around the building safely.

Direct lighting, which is also referred to as down lighting, sends 90 per cent of its light towards its goal. The most common way of providing direct lighting is from fittings attached to ceilings. Unfortunately, this form of lighting can create a marked contrast between those areas which are lit and those which fall into shadow. In order to reduce this contrast, it is necessary to ensure that the general level of lighting is sufficiently high. This can create its own problems, as we have already seen.

Indirect lighting, or uplighting as it is known, projects most of its light towards suitably coated ceilings and walls which reflect much of this light back into the room. The resulting effect is much closer to that brought about by natural light and is diffuse, causing practically no shadows. However, the bright ceilings and, especially, the walls can produce reflections on the screen and give rise to contrast glare between the VDU and the wall. A number of suppliers now offer vertical lights of some kind: the free-standing pedestal variety; wall-mounted lights and those which can be affixed to the top of a divider, cupboard or bookshelf.

Indirect lighting is most suitable for rooms over 2.5 m high. If the height of rooms is above 3.5 m then suspended or wall-mounted uplighters or direct lighting should be used (CIBSE, 1989). Where a large number of pedestal uplighters are required, a separate floor power system must be installed for safety reasons. This means that there must be adequate space beneath floors for the necessary trunking of cables. It must also be remembered that pedestal uplighters take up floor space, need to be evenly spaced to obtain an even light and depend on the reflectance of the fittings and surrounding surfaces. Regular maintenance of fittings and decor is, therefore, essential.

Uplighting installations give the impression of a light, airy interior but, because the lighting is very diffuse, room surfaces may lack visual variety. For this reason it is important to add small areas of colour and texture to the interior decor.

Direct and indirect lighting can be combined (see Figure 11.1). Uplighters can be obtained with a translucent shade, allowing 40–50 per cent of the light to radiate to the ceilings and walls while the rest is thrown directly downwards; this results in the whole environment being evenly lit.

TASK LIGHTING

Overall, general room lighting should provide about two-thirds of the required illuminance (CIBSE, 1989). Task lighting should provide the remainder. This gives a good balance between the general and the task lighting and also avoids highly contrasting areas of light and dark. The most common way of providing task lighting is with the use of adjustable desk lights. These are available as free-standing units or may be fixed to the desk by screws or a clamp fitting.

Wherever possible, lamps should be positioned so as to throw light from the side of the task area. The spread of light should cover all the desk area used for reading and writing and the user should be able to switch it on and off when required. The user should also have some control over its

Figure 11.1 Different light types used in one work area

position and light level. This means that the luminaire should have a limited range of adjustment and a dimmer control.

Care must be taken not to position the light at too shallow an angle. The CIBSE (1989) recommends that the height be 0.5 times the width of the area being lit, which means that the light needs to be at least 500 mm high to light across 1 m of desk. Because the height of a task light is low, it is necessary to ensure that it is shielded to prevent it becoming a glare source for the operator and to position it to avoid causing glare to others. Another problem is that lamps give off heat which may, if poorly positioned, be a source of thermal discomfort to the operator.

The most suitable types of lamps for task lighting are linear and compact fluorescent lamps. Low wattage discharge lamps could be used but they have the disadvantage of being slow to react and difficult to dim.

LIGHT SOURCES

There are many different types of light sources and the choice depends on the level of illumination required, the task being undertaken, the type of light required (direct or indirect), colour rendering properties, the initial cost of installation, the running costs, how frequently they are used, lamp life and ease of maintenance. Three types of light sources have been recommended for use in a library: linear tungsten halogen lamps, tubular and compact fluorescent lamps, and high pressure discharge lamps (CIBS, 1982).

Linear tungsten halogen lamps have a filament or coil of wire, which when heated gives off a warm reddish, yellowy glow. The result is a pleasant and restful light that can be easily controlled and dimmed. Because these types of lamps are small in size and give instant light output they are often used for task lighting. They might also be used in a library for spot lighting, to highlight an exhibition for example, where emphasis and stimulation are required. However, electric filament bulbs can be costly to run, about five times as expensive as fluorescent lamps, have a short life, and may throw insufficient light. The heat output may also cause discomfort and even headaches. For these reasons they are not recommended for general purpose room lighting in libraries.

In modern buildings, fluorescent tubes are more common. Electricity passes through a gas (usually argon or neon) causing it to glow. These tubes are more efficient than electric filaments; they give a high output and last a long time. They also tend to be less troublesome as sources of glare, having low levels of luminance when adequately shielded.

The colour of the light varies, depending on the chemical composition of the fluorescent material, and can simulate natural light. Flicker, however, can be a problem. Fluorescent tubes emit a flickering rather than a constant light – 100 Hz in Europe, 120 Hz in the USA (Grandjean, 1987). Usually, this is too fast for the human eye to perceive, but research undertaken by Wilkins (1985) found that eye movement was affected by intermittent lighting at 100 Hz, even though it appeared continuous.

One operator interviewed by the authors reported that a particular VDU and fluorescent light combination had caused her headaches. When the lights had been replaced with a different kind, her headaches ceased. It is possible, therefore, that certain combinations of fluorescent tube flicker and screen refresh rates cause discomfort to some operators. Removing the tubes near the offending VDU and replacing that light with a task light (see below) may be a solution. In Europe, it is now accepted practice to install two or more phase-shifted tubes within one lighting unit to give almost constant light.

High pressure discharge lamps are sodium or mercury based sources. They are similar in efficiency to fluorescent lamps but take several minutes to produce a full light output, require control gear and cannot easily be dimmed. They do have the advantage of being small in size, of lasting longer than filament bulbs, and are available in a wide range of colour temperatures. The lighting colour offered varies from a cool blue-green colour to a warmer yellow colour according to the type of bulb which is used. Sodium lamps, however, have poor colour rendering; they appear slightly orange in appearance. High pressure discharge lamps are most often used for uplighting and floodlighting.

Running costs of all lamps can be reduced by using low energy bulbs. Economies can be made, too, by operators switching off individual lights when they leave their desks. It should be remembered, though, that if lights are switched off or moved, the overall effect may be destroyed. Installation costs vary with the type of lighting. The capital outlay required for uplighting is not low, for example, either in terms of the new light fittings or the lowering of ceilings and the redecorating which may be required (especially in older buildings). It is possible, however, that a combination of reduced energy consumption and increased productivity would repay investment in new lighting schemes in the long term. Hambro Life installed vertical lighting for their data preparation staff and reported a 15 per cent productivity increase as a result. Before installing the lights, the company had tried using blinds at the windows, but this had the effect of cutting out the natural light without reducing VDU glare. After the uplighters were installed, the operators reported fewer headaches (Hunter, 1983). It may not be the uplighters, however, which caused the improvement. Employees singled out for special treatment have been known to respond favourably even when there was no measurable improvement in conditions – a phenomenon known as the Hawthorne effect.

LIGHTING COLOUR

Whatever form of lighting is used, the colour of the light is important. Light colours have been referred to in the previous discussion and can range from a bright, dazzling white to a warm, cosy yellow or, more precisely from cool, through intermediate to warm. The colour chosen will depend upon the tasks which are to be performed in the room.

A cool light is good for visually intensive work, but can make a room look gloomy, especially if the decor is drab. Intermediate lights are good for clerical work and moderate VDU use, while a warm light is more restful, suitable for the browsing areas of a library. Problems arise when different types of tasks are performed in the same area, as lamps of different colour appearance should not be mixed in one room (Damodaran *et al.*, 1980). The best solution is to choose an intermediate colour if in doubt.

LIGHTS AS GLARE SOURCES

Glare occurs when there are excessively bright contrasts in the workplace. In severe cases it may interfere with performance (disability glare) or, if

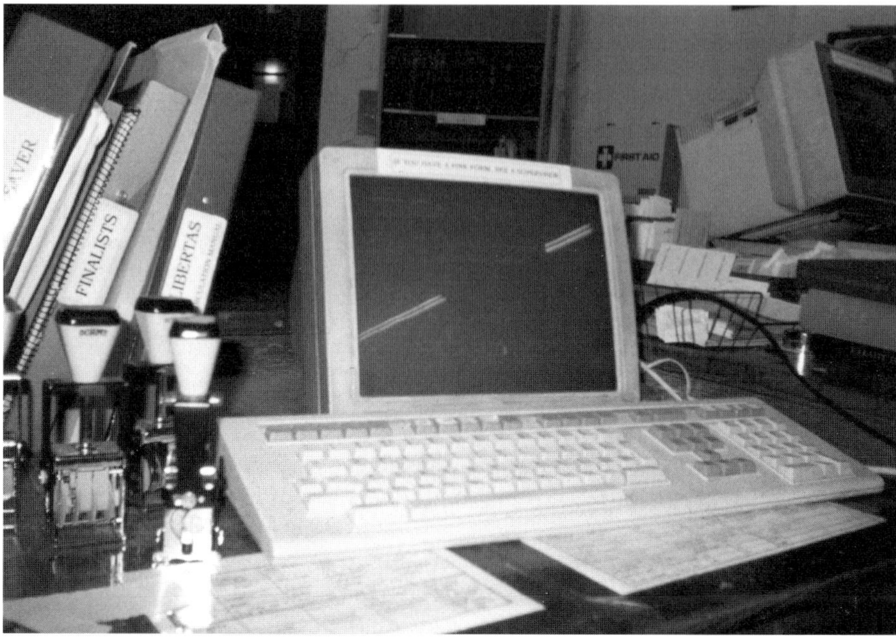

Figure 11.2 The effect of glare from lights on the VDU screen

less severe, it may cause discomfort (discomfort glare) without affecting work output.

As mentioned in Chapter 4, the most common sources of glare are windows, light fittings and reflections from polished surfaces such as VDU screens, glossy paper, book covers, shiny desks, and so on. The degree of glare resulting from these sources depends on factors such as 'the brightness and area of the sources, their position in the person's field of view, and the average brightness of the surroundings against which they are seen' (Galer, 1987). As a general rule, however, the closer the source of glare is to the centre of the visual field, the more performance is impaired, errors increase and productivity is reduced (Grandjean, 1987). The angle of a source of light relative to the operator is therefore important. An overhead light may not be a problem to an operator sitting directly underneath, but may be directly in the line of sight of an operator at the far end of the room, or reflected in a distant screen (see Figure 11.2).

Reflected light bounces in much the same way as a tennis ball, that is in an equal and opposite direction, therefore it is important to position the VDU so that reflections do not bounce into the operator's field of vision. The best way to ensure this is to site the VDUs so that the operator's line of sight is at right angles to light sources – both windows and artificial lights.

However, as Damodaran *et al.* (1980) point out, this 'may, in practice,

Figure 11.3 The effect of reflected light from windows on the VDU screen

result in too uniform a workroom layout. So, provided that the arrangement of the lamps minimizes the risk of glare, these recommendations need not be rigidly adhered to.' Even if the recommendations are adopted, other factors, such as a long window stretching behind the operator, can result in reflected glare (see Figure 11.3).

It is also recognized that because of the arrangement of lights and windows in many existing library buildings, it is not possible to achieve the ideal. It may be necessary, therefore, to use one of the devices listed on pp. 60–63 to help reduce reflected glare. Glare can be also reduced by:

- Ensuring that there are no highly reflective surfaces in the line of sight of the operator (it should be noted that even the protected covers of books awaiting processing may be sources of reflected glare, and these may need to be positioned out of the line of sight).
- Providing all lights with shades or glare shields (see Figure 11.4).
- Keeping the levels of illumination as uniform as possible. As a rule, it is better to use several lamps of lower power than a few, high powered lights to increase the overall level of lighting and to reduce excessive contrasts and heavy shadows.
- Painting walls and ceilings a light matt colour. Glare can be relative

Figure 11.4 The effect of diffuse lighting

and the perceived glare from a bright light may be reduced by painting the surrounding ceiling a bright yellow or white.

- Recommending operators not to wear white clothing.

FLOORS, WALLS AND CEILINGS

Light is reflected from the surfaces of the working environment. The amount of light which is reflected, the size and colour of the glare source and its distance from the operator will all play their part. The reflective properties of a surface will depend largely upon its finish, shiny surfaces reflecting considerably more light than matt finishes. In order to reduce reflected glare, matt surfaces are to be preferred. In general, surface reflectance should decrease from ceiling (usually about 75 per cent) to floor (20–40 per cent) (BS, 1990).

The colour of the surrounding environment will also affect the overall illumination and the emotions people feel towards the working environment. Grandjean (1987, p. 42) lists the following qualities:

- Blue and green seem further away, rooms appear larger, slightly colder and more restful.
- Red and orange seem close, are warm and rather exciting.
- Brown seems very close, rooms appear small, and many people find brown colours to be cosy.

Roberson (1992) summarizes research on colour. She points out that bright, light, and warm colours have been shown to initiate outwards activities and actions; cooler, softer colours to foster withdrawal; red backgrounds to generate ideas and actions; and green to aid meditation and performing tasks.

High contrasts between colours should be avoided and, for the most part, restful neutral colours, such as light, soft pastels prove the most popular for interiors of libraries (Niedzwetzki, 1991). It is advisable to select colours of similar brightness for the different surfaces and to avoid eye-catching effects. White ceilings are recommended. Flooring should have very low reflective properties, carpets being good from the point of view of sound reduction as well as light absorption. Strong desk colours should be avoided where possible, but it is appreciated that in heavily used areas, light colours are impractical.

Paznik (1986) describes the dramatic effects of colour on one company which had darkened the walls in order to reduce glare. Unfortunately, the darkened walls also reduced productivity and energy levels; people were inhibited from communicating with each other and became more irritable. Once the operators were moved to a lighter environment, productivity increased. Mental capacity can be consumed in having to compensate for an unpleasant environment. If the environment is improved, therefore, spare capacity is made available for work and, hence, productivity increases.

EMERGENCY LIGHTING

Employers in most countries are compelled by law to provide emergency lighting in case the electricity supply fails. In the UK this is covered by the Provision and Use of Work Equipment Regulations 1992 (HSE, 1992c) and by the Standard BS 5266 Part 1 (BS, 1988). It is recommended that escape lighting should reach the required illuminance (not less than 5 per cent of the normal illuminance) within 5 seconds of the failure of the main lighting system. This time may increase to 15 seconds if the occupants are familiar with the building. The need for immediate light output when the main supply fails means that the only lamps suitable for emergency lighting are tungsten and tubular fluorescent lamps. To ensure safety, emergency lighting needs to be tested and maintained regularly.

RECOMMENDATIONS

In most libraries, technological change has overtaken architectural change. For example, in one old library, in order to automate at all, it was necessary first to install a ring main. The lighting systems in other libraries may be equally archaic. For those librarians with access to sufficient funds to rectify the situation, the following steps should be taken (Damodaran *et al.*, 1980):

- Review the existing lighting scheme, position and type of lighting, position and type of plug sockets and cable runs, position and size of windows, and ask the staff their opinions on the lighting for their work.
- Identify the activities to be undertaken and where they will take place.
- Establish the recommended lighting levels for the different tasks (see Tables 11.1 and 11.2).
- Measure the existing lighting levels using a photometer.
- Consider how to adjust the existing lighting levels to the recommended lighting levels.
- Assess any glare problems, whether direct or reflected, and consider how to eliminate them.
- Install the new lighting system and review it by repeating the above steps after a period of approximately one to two months.

Where lighting systems are relatively modern but giving rise to complaints, specific remedial measures may be needed. In these cases the complaints need careful investigation. If the complaints are made by several people in various parts of the building doing different tasks there is likely to be something fundamentally wrong with the general lighting system. If the complaints are made by people doing different tasks in one area then the lighting in that specific area is likely to be at fault. If the complaints are made by people doing the same task but in different areas of the building then it is likely that the lighting is inappropriate for that particular task. Finally, if complaints are made by a just a few people doing different tasks in different areas then it may indicate that the problems are specific to individual workstations or related to an individual's visual abilities. In all the above cases more detailed investigations may be necessary to identify the problems. This will undoubtedly involve discussions with the operators and an assessment of the workplace.

This chapter has focused on many issues concerned with the visual environment and has given recommendations on how it may be improved. For further advice on remedial measures readers are referred to the Health and Safety Executive guide *Lighting at Work* (HSE, 1987).

CHAPTER 12

NOISE

INTRODUCTION

Noise is simply a sound which the hearer considers to be unpleasant or disturbing. Psychologically and physically, people differ in their reactions and receptivity to, and perceptions of, sound. What may be wonderful music to a teenager may be regarded as an irritating noise to a parent, for example. High noise levels can be not only annoying but also can cause difficulty in understanding speech, may result in hearing loss and, in some cases, may interfere with working efficiency. While it seems paradoxical to talk of noise in respect of libraries, the noise level will affect workers' concentration and stress levels, even if their hearing is not directly threatened. Indeed, there have been some reports of noise causing problems, particularly in modular designed libraries (Shoham, 1991).

UNITS

Sound is caused by a fluctuation in air pressure. The perception of sound depends on two factors: the frequency of the fluctuations in the air pressure and the pressure amplitude. The frequency at which the pressure fluctuates is measured in cycles per second (Hz); the higher the frequency the higher the pitch of the sound. The loudness of the sound is a function of the size or amplitude of the air pressure fluctuation and this is measured in decibels (dB).

Pitch and loudness together affect the apparent 'loudness' to the human ear. Our sensitivity to sound depends on the frequency. We are less sensitive to low frequencies and more sensitive to higher frequencies. Thus, tones with higher frequencies appear louder. Various scales have been devised for measuring noise to reflect this variation in sensitivity. The most common one in use is the dB(A) scale, which gives the best correlation with subjective tests of perceived loudness (see Figure 12.1). The scale is logarithmic because as the sound levels increase, an even greater increase in sound is required for a noise to appear louder (Grandjean, 1987):

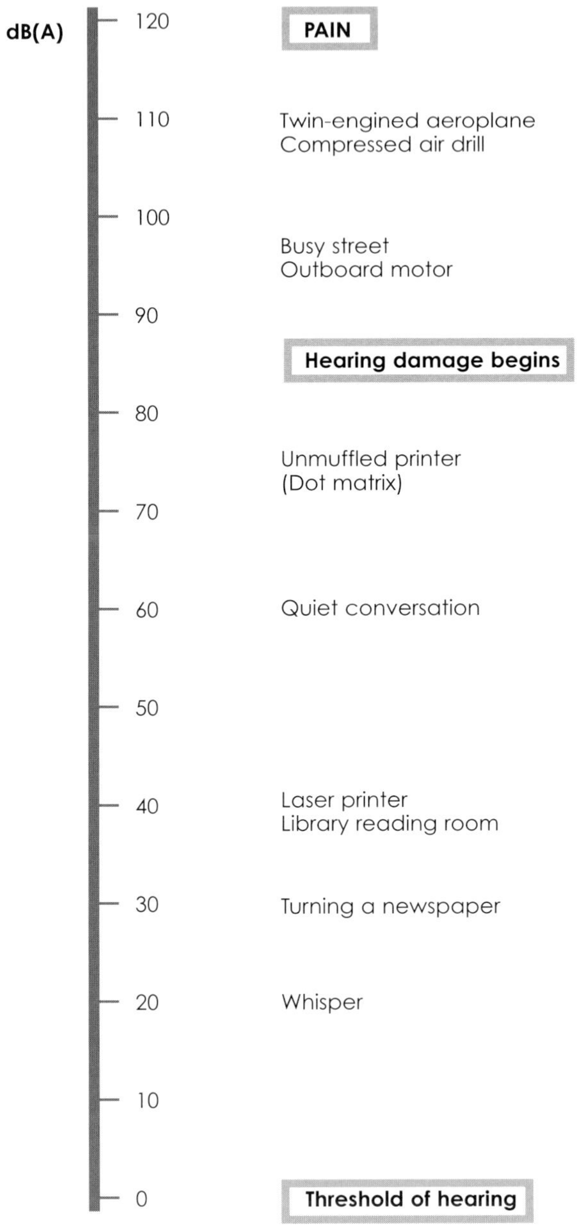

Figure 12.1 The dB(A) scale

> 1 dB is the smallest change the [human] ear can distinguish; a 6 dB increase is a doubling of the sound pressure level, although a 10 dB increase is required to make it sound twice as loud

GUIDELINES ON NOISE LEVELS

Employers have general obligations to safeguard workers' health including their hearing. They must also comply with the Noise at Work Regulations 1989 (HSE, 1989). The Regulations are based on a European Community (EC) Directive requiring similar basic laws throughout the Community on reducing the risk of hearing damage as far as is reasonably practicable. In the USA, employers must abide by the noise exposure limits for both continuous and intermittent noise at workplaces set by the Occupational Safety and Health Administration (OSHA, 1981, 1983). For the purposes of the UK Regulations any audible sound is treated as noise. Briefly, the UK Regulations require employers:

- to take action when people are exposed to a daily personal noise exposure ($L_{EP,d}$) of 85 dB(A). $L_{EP,d}$ is defined as the total exposure to noise throughout the day, taking account of the average noise levels in working areas and the time spent in them;
- to arrange for a noise assessment, by a competent person, wherever an employee is likely to be exposed at or above a ($L_{EP,d}$) of 85 dB(A). As a rough guide, assessments will have to be made wherever people have to shout or have difficulty being heard clearly by someone about 2 metres away, or they find it difficult to talk to each other;
- to keep records of assessments made; these will need to include details about the workplace, areas or jobs assessed, the results and when the assessments were made;
- to provide adequate ear protection, where necessary, and encourage use;
- to mark ear protection zones with signs;
- to make sure all equipment is maintained properly;
- to provide information, instruction and training for employees where they are likely to be exposed to excessive noise levels.

The Health and Safety Executive's publication *Noise at Work* (HSE, 1990b) gives detailed guidance on equipment and procedures for noise surveys, how to reduce noise, the types and selection of personal ear protectors, the training of noise assessors, and the procedures for testing noisy machinery.

These Regulations are obviously vitally important in industry where environments can be extremely noisy, but what about libraries? Since it is

Table 12.1 Maximum noise levels for various activities (adapted from Dul and Weerdmeester, 1993)

Activity	*dB(A)*
Unskilled physical work (e.g. cleaning)	80
Routine administrative work (part-time)	70
Simple administrative work with communication (e.g. typing pool)	60
Administrative work with intellectual content (e.g. drawing and design work)	55
Concentrated intellectual work (e.g. working in office)	45
Concentrated intellectual work (e.g. library reading room)	35

highly unlikely that any librarian or library user will be exposed to a daily personal noise exposure ($L_{EP,d}$) of 85 dB(A) should we be concerned about noise levels? The answer is, of course, yes. Levels of 85 dB(A) and over are likely to cause hearing damage which is why the Regulations require action to be taken over this level. However, levels considerably below this are recommended for office tasks and library reading rooms (see Table 12.1).

There is some variation in recommendations for office noise levels. Apex (1985) suggests an office noise level of 65 dB(A) but argues that a lower level should be aimed for. The ANSI standard cited by Czaja *et al.* (1987) recommends a level of 55 dB(A) and the British Standard BS 7179: Part 6 (BS, 1990) suggests that 'the noise level in VDT workplace should not exceed 55 dB(A) for a task requiring a high degree of concentation, and 60 dB(A) for other VDT tasks'.

In an open plan office or library, 50–55 dB(A) is to be aimed for. Ideally, this should be made up of steady, unchanging sounds such as the air-conditioning system and the hum of a computer. Such sounds help to mask other people's conversations, which can be distracting. Less than 45 dB(A) allows other sounds, such as speech and the rustle of paper, to travel further, and this can be annoying. Apex (1985), for example, reports that in open plan offices in Sweden, considerable attention was given to eliminating all sources of noise. The result was that the staff were found to be suffering from a state of shock and it was necessary to reintroduce an acceptable level of background noise. 'White noise' generators, which produce a low level of background noise, can be used for this purpose.

An investigation undertaken by Czaja *et al.* (1987), however, indicates that noise level need not be the biggest problem. Czaja and her colleagues studied 105 female workers in both traditional and automated offices. The average noise level was found to be 51 dB(A); the range was not stated. Of those studied, only 31 per cent found the office to be too noisy, yet 66 per cent were annoyed by talking, 42 per cent by telephones ringing and 28 per cent by printers and keyboards. So, although only about one-third of the workers felt that the overall noise level was too high, two-thirds were irritated by a single type of noise – talking.

Thus, rather than simply reducing the overall level of noise, it may be advisable to obtain employees' views on the various acoustic distractions which they may experience. This is best carried out over a period of time to allow for the daily and even hourly fluctuations.

THE CAUSES OF NOISE

Noise is cumulative, and may be made up of a number of contributing factors: people talking on the telephone; people talking across the office; constant movement of people; ringing telephones; typewriters in use; printers; air-conditioning systems and heating systems; cooling fans in VDUs and acoustic hoods – especially if they become worn and/or faulty; high frequency noises from the cathode ray tube; photocopiers; office fans (in summer); and unrelated noise such as traffic, not all of which relate directly to automation but which are all likely to be present in a library or workroom and will add to the overall noise level.

In addition, noise, like light, can be reflected or absorbed. Noise bounces back from floors, walls, ceilings, work surfaces and metal office equipment such as filing cabinets. Hard surfaces, such as metal and stone, cause the sound to be reflected much more than do absorbent surfaces, such as carpet and hessian wall coverings. The shape and layout of the room also affects its acoustic qualities. A long narrow room will seem louder than a square room of similar area, because the sound has only a short distance to travel before being reflected and, as noise decreases over distance, it hits the wall of the narrow room at a greater volume than it would in the square room. Low ceilings have a similar effect.

The layout of the room dictates whether sound will be encouraged to carry or to be blocked. In one library, the room was arranged so that the bookshelves fanned out from a central issue desk. The purpose of the arrangement was to ensure visibility throughout the library to increase security. Where light could travel, so also could sound, and noise from the busy counter ran unchecked to the 'quiet' study areas. In another library the non-fiction and reference material was arranged as a gallery above the issue desk. Consequently, the noise from the busiest part of the library travelled upwards and disturbed the library clients on the first floor (Saunders, 1993).

PROBLEMS CAUSED BY NOISE

ANNOYANCE

Everyone has experienced noise which is annoying. Generally, the louder the noise is, the more people complain about it. However, there are various other factors which can influence whether we become annoyed by noise in addition to its level. These include:

- *The nature of the noise* – unexpected and irregular noise being more disturbing than regular noise, and high-pitched noise more annoying than low-pitched noise.
- *The nature of the work of the hearer* – if a person 'approves' of the source of the noise (for example, if it comes from the machine he is operating) it will be less annoying than if it is from an outside source.
- *One's involvement with the noise* – other people's conversations can arouse curiosity and therefore be distracting; the mention of one's own name calls attention more than the rest of the conversation.
- *One's control over the noise* – a sound which is imposed on a person is more annoying than one which can be controlled.
- *The source of the noise* – sounds whose sources are unknown are especially irritating.
- *The degree to which the noise is necessary* – noise which is thought to be unnecessary and a result of thoughtlessness is much less acceptable.
- *The time of day and previous exposure to noise* – frequently occurring noise late at night such as a neighbour's dog barking can be particularly irritating.
- *The degree of acceptance of and adaptation to the noise* – people vary in their ability to concentrate and to shut out noise, also different kinds of noise affect people in different ways: a ticking clock, for example, can be calming to some, ignored by others and annoying to a third group of people.

It is well known that people can adapt to some types of noise. Residents living on main roads, for example, often say they are unaware of the traffic noise most of the time. However, it is dangerous to make the assumption that employees will adapt to working in noisy environments.

COMMUNICATION DIFFICULTIES

As with our other senses, hearing is for a purpose: to enable communication between us and other people and to act as an alerting device. Since the human brain responds particularly well to sudden or intermittent noises, such as the ringing of a telephone (see Chapter 1), such noises can be a source of distraction to us which can, in turn, lead to stress, especially if the work being performed requires continuous attention.

In order to communicate effectively, we need to be able to hear and to distinguish what other people are saying. In order to hear speech clearly, it must be 10 dB higher than the level of the background noise and, if the subject is new or contains unfamiliar words or if the speaker has a difficult accent, the voice needs to be 20 dB higher. This is likely to be especially true, too, of conversations with people who cannot be seen, such as telephone conversations. The apparently illogical but surprisingly common statement, 'I'm sorry I can't hear you properly, I haven't got my glasses on', may be due to the fact that, when hearing is difficult, the listener often unconsciously supplements the hearing loss by lip reading.

HEARING DAMAGE

Continued exposure to certain levels and types of noise – for example, working with a compressed air drill – can cause hearing damage. Work in libraries and offices is unlikely to cause deafness but, if much noisy equipment is used in poorly designed acoustic environments, there may be some risk. For the most part, though, the problems of noise experienced in automated libraries are interference with verbal communication, loss of concentration and stress.

Hearing loss occurs naturally as we get older. Loss of hearing is greatest in the higher ranges of frequency and is more pronounced in men than in women (Grandjean, 1988). Initially, hearing loss may go unnoticed as speech mostly consists of low frequencies, generally between 300 and 700 Hz. However, when the loss of sensation spreads to the lower frequencies the perception of consonant speech sounds will be affected. Age deafness is one of the reasons why older people have difficulty hearing what someone is saying when numerous conversations are taking place at once. Hearing loss is measured using an audiometer which can detect the smallest sound pressure perceived at a number of frequencies.

EFFECTS ON WORK

Studies have shown that noise can affect the performance of an individual. In some circumstances where the work is boring or repetitive the presence of noise may actually help to keep a person alert and therefore improve performance. However, noise adversely affects performance when work is more demanding such as when the job requires a lot of concentration or it is done under fairly stimulating conditions. Weinstein (1977), for example, investigating the effect of noise on proof reading tasks, found that the detection of grammatical errors was impaired when subjects were exposed to noise levels of 68–70 dB(A). In another study (cited in Grandjean, 1988), Wisner obtained a reduction of 30 per cent in typing errors when he reduced the noise level of an industrial typing pool by about 25 dB. Such results, however, must be viewed with caution. It is possible that the typists involved may have been motivated to do well because someone was taking an interest in their work.

PHYSIOLOGICAL STRESS FROM NOISE

Several studies have shown that noise can stimulate the autonomic nervous system and produce symptoms such as raised blood pressure, increased heart rate, contraction of the blood vessels of the skin, increased metabolism, increased muscular tension and a slowing down of the digestive organs (Grandjean, 1988). The stimulation of the autonomic nervous system is a natural defensive mechanism which prepares the body for action when facing danger. Over-stimulation by exposure to excessive noise levels over a period of time may lead to stress and other associated health problems.

RECOMMENDATIONS

The effects of noise can be reduced in several ways or by a combination of approaches. These include reducing the noise at its source or on its way to the hearer, and minimizing an operator's exposure to the noise.

REDUCE THE NOISE AT SOURCE

The best solution, if possible, is to reduce the noise where it is generated. This may be done in one of the following ways:

- Ensure that equipment is regularly checked and maintained so that cooling fans and so on do not become noisy.
- If buying new equipment such as a computer or a printer, find out from the manufacturer the working noise level, but be aware that it is likely to be the average rather than the peak level and will not take into account the environment in which the equipment will be situated.
- Isolate noisy equipment by, for example, putting the printer in a separate room. However, this may be inconvenient, if people need frequent access to the printer. Problems may arise, too, if an operator is forced to work in isolation, not so much because of the equipment noise, but because of loneliness.

REDUCE THE NOISE ON THE WAY TO THE HEARER

This means, in effect, putting a sound absorbent barrier between the source of the noise and the operator(s) who will be affected. Double glazing can cut down external noise such as traffic but, in summer, especially, it will be important to ensure effective cooling systems as opening the windows for air could also admit noise.

In addition, the environment can be made to absorb sound generated from inside or outside the building. Any of the following sound-absorbing materials could be used: carpets on the floors; acoustic ceiling tiles; partitions (shelves of books are extremely effective but care must be taken to ensure that operators do not feel isolated); and absorbent paint or materials to coat the walls. The treatment is more effective when sound-absorbing material is applied to the area immediately behind noisy equipment (HSE, 1990b).

REDUCE THE PERIOD OF EXPOSURE TO THE NOISE

While some work may have to be done in a noisy area, it may be possible to allocate times away from distractions to undertake work requiring greater concentration. Reduced noise levels and fewer interruptions may enable the operator to work more effectively, thus attaining a greater sense of job satisfaction.

CHAPTER

13

HEATING AND VENTILATION

INTRODUCTION

Studies of conditions in offices have shown that poor ventilation and temperatures that are either too hot or too cold can result in 'loss of efficiency, discontent, and increased rates of accident and sickness' (Galer, 1987). Poor thermal conditions may also result in a higher turnover of staff.

There are six important factors that will determine whether staff and users will feel hot, cold or comfortable in a library. These are:

- air temperature
- radiant heat
- relative humidity
- air movement
- type of activity
- clothing.

All these affect each other and cannot be considered in isolation; nevertheless, for ease of presentation, they will be discussed separately and then combined in a later section. First, however, it is useful to discuss how the temperature of the body is controlled.

BODY TEMPERATURE REGULATION

The body tries to maintain itself at a core temperature of around 37.0°C (98.6°F). The brain receives messages from heat-sensitive nerves located throughout the body, but especially in the skin, and then sends out regulatory messages to maintain the core temperature. Temperature is monitored by the control of blood flow, the secretion of sweat and, if the body is too cold, by shivering.

Heat is continuously generated inside the body by a metabolic process which converts the food we eat into heat and energy. Under most circumstances, we generate more heat than is needed, which means that the body usually has to find ways of losing heat. In cold conditions, however, the body has to find ways of generating heat.

Being too hot or too cold may affect the whole body. Overheating can cause lethargy and lack of concentration, while overcooling can induce restlessness. Both conditions may lead to increased errors and poor performance. The maintenance of a comfortable thermal environment is therefore essential.

AIR TEMPERATURE

Individuals vary in their preference for heat and cold; older people, for example, generally prefer hotter environments to younger people. Even within age groups, perceptions of comfort differ which means that it is impossible to please everybody. However, recommendations have been made that the air temperature should be maintained between 19°C (66°F) and 23°C (73°F) for office work using a VDU (BS 7179) and a legal minimum of 16°C (61°F) has been set, below which it cannot be expected that sedentary work will be undertaken (Damodaran *et al.*, 1980).

The British Standard, BS 7179 (BS 1990), also states that, for jobs which involve extensive periods of seated work, the difference between temperature at floor level and that at head level should not exceed 5°C (9°F). Galer (1987) suggests that the temperature difference should be lower than this. He recommends that the temperature difference should be no more than 2°C (3.6°F); above this people feel uncomfortable.

Air temperature can be measured using a mercury in glass thermometer or with electronic instruments. Mercury (in glass) thermometers are simple to use, but electronic instruments have the advantage that values can be continuously recorded and analysed by computer.

In most modern academic library buildings, thermostatically controlled air conditioning is installed, bringing hot or cold air into the library as required to maintain a constant temperature. However, variation in air temperature within a library can occur even with air conditioning. Both VDUs and people generate heat (approximately 100–300 watts per unit) and it is possible that, if VDUs and their operators are concentrated in one section of the library, as online public access catalogue (OPAC) terminals often are, a 'hot spot' will occur in that area. Since it is recommended that the heat built up in areas around equipment should not exceed 3°C (5.4°F) above the ambient level (BS 1990), VDUs should be well spaced throughout the workroom or library. When positioning the VDUs throughout the library, it is also recommended that heat output from the vents should not be directed towards the operator or their neighbour.

RADIANT HEAT

In non-air-conditioned libraries, heating may be an even greater problem than in air-conditioned buildings. Cakir *et al.* (1980) found that 50 per cent of people in traditionally heated rooms complained of problems relating to heating, compared with 30 per cent of those in air conditioned environments. Certainly, it is in the non-air-conditioned libraries that many of the problems of radiant heat occur.

The term 'radiant heat' is used to refer to heat emitted from a surface or a body, such as a central-heating radiator, a light bulb or a VDU. Radiant temperature can be measured with the use of a black globe thermometer. This comprises a thermometer inside a globe, approximately 150 mm in diameter, which is painted black to absorb the radiant heat. The globe temperature has to be corrected for air temperature and velocity to give the mean radiant temperature.

Ideally, the radiant temperature should be the same as the air temperature. Thermal discomfort may occur if the heat (or cold) emitted from the item is considerably different from the air temperature (± 50 per cent). It may be necessary to rearrange the furniture and equipment to ensure that the operator, for example, is not situated uncomfortably close to a radiator or a window.

Another form of radiant heat is the sun, especially through south-facing windows. Effective curtains or blinds must be available to cut out the unwanted sunlight when required. A combination of a radiator and sunshine could be particularly uncomfortable.

RELATIVE HUMIDITY

Relative humidity is a measure of the amount of moisture in the air and is expressed as a percentage. It is now generally recommended that at temperatures of 19–23°C (66–73°F) the relative humidity should be between 40 per cent and 60 per cent for office workers using VDUs (BS, 1990). It has been suggested by some that the comfortable range is between 30 per cent and 70 per cent relative humidity. However, two-thirds of operators in a survey reported by Cakir and his colleagues recommended that a level between 30–40 per cent was too dry (Cakir *et al.*, 1980). Thus a level higher than 40 per cent is recommended with 50–60 per cent being preferable (Bridger, 1995). Air humidity can be calculated from the wet and dry bulb measurements of a whirling hygrometer.

If the atmosphere is too dry and, especially, if it is also too hot, staff may experience irritation of eyes and dry noses and throats. The relative humidity can be increased by reducing the temperature or by improving the air-conditioning system so that it also humidifies. A simpler and cheaper solution is to use humidifiers which clip onto the radiator or to distribute plants around the library. Electric humidifiers can also be purchased.

AIR MOVEMENT

The rate of air movement should be less than 0.25 metres per second (m/s) (BS, 1990). Provided the air and radiant temperatures are correct then the ideal air movement should be at 0.15 m/s (Galer, 1987). Environments with air speeds of less than 0.1 m/s, are usually perceived as airless whereas those with air speeds of more than 0.25 m/s are usually regarded as draughty. In one study by Cakir *et al.* (1980) draughts were found to cause approximately 25 per cent of neckache complaints. They were also found to be more prevalent in air-conditioned environments, especially during hot weather when air currents are used to cool the atmosphere. Air movement is more readily perceived if the air is cool, so if the air temperature is raised, its movement is less noticeable.

In hot weather, however, fast air movement may be desirable. A rapid flow of air over the body encourages heat loss by evaporating perspiration; hence the use of fans in summer. Opening windows is another way of encouraging increased air movement but brings with it the hazards of admitting noise and fumes as well as causing disturbance to piles of documents.

Air movement can be measured using instruments such as Kata thermometers which take into account air temperature.

TYPE OF ACTIVITY

The different kinds of work undertaken in a library create their own problems for the heating system. Active work, such as shelving, requires moderate physical exertion, causing the person to generate metabolic heat (see Table 13.1). The shelver will therefore require a lower air temperature than the sedentary worker. Variations in the routine of the VDU operator need to be made for postural reasons (see 'How to minimize postural problems', Chapter 3). It may also be necessary to consider the effects of long-term sedentary work on the operator's thermal comfort.

Table 13.1 Estimated metabolic heat generated for different tasks

Activity	*Metabolic heat Wm^{-2})*
Seated, no activity	58
VDU use	70
Standing, issue desk	100
Standing, shelving	115

CLOTHING

The type of clothing worn can affect whether we feel hot or cold. A person wearing a heavy outdoor coat will obviously feel more comfortable at lower temperatures than someone wearing lightweight clothing. This does give rise to problems in public libraries. Staff will need to have air temperatures between 19°C (66°F) and 23°C (73°F) but the general public may feel this is too hot, especially if they keep their coats on. Ideally, public libraries should provide cloakroom areas to allow users to remove and deposit coats.

The measure of clothing insulation is the CLO. One CLO represents the amount of thermal insulation necessary to maintain in comfort a sitting, resting person in a normally ventilated room at 21°C (70°F) and 50 per cent relative humidity. Typical CLO values are given below:

Light summer clothing	0.3
Light suit	1.0
Heavy suit	1.5

The concept of needing to wear different types of clothes for different types of temperatures is not new. Sanders and McCormick (1987) point out that the Chinese have for many years described the weather in terms of the number of suits required to keep warm. In China, a one-suit day is reasonably comfortable, a two-suit day is a bit chilly and a 12-suit day (the limit) indicates very bitter weather.

THERMAL INDICES

Several studies have been made to examine the relationships between the six variables and a number of indexes have been derived which attempt to

Table 13.2 The relationship between PMV values and PPD values (adapted from Parsons, 1990)

Sensation	*Cold*	*Cool*	*Slightly cool*	*Neutral*	*Slightly warm*	*Warm*	*Hot*
PMV	−3	−2	−1	0	1	2	3
PPD (%)	99	75	25	5	25	75	99

give an overall measure of thermal comfort. One of the most useful indexes is the predicted mean vote (PMV), derived by Fanger (1970), which is used in ISO 7730 (ISO, 1984).

A PMV is calculated by combining air temperature, mean radiant temperature, air velocity, humidity and values for clothing and the type of activity undertaken. This results in the prediction of the mean thermal sensation vote of a large group of people on a seven point scale from −3 to +3. A PMV of −3 would mean that the environment is too cold, whereas the other extreme of +3 would suggest that it is too hot. A PMV of 0 represents ideal comfortable conditions. PMV values can be used to calculate the percentage of people likely to complain about thermal conditions. This is represented by a PPD (predicted percentage of dissatisfied) value (see Table 13.2).

SICK BUILDING SYNDROME

Even if thermal conditions are within reasonable limits some employees can feel ill when working in particular buildings. Terms used to coin this include 'building sickness' and 'sick building syndrome'. Building sickness is 'recognized by the World Health Organisation as a syndrome of complaints covering non-specific feelings of malaise, the onset of which is associated with occupancy of certain modern buildings' (Wilson and Hedge, 1987). A syndrome is defined as a 'collection of signs and symptoms fitting a recognizable pattern' (World Health Organisation, 1983).

In 1993 the World Health Organisation listed the symptoms characterizing sick building syndrome. They include eye, nose and throat irritations, dry skin, headaches, coughs, wheezing, nausea, dizziness, lethargy, hypersensitivity and erythema (World Health Organisation, 1983). All of these symptoms are commonplace in the general population, but it is the pattern of their expression that points to the diagnosis of sick building syndrome. In sick building syndrome symptoms are associated with being in a particular building and are relieved by leaving or staying away from that building.

Research investigating sick building syndrome in offices indicates that:

- air-conditioned buildings are likely to be more affected than naturally ventilated buildings (Finnegan *et al.*, 1984);
- women are more prone to the syndrome than men (Vince, 1987);
- clerical workers are more prone to the syndrome than professional or managerial staff (Vince, 1987);
- workers who perceive their workplace to be hot and stuffy and the air to be humid and stale report more health problems (Hedge *et al.*, 1989).

Pinpointing specific causes of building sickness is difficult because the symptoms reported are often non-specific and diverse, not all employees in a 'sick' building are affected, and causes appear to vary from building to building. Nevertheless, it is generally thought that the main causes of sick building syndrome is physical, and it is probably caused by the presence of air-borne volatile organic compounds plus interactions with other physical factors such as ventilation, humidity and temperature (Lundin, 1991). Psychological factors such as stress may also contribute to the perception of a building being 'sick'.

Like offices, libraries can be affected by sick building syndrome (Morris and Dennison, 1995). Evidence of sick building syndrome was found in a survey of 147 academic libraries and 79 public libraries. The results also supported previous research that suggests air-conditioned buildings are more likely to suffer from sick building syndrome than those that are naturally ventilated. Staff in air-conditioned libraries experienced more health problems, had more days off sick and rated their environment more stale than staff working in naturally ventilated libraries. The research has also confirmed that psychological factors, such as allowing staff control over environmental levels, may also contribute to the perception of a building being 'sick'. Significantly more health problems were experienced by staff working in libraries where they did not have control over the opening of windows and level of lighting and air temperature.

It is important that if sick building syndrome is thought to be the cause of staff illness, matters are investigated quickly. Climatic conditions should be checked (see recommendations below), ventilation systems serviced and stress levels should be monitored and reduced if necessary. If this fails then the only option may be to move the employee(s) affected.

In one particular case known to the authors, an employee asked to transfer from one library to another which was closer to her home. After a few days, at the new library, she began to suffer all the classic symptoms of sick building syndrome. Within two weeks her eyes had become so sore and bloodshot and she had become so listless that she sought medical help.

Fortunately the doctor recognized the problem. Unable to locate the source of the problem within the library building her employers eventually moved her to another site where she quickly recovered.

RECOMMENDATIONS

A comfortable climate is necessary to create a good working environment, and the following actions are suggested.

- Elicit from the staff their feelings about the climatic conditions.
- If there are a lot of complaints, employ an ergonomist to take measurements at ankle, chest and head level throughout the library but especially at trouble spots.
- Look, in particular, for 'hot spots' and sources of radiant heat and reorganize the area if necessary.
- Where possible, obtain VDUs and other equipment with low heat output, and position the equipment so that the operator, neighbouring operator or library user are not adversely affected by the heat output vent.
- Ensure effective curtains or blinds are available to shut out unwanted sunlight.
- Regulate the relative humidity by the use of humidifiers or dehumidifiers as necessary.
- Allow people some control over their own environment. However, this may be feasible only in individual non-air-conditioned offices where windows can be opened and radiators regulated.

PART

V

WORKPLACE DESIGN: WORKSTATION DESIGN

CHAPTER
14

WORKSTATION LAYOUT AND DESIGN

INTRODUCTION

A workstation is the combination of furniture and equipment used by an operator to carry out their job. The aim of good workstation design is to select and arrange the appropriate components in order to facilitate efficient operation and employee comfort, well-being and safety. Consideration has already been given to how to adjust a workstation to attain a good posture and minimize postural discomfort. This chapter considers how best to organize the various components with a view to efficient operation of tasks. In particular, it centres on analysing a task to assess its design requirements, equipment considerations, the physical limitations of the operator, the principles of workstation design and its effect on productivity.

TASK ANALYSIS

Before a workstation can be designed, it is necessary to understand how it will be used. This implies the need to study the various functions of the workstation in some detail, a stage known as task analysis. The objective of the analysis is to ascertain how frequently and for how long a particular task is to be performed. The more often a task or sub-task is performed or the longer it takes, the more consideration needs to be given to exactly how it is performed and to design the workstation to maximize comfort and efficiency in its operation. The nature of libraries is such that no rules can be laid down as to how a particular workstation should be organized. Libraries vary both in size and the way in which tasks and sub-tasks are allocated. In larger libraries, for example, it is possible that a workstation can be specialized for a particular task: cataloguing, for instance, may always be done by one individual at one terminal. In smaller libraries, however, many different functions will need to be performed by the same person at the same desk, and so a compromise may need to be reached. A workstation to perform the same function may also have different requirements in different libraries because of varying library policies. Rea (1987, p. 20) gives the following example of the effect of the University of Bath's library policy on the issue desk:

> In our library there is no restriction in the number of books a reader may borrow. Because there is no restriction, there is no compulsion for readers to wait at the desk while their books are discharged before they can borrow further books. Space, therefore, needs to be set aside where books can be piled up; hence the larger than normal size of our returns counter.

Task analysis considers how the workstation is used and the various tasks which are to be performed. The operator's posture, their interaction with other people and the amount of movement needed are also examined. In addition, the use of ancillary equipment, notices or operating instructions, and the changes brought about as a result of automation or upgrading are studied. Each of these is discussed below.

HOW THE WORKSTATION IS USED

While it is possible to analyse how the workstation is currently used, it should be noted that the introduction of a new computer system may bring about a change in the task. This may be because it will need to be performed in a different way with the new system, or the job may need to be redesigned (see Chapter 22) to optimize operator performance. The library manager, in analysing the task, needs to consider a number of factors:

- The number of users. In the libraries studied by the authors the number of users using a single workstation varied from one to six.
- The age, sex and fitness of the users. The anthropometric data in Table 1.2, p.24 indicate the range of sizes to be catered for. Statistically, the variation of body sizes across a population fit within a normal bell-shaped curve (see Figure 1.4, p.23). Ideally, workstations used by both males and females should be designed to accommodate the needs of 90 per cent of the adult population. The 5 per cent of cases at each tail end of the curve are usually disregarded as being extreme cases. Targeting 90 per cent of adults means that we have to design for both the 5th percentile female (a small woman of 1 505 mm) and for the 95th percentile male (a tall man of 1 855 mm).
- Whether the users are right- or left-handed.
- The length of time spent by each operator and on each task. The time spent using computers may vary from a few minutes at infrequent intervals in one library to several hours at a stretch, several days a week in another library.

- The frequency with which each task is performed.
- The frequency with which the workplace is used.
- The number of different tasks performed.

THE TASKS PERFORMED

The tasks to be performed at a workstation will differ according to individual circumstances. It is, therefore, essential to examine them in detail, bearing in mind such factors as the proportion of VDU and non-VDU work, the category of VDU work and the sequence of activities.

The proportion of VDU and non-VDU work

In small libraries, for example, the operator may fulfil all the functions of the library, creating catalogue entries, ordering books, answering reader queries, doing online searching and so on. Such a range of tasks will inevitably include a mix of VDU-based and non-VDU work. The proportion of time spent on each will influence the amount of space required for each type of task. The use of documents or manuals in conjunction with the VDU will require adequate space, especially if two or more need to be read in conjunction with one another as may be the case in online searching. Space may be needed to sort documents, such as overdue notices prior to posting, to make piles, such as books for cataloguing or serials for check-in, or to make notes. In larger libraries, jobs tend to be more specialized. However, while specialization makes it easier to design the workstation it must be remembered that jobs should have a mix of VDU and non-VDU work (see Chapter 22).

The category of VDU work

This may be loosely divided into the following: data entry; data enquiry; interactive or conversational; and word processing. The distinction is drawn between each according to the relative use made of the three main components in VDU work, namely the screen, the keyboard and the source document.

Data entry is the term used to describe continuous input of data, such as retrospective catalogue conversion from the catalogue cards or the input of a document to the word processor. The operators, for the most part concentrate their gaze on the source document with an occasional glance at the screen and keyboard. Studies have indicated that the percentage of time spent looking at each of the three main components is: 70–80 per cent on the source document; 15–20 per cent on the display screen and 3 per cent on the keyboard (Grandjean, 1987). Keying is usually rapid and operators tend to adopt a fixed posture. Ideally, the items which are to be viewed most frequently need to be situated near to

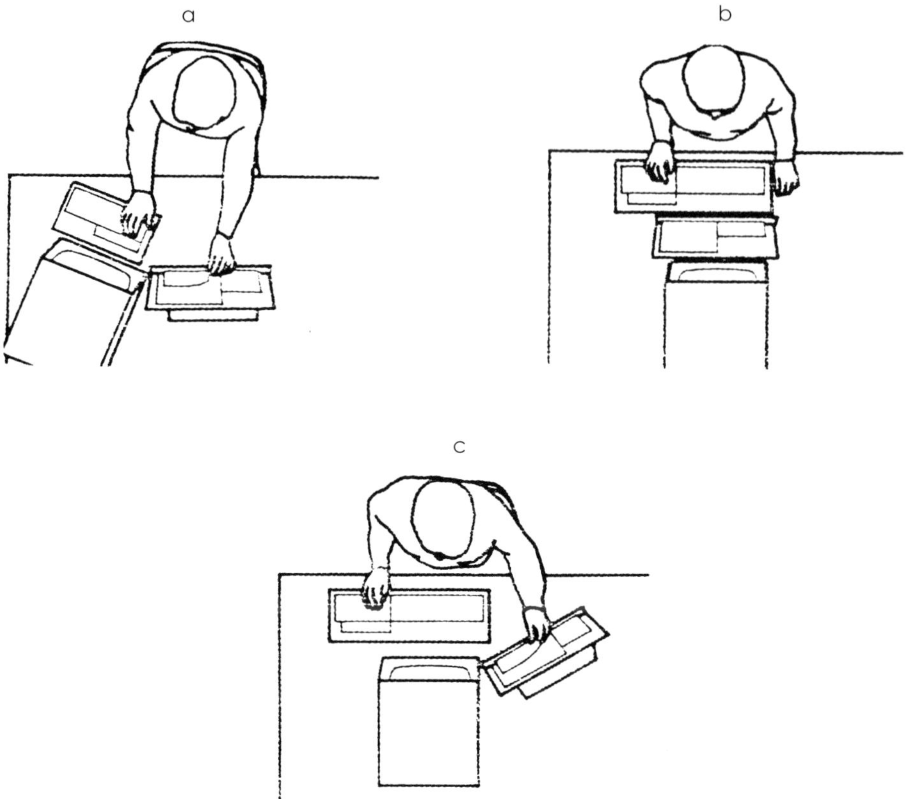

Figure 14.1 Different workstation layouts. Adapted from Health and Safety Executive, 1993

and at the same angle as, one another. In the case of data entry, the source document needs to be in the prime position with the screen nearby and at a similar angle (see Figure 14.1a).

It has been found that, for data entry, operators using a straight-line VDU arrangement, similar to that illustrated in Figure 14.1b, were more efficient and more comfortable than with a lateral arrangement as in Figure 14.1c (Marek *et al.*, 1984). Grandjean (1987), however, argues in favour of positioning the source document in front of the operator with the screen to the right-hand side and the input pile to the left (for right-handed operators) as shown in Figure 14.1a. Operators with personal preferences should have the last word.

Data enquiry involves calling up information from a screen and reading it. This is the kind of operation which is performed on an OPAC terminal or when using the Internet. Little use is made of the keyboard, the operator spending most of their time reading from the screen, possibly searching for specific items of information, such as exchange rates or flight times on the World Wide Web, rather than reading the entire screen. It may be necessary to note down what appears on the screen.

An interactive, or conversational, task combines a mixture of data input and data enquiry, further input often being required in response to the screen display. A good example is online searching. As it is likely that the operator is not a touch typist, keying speed will be low and the operator may have to look at the keyboard while typing. Grandjean (1987) reports the following percentages of time spent looking at the three elements: 35 per cent at the source document; 30–40 per cent at the display; and 20–35 per cent at the keyboard. Thus the three elements should be positioned so that they are equally easy to see. Grandjean recommends placing them in a straight line, as in Figure 14.1b, but if, as with online searching, it is necessary to write notes, it is preferable for the form to be positioned on the right (or left) rather than behind the keyboard.

Word processing, or to be more precise, the amendment of text on a computer, demands a considerable amount of time looking at the screen, with less time each at the keyboard and the source document.

The sequence of activities

If there is a logical flow to the work, items need to be arranged to facilitate ease of movement between them.

POSTURE

Posture is discussed in depth in Chapter 9. For the purposes of task analysis, it is essential to ascertain whether the operator(s), for whom the workstation is to be designed, will be sitting, standing, or a mixture of both. (The last is to be encouraged where practicable.)

INTERACTION WITH OTHER PEOPLE

A variety of forms of interaction may occur. There may be interruptions from colleagues, readers may be present at an online search and work at the issue desk will usually involve interacting with readers. The direction from which these interruptions come is important. If interaction is to be encouraged, it is desirable for the operator to be facing towards the likely source of the interruption.

Interaction with users for online search purposes requires a position which facilitates communication between the user and the searcher but which excludes unnecessary interruption. Alternatively, it may be necessary to minimize unproductive interaction between staff, but without inducing a sense of isolation.

MOVEMENT

If frequent access or egress is required, it is necessary that equipment is designed and space is available for ease of movement; chairs on castors, for example, are easier to get on and off than chairs on legs, and armrests can be obstructive. One of the features of library work, particularly at the issue desk, is the mobility of staff.

ANCILLARY EQUIPMENT

Frequent telephone calls dictate easy access to the telephone. Similarly, other ancillary equipment such as wastepaper baskets, book trolleys, cash registers (for fines on the issue desk) need to be easily accessible. The frequency of use of the ancillary items can be used to rank them in order of importance if there is competition for space.

NOTICES OR OPERATING INSTRUCTIONS

Instructions for use of an OPAC or for end-user searching on CD-ROMs or the Internet need to be given prominent space and should be eye-catching. Less frequently used or more confidential instructions, such as identification numbers and passwords for online searching, need to be secure but easily accessible by authorized users.

CHANGES CAUSED BY AUTOMATION OR UPGRADING

A change from a manual to an automated system may itself impose a change of sub-tasks. The use of laser pens and inset sensitizers and desensitizers at the issue desk, for example, reduces the time spent issuing and discharging books. Rather than allowing periods of enforced idleness for staff, other work may be introduced as an alternative task, bringing with it its own requirements for workstation layout.

The introduction of self-issue and discharge service points, such as the one shown in Figure 14.2, will further affect the nature of tasks undertaken at the issue desk.

EQUIPMENT CONSIDERATIONS

As well as understanding the nature of the work to be performed, it is necessary to consider the computer equipment which will be used. Consideration needs to be given to:

Figure 14.2 Self-issue, renew and return system. Photo courtesy of Automated Library Systems Ltd

- *The ergonomic design* of the equipment (see Chapter 10).
- *The installation requirements*. (Will there be a need for new electrical connections, network points, telephone points, new furniture and so on?)
- *Obsolescence*. (Will the equipment get out of date quickly and, if so, what is the likelihood of a replacement being a different size?)
- *Security*. (Will the equipment be easy to remove? What security measures need to be put in place?)
- *Magnetic fields*. (Will VDU screens need to be placed close to sensitizers/desensitizers or security barriers? Some VDU screens are more affected by these than others.)

OPERATOR LIMITATIONS

It is necessary to place things so that the operator can reach and see them comfortably. An operator's physical limitations need to be appreciated when designing the workstation.

Table 14.1 5th percentile measurements relating to reach (adapted from Grandjean, 1980) (measurements in mm)

	Small men	*Small women*
Standing shoulder height	1 300	1 200
Sitting shoulder height	540	490
Arm length (closed hand)	650	580
Maximum reach from sitting	1 190	1 070
Maximum reach from standing	1 950	1 780

REACH

If an employee is forced continually to reach too far to operate a control or to pick something up, this can lead to pains in the back and shoulders, especially if the object being retrieved is heavy, like a book. As Table 1.2 (p. 2) shows, people vary in stature and consequently in their reach. Reach is primarily in two directions, vertical and horizontal.

Vertical reach

The limit of upward or downward reach is the radius of the arm, up or down, with the hand in a clasped position, that is, the distance from the shoulder to the fist. Unlike the desk heights, which cater for the tallest employee, reach requirements should be based on the smallest employee. Table 14.1 gives the relevant measurements for the 5th percentile (the measurements of the smallest 5 per cent of the population).

The highest position that can be reached comfortably from a sitting position is 1 070 mm for small women and 1 190 mm for small men. Allowance may need to be made for children, if appropriate. This applies only to vertical reach, however, and does not take into account any forward movement. If a shelf is in front of the operator and the required item at the back of the shelf, the distance from the operator's shoulder to the item should not exceed 1 070 mm, which would mean that the shelf should be lower to compensate for the forward distance.

Horizontal reach

The various reach distances are as follows (Grandjean, 1980):

350–450 mm – relaxed arm – normal reach

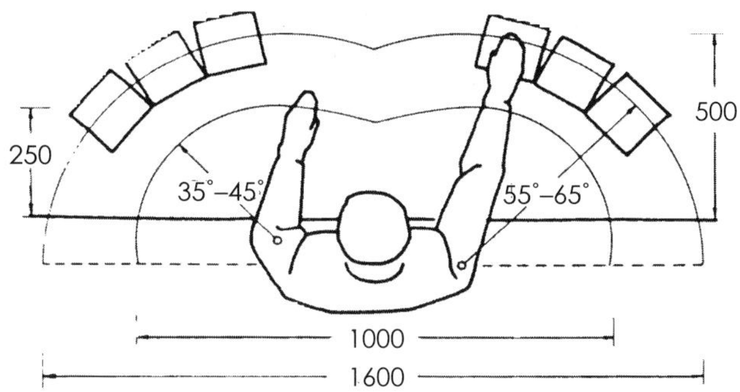

Figure 14.3 Maximum and optimum lengths of horizontal reach (measurements in millimetres). Source: Grandjean, 1980

550–650 mm – straight arm – maximum reach
700–800 mm – stretching, using the whole body

They are shown diagrammatically in Figure 14.3.

Forward, horizontal reach may be divided into two categories, normal reach and maximum reach (Das and Sengupta, 1996). Normal reach is defined as being the area which is easy to reach with the elbow resting on the desk – that is, the arc from the elbow to the hand. Writing, for example, is usually performed within normal reach. Maximum reach describes the arc of the outstretched arm – that is from the shoulder to the hand, in much the same way as for vertical reach. Beyond the maximum reach envelope, it is necessary to stretch the body.

The least tiring arm position is within the normal reach envelope where the elbow can give support. Maximum reach is unsupported, the shoulder and arm muscles carrying the load, not only of the outstretched arm but also of any items which may be being carried: thus a weight in the hands diminishes reach (Pheasant, 1991).

Ideally, the most frequently used items should be within the optimum reach envelope. The following equipment, among others, should be considered: the VDU screen, so that its angle can be adjusted if necessary; the controls – ON/OFF, brightness, contrast; source documents, such as books; paper for writing notes, etc.; telephone; pens/pencils, etc.; CD-ROM or floppy disk drive; waste bins; and personal belongings (Figure 14.4).

The position of the equipment can be laid out according to four methods (Damodaran *et al.*, 1980, p. 86):

- *by importance* – where the controls and items of equipment considered the most important occupy the prime positions;
- *by function* – where equipment relating to one function is grouped together separately from that relating to other functions;

Figure 14.4 Reach problems at the issue counter

- *by frequency of use* – those items used most frequently are located in the normal reach envelope, less frequently used items are situated in the maximum reach area;
- *according to sequence of use* – such as stuffing envelopes for a mailshot or collating papers – in order to reduce the risk of error. A regular order, preferably from left to right and from back to front or top to bottom, is advised.

VIEW

The distance that an operator can see will depend on several factors, such as the quality of their eyesight, the size and clarity of the characters on the VDU screen or source document, the quality of the lighting and the effect of any glare. The recommended distance between screen and eye is 610–930 mm (Grandjean, 1987).

The angle of view will depend on eyeball movement and the tilt of the head and neck. The eye can move within 15° above and below the normal line of sight. In order to see something at a greater angle, it is necessary to move the head and neck.

The normal, or preferred, line of sight is 10–15° below the horizontal plane (see Figure 14.5). The arc of view made possible by eye movement gives a comfortable range of sight from 5° above the horizontal to 30° below it, without moving the head or neck. Thus to reduce the risk of painful neck and shoulder muscles (outlined in Chapter 3), all items which are to be viewed frequently should be positioned so that they may be seen with minimal head movements. Ideally, therefore, items should be at the same angle and the same distance from the operator and situated close together where possible.

The angle of the screen and the source document should be the same and recommendations are that the screen angle should be between 75° and 110° to the horizontal (Grandjean, 1987). The precise angle of the screen will depend on the operator's position relative to it. Ideally, the screen should be at right angles to the operator's normal line of sight.

Incorrect positioning of the screen can cause the head and neck to be held at an awkward angle. The angle between the neck and head and the rest of the body should be between 0° and 15° and should not exceed 20–30° for a prolonged period. Looking up for too long, causing a backward tilt of the head beyond 30°, is extremely uncomfortable, and discomfort has been found to be greater when the screen is too high, rather than too low (Grandjean, 1987). Rea (1987, p. 18) describes the effect of this problem on the staff of Bath University when automated equipment was placed on an issue desk designed for a manual issue system. The VDUs were perched on the ledge which was at standing height for readers and the staff, who were sitting below the level of the

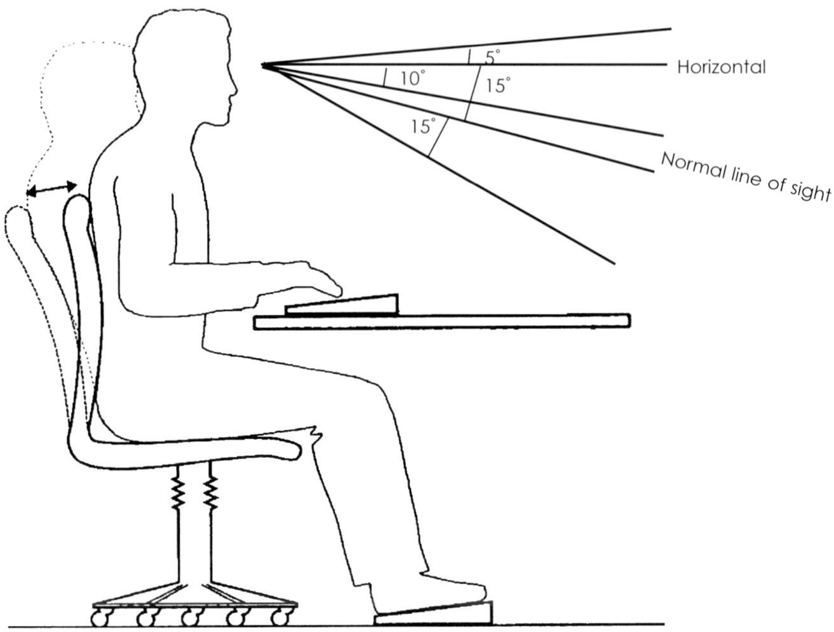

Figure 14.5 Operator viewing angles

VDU, had to look up to see the screen. This 'caused stiff necks and eye-strain (the alternative of standing and leaning over was just as bad)'.

Some manufacturers have brought out workstations which inset the VDU screen at an angle into the desktop. Typically the screen is positioned at a distance of 1 000 mm from the eye and 20–40 degrees below the horizontal (Nova Solutions, 1993). It has been argued that such an arrangement is less tiring for the eyes (Jaschinski-Kruza, 1990, 1991). However, the configuration requires operators to either droop their eyes to view the screen (the recommended way) or bend their neck more than 15 degrees. Because it is debatable whether operators can always be relied upon to do the former, some sceptics believe that any benefits to the eyes may be outweighed by the postural problems that may be caused, particularly if the angle and distance of the screen is not adjustable to accommodate the needs of different sized people.

More effort is expended when the head is moved in the vertical plane, such as when nodding than when it is moved horizontally (Cakir *et al.*, 1980). Ideally, therefore, the source document should be held on the same level as the screen to permit side to side, rather than up and down movement. In order to position the source document at the same angle and distance as the screen, a copyholder may be used. This is particularly useful for data input such as word processing or cataloguing from a source document. It should be noted, however, that the more the angle between the lighting and the document is reduced, the less clear that document may appear. However, Damodaran *et al.* (1980) recommend the use of a document holder to reduce the luminance contrast between the screen and the source document. In support of the use of document holders adjacent to VDU screens, Brand and Judd (1993) found they reduced text-editing time by approximately 15 per cent.

A range of copyholders is now available:

- Those that can be attached to either side of the VDU screen, offering a narrow viewing distance between it and the source document and a similar angle.
- Free-standing models, made of perspex or acrylic. These may be able to accommodate A4 or even A3 paper, but are not designed for books.
- Clamp-supported or heavy-based holders with an angled arm bracket. Once again, these are not appropriate for books, but will hold A4 or A3 paper.
- Electrically operated copyholders with foot pedal adjustment.

With such a range, it is useful to know what features to look for. The following criteria may be used:

- There should be a low reflectance surface to minimize reflected glare.
- It is useful to be able to alter the rotation, distance and slope of the holder.
- A reading guide or cursor line is helpful.
- It should be possible to adjust for varying thicknesses of source documents.
- It should be adjustable by the operator to the right or left of the screen.

Most copyholders are designed for single sheets or small pamphlets and few offer the necessary support or allow for the thickness of books. This may be overcome by the use of a lectern but, in one library, it became necessary to have a copyholder specially made. This was to assist in the cataloguing of heavy reports, many of which had long titles or large numbers of authors, and so required the assistance of a cursor line.

The type of work performed will also affect the type and position of copyholder which is best suited to the task.

Three main items need to be comfortably in view: the screen, the keyboard and the source document (or paper for notes). The relative position of these to one another needs to be such that the correct angle of view is maintained and unnecessary head and neck movements are minimized. As with reach, view must take into account the relative frequency with which each element is observed.

WORKSTATION DESIGN

In order to permit efficient operation, workstation design must take into account factors such as the task to be performed and the physical limitations of the operators. Other factors which must also be considered are: the need to adjust the system should requirements change; storage requirements; ease of cleaning; the amount of work surface required; the need for cabling; maintenance and security; the preferred construction material; and the effects of static electricity. These are discussed below.

FLEXIBILITY

Beckman (1987, p. 16-2) argues: 'There must be flexibility in the design of furnishings which house the equipment because the equipment size and communications connections may change.' Flexibility for the different sizes and shapes of employees is also important. However, they may need to be educated in how to adjust a workstation to suit individual requirements as experience indicates that staff do not usually make adjustments to furniture (Rea, 1987; Ralli, 1988). Flexible, modular and adjustable furniture is now easily available (see the list of suppliers in the Useful WWW Sites and Addresses section). The layout must allow for operator preference in terms of posture, arrangement of items, and performance of the task.

STORAGE REQUIREMENTS

There should be enough storage space to hold all the items needed to perform the task, including manuals, spare paper, forms, trolleys or shelves for books awaiting or following processing, and personal items, such as handbags. The need to be able to access storage easily should be taken into account, as should the need for locks. In many libraries, lack of storage space is a problem.

EASE OF CLEANING

A dirty library or workroom is bad for public relations and staff morale. Dirt traps, such as poorly designed drawer handles should be avoided. The use of dust covers for VDUs and keyboards is also recommended for when the equipment is not in use. They can also protect the equipment from other hazards, such as coffee spills.

WORK SURFACE AREA

If a VDU is placed on an existing desk the usable space on that desk is reduced. Space needs to be available in front of the keyboard to allow the operator to rest their hands and forearms. It may be necessary, too, to be able to push the VDU backwards or forwards to cater for different preferred viewing distances or to give more space when the VDU is not in use.

The use of a 'buffer' area may also need to be considered for incoming and outgoing documents or books. For example when checking in journals, there needs to be space for the waiting pile, the item being processed and the output pile. Similarly, when cataloguing, a small pile of books awaiting processing should be within easy reach of the operator to avoid having to twist and reach for a new item from the trolley or shelves each time. The need for a buffer area within comfortable reach may be met effectively by curving the work surface around the operator. This may be done using specially designed or modular furniture (Figures 14.6 and 14.7), or by positioning trolleys close to the operator. The latter arrangement offers little space for access or egress, however, and so is unsuitable for tasks requiring constant movement to and from the workstation.

In order to increase the available working area, various space-saving devices can be used, such as:

- Folding arms which support the VDU screen. These can be adjusted at an angle to suit the operator but tend to position the screen too high relative to the operator. They do, however, make much more desk space available.
- Keyboard holders which enable the keyboard to be slid out of the way

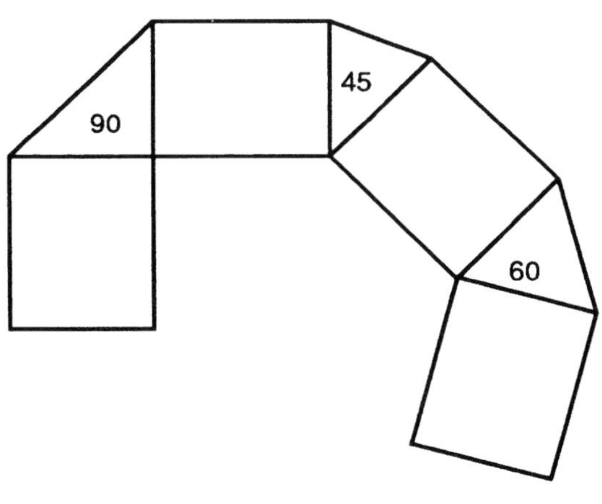

Figure 14.6 The increased work surface offered by modular furniture and a range of angles between adjacent surfaces

Figure 14.7 The use of modular furniture to extend the working surface within reach of the operator. Photo courtesy of Project Office Furniture plc

when not in use. Again, the result may be that the screen is too high if the keyboard holder and system unit are situated underneath.

- Placing the system unit on its side on the floor to one side of the desk. However, this is not without problems (see Figure 14.8). The disk unit is below the comfortable level of reach of the operator, so it is not recommended for tasks where frequent disk changes are important. Also, the system unit and disk may be vulnerable to knocks and dirt, unless it is well positioned and covered.
- The use of portable computers or 'laptops' which are smaller. These have the advantage that they can be easily stored away when not in use and take up little space on the desk. The disadvantage is that the screens on these types of computers are rather low and, if used frequently, may cause neckache.

CABLING

Loose cables under desks or on the surface of desks can be dangerous since they could either get pulled or cause people to trip (see Figure 14.9). This

Figure 14.8 How system units on the floor can be difficult to access and cause cabling problems

can lead to personal injury, damage to the connections (the pins are surprisingly easily bent), or even pull the equipment off the surface and onto the floor. Cabling is an important issue; indeed, it was considered so important in Hillindgon's new library that trunking was said to be the 'backbone' to their IT effort (Westlake, 1988). Not all libraries, however, are in the position of being able to redesign the cabling from scratch. In addition, it would appear that few librarians are in a position to know how to plan for electrical and communication requirements (Somerville *et al.*, 1986), and it is suggested that the importance of providing adequate, safe cabling warrants the advice of an independent expert.

A few general points are worth noting here, however, particularly since it is likely that 'insurers will increasingly demand that librarians show that their staff have competency in basic electrical safety, given their relatively open access to staff, students and the public' (Casale, 1996, p. 27):

- Flexibility should be built into the cabling system to allow for adding or moving VDUs. Beckman (1987, p. 16-3) argues that: 'Panels built into the workstation which hide the conduit or cables at either the base or work surface level can allow for future change. Equipment

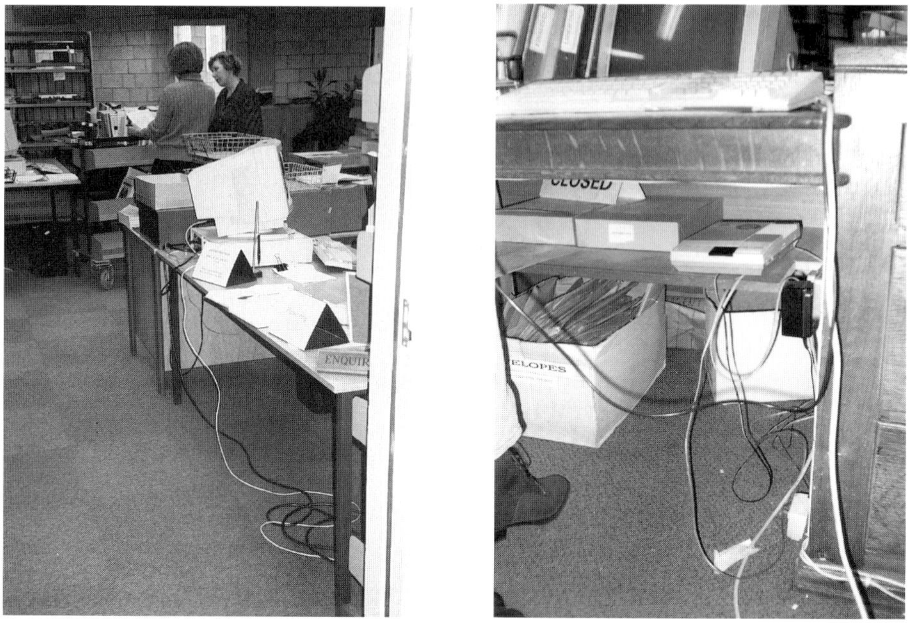

Figure 14.9 Trailing wires and the potential hazards they cause

bars which group a variety of electrical and telecommunications features provide flexibility for groupings of equipment in confined areas.' One library visited by the authors used a slightly different method to hide the cables at the issue desk. A screen was placed at the back of the VDU screen which protected the user from the back of the VDU (and vice versa!).

- Wires should be unobtrusive. Most furniture system units now offer discreet cabling inside desk legs, in channels in the desk and through screens or dividers.

A cheaper, but less attractive, alternative is to drill holes in the existing workstation to feed cables through, although future changes may require more holes to be drilled which may be unsightly.

- Accurate anticipation of future cabling requirements is important, although far from easy. Westlake (1988, p. 8) admits that at Hillingdon mistakes were made causing the 'trunking [to be] opened four times, with the inevitable risks for previously soldered connections'. In particular, it is difficult to envisage the future requirements for public access terminals. Fayen (1983, pp. 41–2) urges libraries to think big when it comes to providing OPAC terminals:

> Studies have shown that users want terminals – lots of terminals – and they want them not only in the library but in other locations. For a public library this might mean placing terminals near the entrance, in the children's room and in the stacks; it might also mean providing dial-up access for patrons who have terminals at home. In an academic setting this might mean putting terminals in all the libraries, in the stacks, in study areas and ultimately in dormitories and other non-classroom spaces.

MAINTENANCE

Periodic equipment maintenance will require that access is available to the engineer or, more frequently, ease of access will be required to the printer for changes of paper and so on.

CONSTRUCTION MATERIAL

The material from which the workstation is made should be matt-finished (avoid chrome trimmings or shiny white surfaces) to reduce the risk of glare. It should not be abrasive or subject to splintering, and it should be easy to clean and hardwearing. It should also be attractive. Opinions on what is attractive will, however, vary. Becket and Smith (1986) report a range of patron reactions to their reference desk which was made in an unusual wood. Comments ranged from 'it's gorgeous' to 'it goes well with your computers' and 'it's got to be the ugliest piece of furniture I've ever seen in my life'.

The construction of the workstation needs to be firm, but heavyweight furniture may hinder flexible office arrangements. Items of furniture for particular functions should be well suited to their task. Printer tables, for example, should be stable, avoiding vibration when the printer is operating and should have adequate space for the storage of paper and the collection of printouts. All exposed corners should be rounded.

COMPUTER SECURITY

Computer security and safety needs to be ensured, especially for smaller computers and laptops which are both desirable and portable. Minimal static electricity should be generated (see below) and air flow through the intake and exhaust vents should not be obstructed. Large temperature and humidity variations should also be avoided.

STATIC ELECTRICITY

The potential for static electricity should be reduced. Static electricity can be generated by people and machines. People build up a store of static by creating friction with synthetic material, such as walking on synthetic carpets. The static is discharged in contact with metal furniture and a shock is received. Static can also be generated by, and transmitted to the computer, causing potential damage to the software, the data and the computer itself. (For methods of reducing static, see Chapter 8.)

EFFECT ON PRODUCTIVITY

The demands placed on the furniture in order to create a comfortable and efficient workstation are greater than most ageing library furniture will tolerate. Although it is appreciated that financial restrictions will preclude many libraries from achieving the luxury of an 'ideal' workstation, it is worth noting the kind of furniture which is available, should funds permit.

Reference has already been made to unobtrusive cabling, adequate leg room, adjustable desks and suitable printer support. Although such ideals may seem remote, they are achievable using the specially designed system furniture offered by some of the suppliers listed in the Appendix.

Dressel and Francis (1987, p. 1) investigated the value of improving the layout of conventional furniture of a group of employees (Improved-Conventional Group) compared with supplying furniture specifically designed for computerized workstations to another group (the Systems Furniture Group). They found that:

> The results of the study supported the hypothesis that improvements to the workstations would significantly, and positively, impact both productivity and satisfaction. The Systems Furniture Group demonstrated a significant increase in both productivity (20.6%) and satisfaction..., while the Improved-Conventional Group showed a non-significant increase (4%) in productivity but a significant increase in workstation satisfaction.

Similar results were found by another study which surveyed 90 office workers before and after workstation adjustment (Halpern and Davis, 1993). The reconfigurations resulted in a significant decrease in bodily discomfort and a perceived increase in efficiency and the usability of office furniture and equipment. It can be concluded, therefore, that an increase in productivity is likely to result from an improvement in workstation design.

CHAPTER 15

WORKSTATION LAYOUTS FOR LIBRARY FUNCTIONS

Because of the variations in library operation, the following layouts for the issue counter, and workstations for cataloguing, serials control and online searching serve only as a guide to illustrate the theory which has been presented. They are by no means definitive.

ISSUE COUNTER

The issue counter has many features which make it unique. Its special characteristics include: the need for staff to interact with users standing in front of them; a (normally) large number of books being handled; returns needing to be stored until they are re-shelved; the requirement for temporary storage for reserved items awaiting collection; the space need for equipment other than VDUs such as security devices, date stamps, tills and so on; frequency of movement; and, often, several staff behind the counter.

The aims of issue counter work have not changed, but the use of computers has altered the way in which the task is performed and the shape of the workspace needed to support the operation. Instead of trays of cards, shared by all staff, it is usual to have terminals or computers. Unfortunately, all too often libraries making the transition from a manual to a computerized system in the past have 'made-do' with the old issue counter often with disastrous results. In one library, visited by the authors, the issue counter was changed only when a 'second-time-around' system was introduced (see Figures 15.1 and 15.2). By this time, many staff had developed health problems and, in particular, WRULDS (see Chapter 4). Fortunately, the new issue counter and system, combined with new working practices, resulted in considerable improvement to staff health and morale.

Rea (1987, p. 18) also describes the problems found at the University of Bath when VDUs were introduced to the existing issue desk:

> There was precious little room for the apparatus, and when Visual Display Units were introduced they had to be balanced precariously on

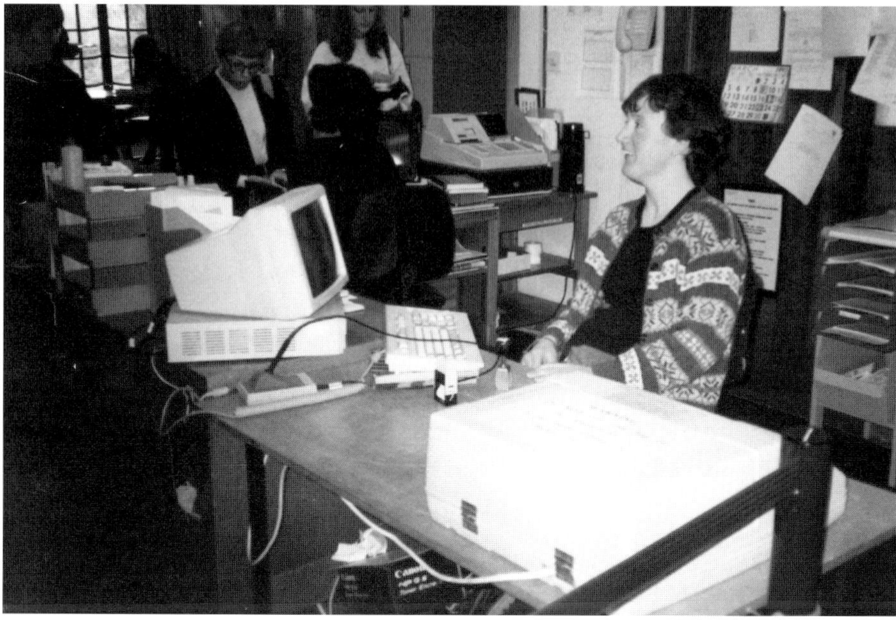

Figure 15.1 Original counter

Figure 15.2 Redesigned counter

> the 9" ledge. This was risky in itself, but additionally the rear of the unit with all its attendant wires was exposed to the public gaze creating a potential hazard. Also, some of the early models had their contrast control at the rear and this proved irresistible to some readers.

A different layout is therefore required for the automated system. The primary consideration is perhaps the need for staff to be able to interact effectively with standing users. Factors which must be taken into account are the need to converse and the need to exchange items. If staff and users are too far apart, conversation can be strained, especially if the background noise level is high. De Jong and Roggeveen (1969) recommend a distance of about 1 000 mm between the two. The need to see the customer and to be able, easily, to transfer books implies that there should be no undue difference in working height between the customer and staff member. Lifting weights up to or down from the raised ledge can be tiring, especially in a sitting position and can impose a strain on the spine. Rea (1987) reports that staff complained of back, neck and arm pains in performing these manoeuvres.

Grey and Wilson (1984) studied a number of issue counters and undertook experiments involving librarians and computer simulated equipment. For busy libraries they recommended the use of sit/stand counters where the height allows staff to either stand or sit using high chairs. These types of issue counters:

- provide more flexibility for staff, those preferring to stand are able to do so;
- allow staff to move easily from a sitting to a standing position or vice versa;
- enable staff to be more at the level of the user (see Figure 15.3);
- reduce the amount of lifting of books.

Ideally sit/stand issue counters should be between 900 and 935 mm high, 775–800 mm deep, flat, and each service point should provide space of at least 700 mm wide and 600 mm deep to provide enough room for chairs and to give leg room. The counter should also be not more than 50 mm thick at the service points; shelves underneath the counter at these areas leave little room for thighs and legs.

Preferably counters should also be fitted with an invisible channel or channels to house cables; fitted with foot space on the customer side to allow them to approach more comfortably; be aesthetically pleasing and modular in design to provide flexibility; and be smooth and rounded at edges for health and safety reasons. Parcel shelves should be avoided where possible especially in very busy libraries. They are not aesthetic and encourage people to delay getting books out ready until they get to the issue desk which slows the whole process. They can also be an obstruction

Figure 15.3 Low desks and the problems they cause clients and staff

and prevent library users getting close to the counter which results in more stretching to pass and retrieve books. This is particularly true if the shelf is not set into the overall top depth front to back.

The area around the counter should allow enough space for all the staff. Grey and Wilson (1984) argue that psychological as well as physical space needs must be considered. They recommend a working width of 1 200–1 600 mm per person with an area behind the counter of at least 1 500–2 000 mm. The latter should provide enough space for a trolley and two people without having to 'breathe in'. The space needs of performing non-VDU tasks – for example, reader registration or alternative clerical work – must also be considered.

The equipment used at the counter needs to be positioned which comfortable reach of all. The security device or cash register should not be situated for the convenience of one staff member at the expense of others, and the positioning of terminals at corners should be avoided. To ensure adequate working space, measurements can be made on the basis of the reach envelopes described on pp. 183–85. Shared equipment may, if necessary, be situated in overlapping maximum reach envelopes but near-reach envelopes should not overlap.

The siting and use of security devices, less formally known as sensitizers and desensitizers, can be problem. Some devices can cause interference on VDU screens if they are within 900 mm of each other.

Using LCD screens may overcome this, however. Staff often use sensitizers and desensitizers incorrectly. If not checked, this may lead to medical problems. Counter inset models are more user friendly than desktop models because they require less lifting of books and fewer awkward wrist actions.

Ideally VDU screens should be placed on stands to enable them to be adjusted to the correct height. For a standing operator the height is 1 400–1 650 mm to the mid-point of the screen and for a seated operator it is 940–1 100 mm. VDU screens are best read at an angle of 10–20 degrees below the horizontal at eye level of the user at a distance of approximately 700 mm. Stands not only allow staff to adjust the height to suit their stature but also allow VDU screens to be swivelled more easily for users to see the contents of screens and frees space on the issue counter.

Tall chairs are obligatory if staff wish to sit at sit/stand counters. An adjustable draughtsman's chair is most appropriate, offering a swivel seat for ease of access and egress, an incorporated footrest, castors and suitable back support. These are available from some of the suppliers listed in the list of Useful WWW Sites and Addresses. A raised chair also eases movement on and off the chair, if it is set at just below hip height.

If an enquiry point is to be incorporated into the main sit/stand issue counter then this should be at a lower sitting height of between 700 and 720 mm. Counters at this height will enable staff to perform clerical tasks more comfortably between possibly infrequent enquiries and have the advantage of being able to accommodate wheelchair users.

In quiet libraries it may be preferable to have issue counters at the lower height of 700–720 mm. This does mean, however, that all tasks performed at the counter should be undertaken sitting down. Staff should not be expected to do any work standing up; this will cause backache and other health problems.

It should be noted that there is a move towards self-operating issue and discharge systems as a means of supporting the issue counter. The implementation of these systems alter the use of issue counters substantially. Much of the repetitive nature of issuing and discharging books can be left to a machine, leaving staff free to deal with more interesting and challenging work.

CATALOGUING

Like all other library functions, the nature of cataloguing varies between libraries and even within the same library in terms of the demands it places on workstation design. Cataloguing may be retrospective or current. If retrospective, machine-readable records may be downloaded by online links to a MARC database or they may be input directly from the stock or the manual catalogue records. Current cataloguing varies similarly between searching for and downloading relevant records and creating original catalogue records at a single terminal. It is possible that a mixture of all these may occur at a single terminal.

The nature of the task will affect the way it is performed and the way in which the workstation must be designed to minimize physical discomfort and increase efficiency. Cataloguing may therefore be divided into the following categories of VDU work.

- *Intensive data input*: for example, retrospective conversion from catalogue cards. This is usually performed by a typist copying from the cards and can induce high-speed input.
- *Original cataloguing*: whether current or retrospective. This is frequently performed by a cataloguer, who may not be a trained typist and who may therefore need to see the keyboard as well as being able to manipulate the item being catalogued.
- *Interactive or conversational*: where MARC records are downloaded from a central source online, for example. This may be done by a senior library assistant who inputs a key (for example, ISBN or author/file key), waits for the computer response, checks the screen for the accuracy of the retrieved item and selects the record for downloading.

The predominating task is the one which should dictate the layout of the VDU (see Figure 15.4). In the last two cases, the following factors also need to be considered:

- Space is needed within the maximum reach envelope for a small pile of books awaiting processing.
- There needs to be room to open and move the item being processed (atlases and some picture books take up a lot of space and it may be necessary to write in the book – for example, to assign an accession number).
- Space is needed for a completed item. It is desirable for this to be close by and on the same level to avoid continuous lifting and twisting.

SERIALS CONTROL

Serials check-in is probably the most VDU-intensive part of the automated serials control process. The requirements of the workstation are: a buffer area for serials awaiting processing; a work area for the serial currently being checked-in with space to manipulate it to obtain details of volume,

Figure 15.4 A workstation layout for cataloguing

issue and so on; an output buffer for items which have been processed and bins for waste paper such as journal wrappers. Additional space may be required for a printer to produce routing (circulation or allocation) slips if required and a date stamp to indicate the date of receipt. The operator will need to see easily the item currently being processed and to be able to check the data with that on the screen. Minimal keyboarding may be required. Some systems have now introduced a barcode reader for inputting bound serial volumes and reports. However, as most single issues of serials do not carry a barcode, a keyboard will still be required (see Figure 15.5).

ONLINE SEARCHING

Online searching is essentially a conversational task requiring constant viewing of the screen and intermittent keyboard use. Printouts of search results need to be easily accessible and preferably within viewing distance during printing to assist in monitoring the search. Space is needed for the storage of manuals and thesauri, both those to which frequent reference is made and those which may only be used occasionally. Consideration also needs to be given to necessary interruption from and interaction with

Figure 15.5 A workstation layout for serials check-in

clients who may wish to be present at the search. Provision may need to be made for personal belongings and any alternative task which may have to be undertaken at the workstation.

RECOMMENDATIONS

The above layouts are only suggestions for workstation designs as it is not possible to legislate for the range of library workstations which exist. In order to design the optimal workstation layout for a particular task or set of tasks, it is necessary to begin by undertaking a thorough task analysis. If other staff are affected, it is wise to get them to participate in designing the layout and to keep all the staff involved informed of the plans. Professional advice may also be helpful.

Once the analysis has been undertaken, it may be best to draw the proposed plans to scale on graph paper, to check that it is workable and to help solicit opinions from other staff if needed. It may even be helpful to set up a mock-up workstation with suitably sized cardboard boxes as these are both lighter and less fragile to move about than computer equipment.

As technology is continually changing, flexibility should be the keyword in choosing and arranging furniture and cabling, in order to cater for present and future needs.

Once the preferred workstation has been chosen, it should be left in use for a month or two, at which stage it is wise to check it over. Ask the opinions of any staff who use it with regard to ease of use, health and safety factors and general attitude. If problems exist, redo the task analysis and check to see if any improvements can be made.

It is in the nature of library and workstation design that no ideal solution will be found and compromises will inevitably have to be made. The objective is therefore to choose the layout which optimizes performance and maximizes operator well being, comfort and safety.

PART

VI

THE HUMAN–COMPUTER INTERFACE

CHAPTER 16

THE SOFTWARE INTERFACE

INTRODUCTION TO SOFTWARE INTERFACE

Although many libraries are now automated to some extent there are many that are still operating manual systems. Automation is not necessarily needed by all libraries. Special libraries which are small and infrequently used may not benefit from an automated circulation system, for example. However, most libraries can benefit from some form of automation.

The decision to automate a particular library's operation can be made for a variety of reasons. The library may need to improve efficiency because of increased demand or reduced staffing levels. Routine and time-consuming clerical procedures may need to be reduced or eliminated. Often, the recognized ability of computerized systems to offer an improved range of services, by making more effective use of existing data and resources, is the instrumental factor. The online public access catalogue (OPAC), for example, offers facilities which improve its usability and functionality over the card catalogue. Before automation, functions such as searching on any field, sorting and combining records, creating custom lists, updating and modifying records were at best time-consuming and labour intensive, and at worst impossible within the constraints of available resources (Henry, 1991). Current awareness and selective dissemination of information (SDI) schemes have only been made viable through automation.

Additionally, libraries are now finding themselves in the position of having to reassess and replace existing systems or add systems, such as CD-ROM or multimedia, to complement their existing computer capabilities in order to meet new needs or offer improved services.

Libraries automating or upgrading an existing system will need to decide whether to purchase a commercially available system or have software specifically developed in-house or by a consultancy firm. Whichever route is taken, the system selected must be thoroughly evaluated both from the point of view of functionality and the usability of the software interface. Functionality of library systems are discussed in detail elsewhere (see Tedd, 1993, for example); detailed here are the human aspects of the software interface.

SOFTWARE INTERFACE

There are many definitions of the software interface, which can also be referred to as the user interface, the human–computer interface and the user–system interface. Bernstein (1989) describes the user interface as the place where the PC and the user meet and the part of the system software that determines how the user and the computer communicate with each other. The latter is often referred to as dialogue design. Borgman (1986a) extends the definition to encompass screen design, the command language, documentation, personal variables of the user, prior computing experience and attitudes among others.

Many writers include the hardware elements of the interaction (e.g. the keyboard, screen, mouse) in the user interface. Above all, the software interface is the method provided by the system for interaction between it and the user. The success of any system will depend on its ability to communicate with the user in a language which is spoken and understood by both the user and the system (Jones, 1988).

The motivation for using an automated system in the library is to find information or to carry out a task. The recent focus on end-user, rather than intermediary or expert usage in libraries requires that the systems employed are transparent to the user, so that they do not need to know the intricate details of how the systems work to use them.

Users do not care about the system they are using unless its use seriously impedes the accomplishment of the task in hand (Marchionini, 1992). Users should not have to conform to a system, rather, the system's job is to help users achieve their goals (Bates, 1990). A well-designed interface will enable the searcher to access systems and navigate them transparently (Hulser, 1993). Hicks and Essinger (1991) state that users should be getting on with whatever they want the system to do and not wasting time trying to make the system work or trying to understand it.

Unfortunately, not all systems are well designed and the costs of poor software interface and, in particular, dialogue design are increased errors in data input and system usage, user frustration resulting in low productivity, employee stress and system under-utilization, and poor user performance because the system is too cumbersome to use or too difficult to learn. Up to 70 per cent of a system's development costs can occur after the initial release because of the need to make the system more usable (Rennie, 1988). Good design, on the other hand, results in increased user satisfaction, usage and productivity, and decreased development and training costs (Johnson *et al.*, 1989). It has been shown that well-designed user interfaces can increase productivity by 25 per cent and reduce errors by 25 per cent.

It is imperative, therefore, for library managers to examine the factors contributing to the effectiveness of the software interface when choosing computer systems. To this end the remaining chapters in this part will examine the human characteristics affecting the use of software, the principles of dialogue design, screen design, interactive feature affecting usability and the process of software evaluation.

CHAPTER 17

USER NEEDS

In his book, *The Psychology of Everyday Things*, Donald Norman (1988) admires the way that the mind handles information gained from artifacts in the world. 'The human mind is exquisitely tailored to make sense of the world. Give it the slightest clue and off it goes, providing explanations, rationalizing, understanding . . . Well designed objects are easy to interpret and understand. They contain valuable clues to their operation.' This has been called the 'latent psychology' of a design.

Dealing with computer systems is one of the most recent skills which many humans have had to acquire in order to cope with everyday life. No longer is interaction with computer systems reserved for the highly educated computer expert or information professional. Many staff at all levels within libraries now have to work with computers and they are increasingly available for library clients. However, the ubiquitous screen/keyboard arrangement at the public access workstation presents no clues, no latent psychology to the library clients as to the types of information available and the way in which the information can be accessed to meet their needs. The users must therefore find clues from the screen display itself, or from their own memory. Additional clues may also be provided in the context/surroundings of the workstation and through instruction or training.

TYPES OF USER

Computer users are not uniform in their experience, skills or expectations. They vary in their knowledge of tasks and of other systems, in motor skills, technical abilities and experience with the system (Card and Moran, 1980). Borgman *et al.* (1989b), have carried out research showing the importance of the need for the interface to be tailored to the expected user population.

Cole *et al.* (1985) identify four groups of users of computer systems (see Table 17.1).

Table 17.1 The knowledge domains of different user types

Type of user	*Knowledge of task*	*Computer knowledge*
Naive user	X	X
Casual expert	✔	X
Associative expert	X	✔
Experienced professional	✔	✔

NAIVE USERS

Naive users have little or no knowledge of the task to be performed or of computer systems. Online Public Access Information Retrieval System (OPAIRS) users will probably include a large number of naive users. Kinney and Pasak (1990) give the following characteristics of the searching behaviour of a naive subject searcher which could represent an OPAIRS searcher (e.g. using a CD-ROM):

- lack of understanding of how information is stored, indexed and retrieved;
- preference for the simplest available mode of interaction with the information system;
- lack of any searching strategy or tactics beyond simple 'subject searching'.

Shneiderman (1992) advises that the vocabulary of systems designed for this type of user should be restricted to a small number of familiar, consistently used terms and the number of choices kept small enabling the carrying out of a few simple tasks to build confidence.

CASUAL EXPERTS

Casual experts understand the task thoroughly but have little computer knowledge. A librarian, well versed in cataloguing, for example, but used to manual systems only, would be a casual expert. Both naive users and casual experts may exhibit anxiety about using computers which inhibit learning. Online tutorials and well-designed hard copy manuals are essential to reduce the distractions of computer concepts and syntax (Shneiderman, 1992). Informative feedback and *constructive* error messages should be provided to encourage exploration of the system by the casual expert.

ASSOCIATIVE EXPERTS

Associative experts have considerable familiarity with computer systems but little knowledge of the particular application. A possible example would be OPAIRS users who are students from a computing studies course but who have little library experience. Associative experts benefit from consistency of action sequences, protection from dangerous actions and a good source of reference of task-related knowledge.

EXPERIENCED PROFESSIONALS

Experienced professionals have skills in both computing and the task for which the computer is used. This class of user needs rapid response times, macro facilities (the ability to pre-programme regularly performed command sequences which can then be actioned with one command) to reduce keying short cuts through menus and the use of abbreviations (Shneiderman, 1992).

Cole and his colleagues point out, however, that these categories are not discrete and that people move between them (Cole *et al.*, 1985). A novice user of an OPAIRS, for example, may become a casual expert in time or an experienced professional may regress to a casual or an associative expert through lack of system use.

The problem of a how a user changes from novice to expert is particularly difficult. The Apple Macintosh and Windows Interface, for example, prints the names of command keys which may be used in lieu of the pull-down menus, actually on the pull-down menus. However, if the pull-down menu has to be accessed to find the relevant command name, it is then much easier to select the command from the options on the menu, rather than to change tactics at this point to use command keys. The user must therefore either rely on incidental learning or make a deliberate effort to remember and use the command keys if they want to graduate to becoming an experienced professional.

If a person is being trained to become a habitual user as part of their daily work, then training should be given in expert performance. Progression from a naive user to that of expert professional without training should not be taken for granted.

ACCOMMODATING VARIATIONS AND CHANGES IN USER TYPES

The type of dialogue best suited to each category of user varies with the user, the task itself and the frequency and duration of the task performed. A naive user, for example, needs prompts indicating what options are available and the implications of selecting a particular option. A casual expert requires guiding through the system, but can comprehend terminology relating specifically to the task. The converse is true of the associative expert who is unlikely to need to be prompted to hit the <Return> key but may require information on the function being selected. The experienced professional will want power and flexibility above all and could become irritated by the distractions of prompts, help messages and lack of speed (Shaw, 1991 and Vandierendonck *et al.*, 1988).

In general, naive or infrequent users benefit from a system-initiated dialogue characterized by simple instructions, menu choices, question /answer interaction or form-filling. This type of interaction requires little training or retention. Frequent skilled users, on the other hand, prefer a 'user-initiated' mode where simple, formatted or natural language queries are entered by the user with no prompts from the system. Command-driven dialogues are flexible and convenient for such users, but require considerable training and memorization (Shaw, 1991; Bernstein, 1989).

The customization of screen displays also needs to take into account the expected user population. For example, the order in which records are displayed, the wording of screens, indexing of fields and default Boolean operators largely determine the effectiveness of subject access for users of the library catalogue (Kinney and Pasak, 1990). Error messages, tutorials and the content of help systems will obviously need to reflect the needs of the users and should vary accordingly.

In choosing software, the nature of the predominant users must be considered. Borgman *et al.* (1990) note the difficulty of satisfying all user types in one interface, and advocate that libraries should lean towards accommodating novices. When interviewing non-users of a library system, they found that this group were unwilling to spend the time to learn the system, even though the same group estimated the learning time to be only 15–30 minutes. This group therefore requires a 'walk up and use' interface which is intuitive to use and requires no investment of learning time.

Complications arise, however, because of the changing nature of users. Resnik (1992), noted the changes in information-gathering habits of aerospace engineers as end users. It was found that engineers in the over-50 age band were unfamiliar with electronic information retrieval, preferring to use the services of librarians if their immediate information needs could not be satisfied from material at hand or by colleagues. Engineering students, on the other hand, took quickly to the use of a menu-driven CD-ROM search system and preferred electronic information retrieval to traditional methods. Subsequently, these same students showed interest and enthusiasm for learning the online database DIALOG command language, thus graduating from naive usage to more expert usage.

Users not only change categories over time, they may change according to the task being performed, so that an experienced professional online searcher may become a mere associative expert when operating the issue system for example. Allowing for the different needs of naive and experienced users makes a difference to the long-term usability of the system (Grossman, 1991). Hulser (1993) notes that no one interface style is the best and only choice for all users and tasks. Personal preferences and style of work will influence the interaction needed.

The development of self-adaptive interfaces which automatically adjust to the user's level of use have been proposed. However, the very adaptation of the system means that the interaction displays inconsistencies to the user's learned expectations, which can inhibit further learning and interfere with the user's existing knowledge.

Belkin *et al.* (1990) note that, with reference to the users of OPACs, the predominant user group is exceedingly heterogeneous on a number of dimensions including age, library and computer experience, tasks and goals, and that individuals consistently, but unpredictably, change their position on these dimensions.

One solution may be to acquire software with an adaptable interface where the user chooses the level of interaction. Some software offers a default menu but with a 'hidden' command-driven interface for experienced users. Norlin *et al.* (1992), for example, discussing the design decisions on a new interface to the BRS/SEARCH software to replace the BRS/MENTOR interface, describe a combination menu selection/dialogue Windows interaction using form fill-in (see 'Form fill-in, Chapter 18, p. 228) and natural language for less experienced library users, with an optional switch to the BRS search mode using the '..' commands for staff and users experienced in this interaction method.

SHORT-TERM MEMORY AND OVERLOAD

Whatever the type of user, interacting with a library computer system will entail the use of the user's short-term memory for information processing. There appears to be a fixed upper limit in a person's ability to process information (see Chapter 1) which cannot be extended. If the demands of the interface exceed the information processing ability of the user, mental overload will occur and user performance will deteriorate rapidly. The age of the user also has a bearing on the amount of information which can be held in short-term memory, which decreases for users over 55 years of age. This is of particular relevance for systems which are to be employed in library environments where significant numbers of older users are expected to use the system.

Orna (1992), reporting on the layout of information on paper, makes an observation that equally applies to using an automated system. She notes that users operate a mental accounting system by which the effort required to gain understanding of what is presented is balanced against the memory load, and, if the effort is felt to be too great, they will quit (reading). Users also tend to trade off performance for minimum cognitive load (Marchiononi, 1992), even though this entails extra physical effort. For example, using a library OPAC, the command <Esc><B> moved the display to the previous screen, and <Esc><M> moved the display back to the start screen. However, repeating <Esc><B> also eventually returned the user to the start screen and users were observed using this tactic, hence only having to remember one command.

PSYCHOLOGICAL CLOSURE

An important factor in the software interface which will alleviate the load on the user's short-term memory is psychological closure and must be remembered when selecting software. When a task is undertaken, the user must make a conscious effort to remember tasks, actions and facts pertinent to the task in hand. This causes some stress until the task has been accomplished, at which time a feeling of relief and satisfaction is experienced, as the factors retained are released from short-term memory. Psychological closure can be enabled by:

- *The information being kept on screen.* The display then acts as an external memory.
- *The user being required to input information only once.* An online information retrieval system which permits searching of several databases, for example, should permit the search to be saved and carried out automatically, if required, on each database rather than requiring rekeying every time the database is changed. Most online databases provide this facility, for example, Dialog and DataStar, in addition to a backtracking facility which retains previous input. Systems are also available now which allow several databases to be searched simultaneously using the same request information.
- *Navigational and status information being provided.* The user needs to know the current position in the system, what actions the system is currently carrying out and what action they should have to take. This can be achieved for example, by page numbering, suitable titles on the display and appropriate screen messages, and instructions for moving to a previous or different screen.
- *Having a screen display which is clear, uncluttered and easy to read.* An overcrowded screen is both difficult to read and to assimilate information from. Only the information which is needed for the task

in hand should be displayed and it is better to have several consolidated screens with sensibly grouped and labelled information. For example, an OPAC's catalogue screen needs to hold an input area, bibliographic, current loan and location information and menu options. One screen containing all this information would be difficult for the OPAC users to deal with. Using different display screens users can choose to navigate through all the screens available, or simply to go directly to the screen containing precise information for their requirements, thus achieving clarity of display with comprehensiveness of information.

- *Having help and instructions provided onscreen*. This includes the provision of clear and consistent labels and headings, navigational instructions and context-sensitive help.
- *Several choices being offered*. Recognition performance is generally better than recall performance. Most naive or infrequent users prefer to have options displayed on screen rather than having to remember them. This means that the choices available do not have to be kept in long-term memory, but can be accessed by recognizing what is required from an offered list. A menu-based library OPAC would clearly be preferable to library users than a command language-based system.
- *The avoidance of abbreviations and acronyms*, unless there is evidence that the use of these is unambiguously understood by the entire user population.

PERCEPTION

The process of visual perception at the software interface involves the user discriminating between the object of interest and the rest of the display (DTI, 1990). The ability to recognize an individual item of information from a crowded screen is therefore affected by the user's experience and expectation and the way in which the item is displayed. An operator who is experienced in the use of the system and is familiar with the data and their position on the screen will be quicker to perceive relevant information than a novice user.

To aid perception, the information on the screen display should be:

- non-flickering and easy to read at a glance;
- distinguishable from the background;
- comprehensible;
- uncluttered;
- meaningfully structured.

Redundant information, which overlaps and reinforces displayed information can be used to increase perception and hence detection of relevant items (Warg, 1988). To assist a novice user, the system needs to use various facilities to highlight specific items to attract the user's attention. Commonly used stimuli include bright colours, bold or flashing text and inverse video, although too much use of these stimuli can induce eye fatigue and headaches.

LONG-TERM MEMORY AND MENTAL MODELS

Geyser and van Brakel (1991) define the mental model as

> the internal representation of the system which the user is assumed to have – it is a model of the system that the user conceives in his head and can be regarded as a qualitative simulation of the model of the system which the user can 'run' to determine his current location in the system, to debug errors and test hypotheses concerning system behaviour.

When confronted with, for example, a CD-ROM bibliographic database system, the user will need to ask, 'Have I used this system before?'. If the answer is 'Yes', then the user will recall the information concerning the system from long term memory into short term (or working) memory to help perform the actions needed to interact with the system. The mental effort required to recall the memory will vary, and the required information may be recalled totally, partially or not at all. Consequently, the mental model they have of the system may be complete, sketchy or non-existent.

If the answer to the question is 'No', or they have completely forgotten how to use the system the user will try to build up a picture of what the system is, how it works and what it does, both for immediate use and for future reference. This picture is the user's mental model of the system and is formed by observation, the context of the system, hearsay, training and comparison with other systems of which the user has knowledge. For example, the CD-ROM system could be compared to the OPAC or an online retrieval system. Furthermore, the user's mental model is constantly reformulated by interaction with the system, each time the system reacts in a way not anticipated by the user, and the model is therefore progressively refined. The mental model formed may be complete or incomplete, correct or incorrect. The degree of completeness and correctness of the user's mental model will affect the understanding and use of the system, and determine whether the user gains full benefit from all the features of the system.

Marchionini (1989) investigated the importance of mental models when making the transition from print to electronic encyclopaedias and found that the mental model which a user builds reflects the user's prior experience and knowledge. Occasional users of a system may realize that their mental model is inadequate to deal with the system, but that the existing exposure to the system, coupled with the 'forgetting' element of infrequent use, inhibits refinement of the mental model. These users need assistance either on screen, or available by pressing explicitly labelled keys.

The mental model built is individual to each person. Van der Veer *et al.* (1990) define mental model builders as being 'imagers' or 'non-imagers'. The former construct mental models according to verbal or visual–spatial metaphors presented during training. Non-imagers, however, need a lot of hands-on experience with examples in order to construct their mental models. Trainers of computer systems need to be aware of these differences and adapt their training material and approaches accordingly.

DESIGNERS' CONCEPTUAL MODELS: TRAINING AND THE USE OF ANALOGY AND METAPHOR

To enable the user to create an adequate mental model of the system in question, the designer provides a conceptual model which is passed to the user through training, documentation and the software interface, to build up a framework on which to base subsequently performed interactions with the system.

Experiments were carried out by Borgman (1986b) into the effectiveness of training in the use of a prototype Boolean logic-based Information Retrieval system. Users were given either procedural (step-by-step) training or conceptual training based on a system model. Although it was found that conceptual training had no effect on simple procedural tasks, users with such training fared much better on complex tasks requiring extrapolation from the basic operations.

Knowledge of the mechanical (procedural) aspects of a system enable a user to use the system while conceptual knowledge enables the user to fully exploit the system. Geyser and van Brakel (1991) point out that it is the job of the system designer to make the system's conceptual model match the user's mental model and therefore, in constructing the system and the conceptual model, the designer must have a clear understanding of the intended users and their needs, expectations and experience. When selecting software, a library manager must assess whether the designer had this clear understanding and therefore whether the system meets the needs of their intended users.

Much use is made of analogies (likening to known systems) in explaining new systems, for example, the desk top and dustbin analogies of the Apple Macintosh and Windows interfaces. An analogical model involves the comparison of two complex systems. One of these is likely to be familiar in an everyday context, and the other system is likely to be less familiar, or even unfamiliar, to users (Halasz and Moran, 1982). However, there is a danger that direct comparison with other objects or systems will disguise the full capabilities of a new system. The analogy 'A word processor is like a typewriter' for instance, or 'An OPAC is like a card catalogue' give only a partial view of the power and capabilities of the systems in question. One thinks of 'Six Blind Men of Hindustan' in the traditional Indian children's poem, who were each only able to gain a partial mental model of an elephant from the part of the animal with which they came into contact (Sake, 1996). Furthermore, some systems violate their own analogies. Take, for example, the dustbin analogy used by the Apple Macintosh interface. Files are deleted by dragging their icons to the dustbin. However, the dustbin is also used to eject a disc. Many new users are fearful of doing this in case all their files get deleted.

However, it has been found that the use of analogy, as a base to explain a particular part of a system in supervised training, is useful in helping to construct part of the user's mental model. This is then augmented by instructions into the unique characteristics of the new system.

Marchionini (1989) observed that new users of an electronic encyclopedia, who transferred only content knowledge rather than organizational and access knowledge from the print version metaphor, were able to begin building mental models of the new system very quickly. Those, however, who transferred all the details from the print version metaphor were unable to accommodate new features of the electronic system.

Designers and trainers of library systems can therefore help users by using analogies and metaphors judiciously to introduce new users to the system, but explicitly highlighting unique features not appropriate to the metaphor.

CONSISTENCY AND TRANSFERABLE SKILLS

Learning and acceptance of a new computer system is easier if the user can link in with already known facts and procedures. Users learn what to expect from a given category of systems because previous products of the type have displayed some standardization by displaying features in particular ways, for example the use of the F1 function key to access Help screens. Users of systems displaying such generic standardization benefit in that they can apply transferable skills from a known system to the new system. Interface consistency between similar applications is therefore an important factor when choosing a new, replacement or complementary system and affects users with prior experience on alternative systems.

Matthews (1987), on the design of online catalogues states, 'As online catalogs proliferate and users move from system to system, it becomes crucial that the users of the online catalog be presented with screens which are relatively similar in layout and content . . . familiar screens mean less time will be needed to (re)learn the use of the online catalog in a variety of library settings.' Matthews points to consistent display formats and labelling as being important characteristics to be incorporated into the screen design to enable transferable skills to come into play. He also gives more specific guidelines for the screen layout of online catalogues.

Snow (1986), writing in the Caduceus column in Online, calls for consistency between the screen displays and syntax employed by database producers and hosts for online information retrieval, where, of necessity, the user needs to access varying hosts and databases continuously. Users cannot hope to retain the details of all the different procedures and displays utilized.

Consistency within an application however, is a more important factor to be borne in mind when selecting a system. Internal consistency affects all users of a system and ensures reinforcement of learned procedures by repetition over all options offered. For example, error handling procedures need to be clear and consistent throughout an application. Spigai (1987), discussing displays in database systems, notes that whereas new database interfaces tend to be internally consistent, there are often changes made as databases grow in size and evolve to include additions in scope, which leads to internal inconsistency. Generally the older the database is, the more inconsistencies it will display.

Unique application interfaces lead to a decrease in usability and to difficulties remembering the unique methods used by different systems. Polson (1988) quotes a 100–300 per cent reduction in training times for subsequent task training as a result of employing consistency within and across applications. Inconsistency, however, blocks the transfer of skills, and increases both forgetting and learning time.

Grudin (1989) discusses three types of consistency in software interfaces, internal, external and correspondence to familiar features in the real world. He makes the following observations regarding the different types of consistency and expert usage. Analogy to a familiar domain facilitates immediate learning but can have negative consequences for subsequent performance. Ease of learning can conflict with ease of use, as built-in consistency often breaks down when expert users' tasks are better understood. There is often a trade-off of the conceptual model against the users' work context with a result that the elegant, consistent logical designer's conceptual model has to be altered to fit in with end user requirements. The best type of consistency for a skilled user interface therefore, is that which maps to the user's perceptions rather than that which blindly copies other systems or follows the designer's conceptual model.

SUMMARY

Users are by no means uniform. Thus, in choosing software, the nature of the predominant users must be considered. Complications arise, however, because of the changing nature of users. Users not only change categories over time, they may change according to the task performed. All people have characteristics which also affect the way in which they process information which, in turn, has implications for the software interface.

CHAPTER 18

DIALOGUE DESIGN

PRINCIPLES OF INTERFACE DESIGN

From a human factors viewpoint, the way in which the system achieves a function is just as important as what it achieves. A system which is cumbersome in use will offer few benefits for the operator (if any) over a manual system, and will be resisted by end users, even if advantages such as increased management information statistics are available.

The interaction between the end user and a computer system is a two-way process. In most library systems, the end user gives instructions by keying in at the keyboard or by making selections using the mouse, and receives feedback through the VDU screen. However, there are a variety of different input methods and dialogue styles.

INPUT METHODS

The dialogue between system and user relies on there being a means for the user to input a request, command or reply into the system. There is a vast array of input devices in use and many more under development. Not all input devices are suitable for library systems, but some of those which are commonly used are now discussed from the point of view of the dialogue between the user and the system and their ease of use.

KEYBOARD

The keyboard – the most commonly used input means – is used to enter textual and numeric information, to move around the screen display using cursor or control keys, to give certain programmed commands using labelled function keys and to send information to the computer using the <Enter> key. The use of the keyboard presupposes some typing ability and keyboard familiarity. Unfortunately, all too often staff are expected to use keyboards without any training. Not only is this inefficient but may leave staff being exposed to upper limb disorders (see Chapter 4).

MOUSE

The mouse – the pointing device for control of the movement of the cursor – is becoming increasingly more common as a means of input, and is often used in conjunction with the keyboard. It has been found that the use of the mouse needs less mental translation to match the intended movement of the cursor into the hand movements needed to effect the change compared to using arrow keys. However, effective use of the mouse does require some experience and if used extensively or incorrectly may result in mouse-arm syndrome (see Chapter 4).

Other similar cursor control devices are available although less popular. Tracker balls are an example of these.

BARCODE READERS

A barcode reader is a hand-held or embedded device which reads a printed or magnetic bar code which is passed over the reading head (see also Chapter 10). It is used to input identification numbers rapidly and accurately, such as accession numbers and reader numbers at the issue desk, where the amount of information to be input is limited but unmemorable, and the chance of keying error is high. This input system can sometimes be temperamental, for example where the barcode is physically damaged, and an alternative keying method must always be provided. Laser barcode readers are generally easier to use than a hand-held wand.

DOCUMENT READERS

A document reader is a high speed scanner which reads whole pages and is used to input large amounts of text, or pictures and diagrams. It is used to record the appearance of the hard copy page and can be used for archiving (DTI, 1990). Document readers may also be used in the library to 'read' to visually impaired library clients.

TOUCH TECHNOLOGY

This is a less familiar method of input which removes the need for keyboard or mouse expertise and is therefore useful for 'walk up and use' systems (see also Chapter 10). It is used by touching the screen or a touch pad in an appropriate position in order to select from a number of choices offered on the screen. This input method is popular for public access systems which are accessed in communal places such as shopping precincts or health centres. Users gave the following reasons for preferring touch screens more than the keyboard (Maddox, 1989):

- Concentration is improved and the eyes are less fatigued because there is no need constantly to change attention and focus from keyboard to screen.
- Touch screens seem to be more friendly for those not accustomed to keyboards.
- They are sometimes faster for entering information.
- They are well adapted for multiple choice decisions.

However, the use of touch technology as a text input device is limited, as the use of an on-screen touch keyboard is slower and more awkward than a conventional keyboard. In comparison to the mouse, the touch screen was found to be quicker in selection tasks, but less accurate in tracking tasks (Fernandez *et al.*, 1988). There are also problems caused by differences in finger sizes; users with thick fingers may have difficulty pointing to exact spots on the screen. As stated in Chapter 10, keeping the screens clean can also be a problem, especially during busy times. Some screens are activated by touch and others by release of fingers after pointing. Those activated by touch can be a particular problem to older or younger people who often move their fingers along the text on the screen when reading to keep their place. Every time their finger accidently moves over a 'hot key' the system moves on, which can leave the user exasperated and with no feeling of control.

FUTURE DIRECTIONS

New methods of user input in human–computer interaction are already in the laboratory and being used in innovative computer systems. Data-gloves allow the users to manipulate screen objects by using their own hand movements. Work is also being continued on gesture, speech recognition and eyeball trackers. Virtual reality systems have already been applied in the fields of games arcades, supermarkets, architecture, and training, to name but a few and are being experimented with in a few libraries.

DIALOGUE STYLES

The optimal dialogue style for a system depends on the goals and experience (or expected experience) of the end user. Different applications will employ different dialogue styles depending on the goals and experience of the intended users of the system. When selecting a system, a library manager must, therefore, examine the dialogue style carefully to make sure it fits the purpose of the system, the type, ability and level of training of the intended user, and existing hardware.

Shneiderman (1992) classifies dialogue styles into five main groups, that is command languages, form fill-in, direct manipulation, menu selection and natural language.

COMMAND LANGUAGES

Command languages, as the name suggests, are vocabularies of instructions which the user can issue to the computer system. By their nature they are concise and powerful, and enable expert users to proceed with the task in hand without delay or frustration. Command language interfaces are user-initiated and minimize system overheads and response time (Cole *et al.*, 1985).

Originating in the early days of computing, command languages require the input of precise syntax by the user and are intolerant of errors. The onus is on the user for navigation through the system. The user therefore must know what functions can be performed and what commands are needed. Training and practice is needed before the user can be proficient in carrying out the interaction (DTI, 1990). Command language interaction in online information retrieval is suitable for search intermediaries where domain knowledge is shallow, but there is a good knowledge of search tactics (Resnik, 1992).

Query languages are a type of command language. Gittings (1986) defines a query language as 'A system designed to support an interactive dialogue for the retrieval, display, and sometimes update, of records, using variable criteria.' A query language system is used in conjunction with databases of records for data management and extracting management information. It is particularly useful for displaying different aspects of a database.

Query languages use a form of structured natural language to express the user's requirements, and avoid the need for the systems designer to anticipate all possible end-user requirements in advance. An example of a typical query could be as follows:

```
? READ BORROWER FILE
? SELECT IF BORROWER_NO IS 1265748
? PRINT SURNAME, ADDRESS GO
```

However, using a query language involves some investment in both learning the syntax of the language and understanding the structure of the database. It is therefore only a suitable interaction style for professional users.

Advantages and disadvantages of command languages

The advantages of command languages are:

- *Flexibility*. The commands which can be given are limited only by the user's knowledge and imagination and not by what is presented on-screen.
- *Power*. The use of a command language enables a user to achieve results with just a few key presses rather than the sometimes tedious and long-winded selection of menu options.
- *Speed*. The commands given are more closely related to the machine code which activates the machine's response, than for example, menu choices. Additionally, expert users can condense several commands into one line of input or specify macros which replace frequently used sequences of commands.
- *Control*. The use of command languages gives all the control of the interaction to the user.

The disadvantages of command languages are:

- Substantial training in use is required.
- No assistance is given to the user by the interface in acquiring a mental model and knowledge of the system's functions and syntax.
- Heavy reliance is placed on the user's ability to memorize and recall the exact syntax of the language, and frequent use is needed to maintain familiarity.
- Command languages often have poor error handling mechanisms and can be very unforgiving.
- Feedback is notoriously poor. In addition to the commands issued by the operator to the system, the program will issue instructions which will convey an impression to the user. The National Electronics Council (1983) cites the following command: 'You have been idle too long . . . Please respond . . .' Other common messages include 'Fatal' error, job 'aborted', and error 'xxx'. While people who are comfortable with computers may simply ignore such prompts and messages, it could be off-putting for new or timid users.
- Very accurate typing ability is required.

The use of command languages is therefore appropriate for frequent use by highly trained expert users. If a person is expected to be a very frequent user, then it is important that training is given in the use of the command language right from the beginning. Transition from novice to expert usage is seldom an automatic progress brought about by practice.

Characteristics of command languages

The characteristics of command languages are as follows:

- The vocabulary is limited and may not be representative of or even a subset of the total vocabulary of the user's natural language. Most of the command languages used in automated library systems are designed by computer programmers rather than by librarians and may not always reflect the terminology with which the latter are familiar.
- There is a limited set of grammatical rules for any one language and each language has its own grammar. Command options to achieve the same effect may therefore vary between systems. One system, for example, uses the abbreviated command ^KD (i.e. CONTROL-KD) to finish working and to save the file, while another system uses FD to leave the system and to delete the file. Therefore while an interface may be consistent within itself, it may be dangerously confusing to an operator using more than one system. Each word utilized in a given language has one meaning only to limit potential ambiguities.

FUNCTION KEYS

Function keys are dedicated or programmable keys on the keyboard which can be used to perform commands at the press of one or more keys. The use of function keys is useful in command-dominated interactions and can be appropriate for naive users if combinations are simple. As with commands, however, consistency is important and industry standards are appearing (for example, F2 is commonly used to save a file).

Soft function keys are normal keyboard keys used in conjunction with the Control key to convey commands to the computer (DTI, 1990). Soft function keys can also appear on-screen in direct manipulation systems (see below), and can serve as 'short-cut keys', preferred by expert users who find the mouse and menus to be too long-winded. They usually take the form of labelled boxes or buttons which are activated by clicking on the area using a mouse or other pointing device. On-screen function keys have an advantage over keyboard function keys in that they are context specific (Cole *et al.*, 1985).

When evaluating interaction styles, Borgman (1986a) noted that function keys need clear labelling, which can be a problem with respect to the space available. She also noted that frequently needed keyboard function keys should be in easy reach of the main section of the keyboard to avoid as much as possible having to leave the 'home' position when the command is given.

FORM FILL-IN

Where the same type of information needs to be recorded on many distinct but similar items, form fill-in interaction is a suitable dialogue style, for example for catalogue input or database entries. Form fill-in employs the metaphor of filling in a printed form (Norman and Chin, 1989) and is often accessed as a choice from a menu. This offers the twin advantages of prompts for the user input and an overview of the whole record.

Another usage of form fill-in interaction is that of utilizing an on-screen form to specify a simple search request when searching bibliographic databases. This type of interaction is called 'Query by Form'. The user types in the search term in the required labelled field or fields rather than typing search commands such as 'ti=automation'. Simple combinations of criteria can be accommodated and the interaction is easier to learn than a full search command language as the user does not have to remember prefixes or suffixes for field names (Belkin *et al.*, 1991). In general, form fill-in is most appropriate for frequent or knowledgeable users.

Advantages and disadvantages of form fill-in

The advantages of a form fill-in interaction are as follows:

- Form fill-in is good for structured repetitious tasks with predictable input.
- Data entry is simplified and requires only modest training.
- Form fill-in can support a higher density of information on the screen than menus and many items of information can be entered in one transaction. If, however, the amount of information to be input cannot be accommodated, then logical sections of the form can be displayed successively.
- On-screen guidance can be provided to users at appropriate and convenient locations, for example to give acceptable date formats or explanations of acceptable abbreviations.
- Input can be free text entry, binary choices, numerical information or selections from a limited range of possibilities.
- Ranges of acceptable and/or needed values can be assigned to individual fields to prevent input errors.

Disadvantages of form fill-in are:

- The screen design consumes a large amount of space.
- Typing ability is needed.
- Understanding of the fields, screen labels and permissible values for input is required.

Features of form fill-in interaction

For ease, the order of the initial input for a new record should be from left to right and from top to bottom. Form fill-in should also include full-screen editing for error correction and subsequent amendments to the information. Other important features are a final acceptance of the completed form and the ability to reject the record at any time before that.

A form layout may imply a restriction on the amount of data which can be input to certain fields. In practice, where variable length fields are permitted, this limitation may be only on the amount of data which may be viewed at any one time, additional information being scrolled horizontally or vertically out of sight. While it is therefore possible to input large quantities of text using form fill-in, the difficulty of reading and subsequent editing must be considered.

Form fill-in for data entry may involve the transfer of information from a handwritten form to the screen, and it is important that the language and layout of the screen reflects the source document. If form fill-in is being introduced for data entry, then existing forms should be scrutinized to ensure that they are easy to use and ordered logically. If necessary, the form should be redesigned first and the screen then designed around it (DTI, 1990).

Cole *et al.* (1985) give the following guidelines for form fill-in systems:

- When the screen is initially displayed, the cursor should be in the first character position of the first input field.
- Progress to the next field should be achieved by a simple one-stroke key press. This is usually by use of the <Tab> key.
- The screen display should indicate the maximum field length allowed for any field and/or should show the outline and boundaries of entry fields.
- Screen labels should be protected from keyboard modification and editing only allowed within the area of the screen fields.
- The cursor should jump over fields containing computer-entered data such as dates or serial numbers.
- Default values should be provided automatically in fields which take a certain value in the majority of cases. Any exceptional cases can then be entered by editing the default value.

DIRECT MANIPULATION

The trend in interface design is now towards direct manipulation, which also uses function keys. This type of interface style is usually referred to as a WIMPs environment. WIMPs is an acronym for Windows, Icons, Menus, Pointers (Illingworth, 1991), or sometimes Windows, Icons, Mice, Pull-down menus. It is also referred to as a window or iconic interface or a graphical user interface (GUI). The objectives of this interface are to offer the power of a command-driven system with the light memory load of a menu-driven system. It uses the metaphor of movement to perform operations and manipulate information in the system, and is characterized by the visibility of objects of interest by the use of icons. Direct manipulation exploits the sophisticated visual information processing capabilities of humans, spatial awareness, and the ability to extract meaning from spatial patterns (Cole *et al.*, 1985).

A mouse is used to control the cursor and to select commands from menus. Keyboard alternatives are also available for menu selection. Menus and sub-menus usually appear in blocks of text or windows which temporarily overlay part of the screen. Icons are used to indicate graphically a function which can be performed (for example a dustbin is sometimes used to indicate how a file or record may be deleted), or an object such as a file or directory. Clicking on an icon to make a selection, 'dragging' an icon to move or copy a file to another directory, and double-clicking on applications or file icons to open them are some of the actions which are performed in this type of interface.

Direct manipulation interfaces also permits the use of windowing, a technique where only part of the display screen is used for a particular interaction or task. Two or more mini-screens can be available to contain separate aspects of the task in hand. Therefore, more than one application, file or interaction may be in progress at the same time.

The windows may be sized by the user to be appropriate to the task and may be organized in a tiled format, that is side by side, or overlapping each other, as sheets of paper on a desk. The windows may be moved at will and the needed window brought to the top when work is in progress on that window.

The advantage of the windowing technique is that the user's short-term memory load is reduced, because the windows act as an external memory. At the same time, more information is immediately available to the user. Manipulating windows also produces fewer errors than using non-windowing techniques but, at the same time, can be very time consuming.

Following the general trend in software development, CD-ROM and other library applications are increasingly incorporating direct manipulation (WIMP) interfaces. Bosch and Hancock-Beaulieu (1995) evaluated the appropriateness of such interfaces for CD-ROM users. They concluded that 'although some degree of standardization is achieved by the mere fact that the interfaces are based on a common platform . . . the great flexibility of the Windows environment has led to an uncontrolled diversity of design' (p. 268). They also point out that current GUIs for CD-ROM applications do not provide effective interaction at the task level and favour the professional searcher. In 1992, Hancock-Beaulieu suggested that the WIMP environment provided by CD-ROMs has increased expectations and user awareness in interface design, and that direct manipulation in other library systems would allow a large range of options, reduce cognitive load and aid navigation by retaining previous positions on screen. Clearly, there is still room for improvement in the library software arena.

Advantages and disadvantages of direct manipulation

Shneiderman (1991) and Henry (1991) give advantages of direct manipulation interfaces as follows:

- Novices learn the functionality of the system quickly, usually by demonstration.
- The user's cognitive load is lower than other types of interaction.
- Physical movement is used in the place of a complex command language syntax.
- Experts can work rapidly on a wide variety of tasks.
- Intermittent users tend to retain operational concepts.
- Error messages are rarely needed. Errors are handled well and all actions are reversible.
- Visual feedback keeps the user informed of progress.
- The user feels more in control than with other interface styles by apparently handling data directly.
- Direct manipulation supports the way we normally think about a domain and relates to the real world. Users can transfer spatial management skills from the desk top to the computer and to change focus from one task to another. Direct manipulation supports moving between different documents and applications.
- Actions are rapid, reversible and incremental.
- Direct manipulation interfaces offer a standard interface across applications, which has implications for transferring skills.

The quicker ideas can be transferred from working memory to action the less the likelihood of forgetting (Gould and Boies, 1978).

Disadvantages of direct manipulation are as follows:

- Users' actions need to be precise when using the mouse and some practice is needed.
- Users need manual dexterity when moving from keyboard to mouse and vice-versa. Some users find the constant changing of input device annoying.
- Icons have limitations in meaning because of their size. Additionally, pictorial representations mean different things to different people. Icons therefore need to have meaningful labels (Bonner, 1988).
- Users can spend too much time in rearranging the display and other associated tasks.

Direct manipulation (WIMP) interfaces are available for a variety of computers and several guidelines for them exist (see, for example, Weinschenk and Yeo, 1995). The WIMP interface has influenced the design of many products by giving a consistent look and feel to the user interface and a shared set of system resources for managing applications (Hicks and Essinger, 1991).

MENU SELECTION

Menu selection can be used as a dialogue style on its own, separate from direct manipulation; however, this is becoming less common as WIMP interfaces grow more popular. Smith and Mosier (1986) define menu selection as 'a type of dialogue in which the user selects one item out of a list of displayed alternatives'. It is a system-driven interface which requires no knowledge of command syntax.

Items can be selected from menus in a variety of ways:

- by typing in the first letter of, or mnemonic for, the menu label. The user types in the appropriate letter;
- by using cursor keys or the mouse to identify choice;
- by using function keys to indicate specific choices.

Making menu selections is faster and more accurate with a cursor-based broad menu, because it offers a wide target area. Selection by a meaningful letter, for example, the first letter of the option label, is faster and more accurate than key-controlled cursor movement or keying in a number (Shaw, 1991). However, difficulties in consistency occur when there is more than one choice which should logically be given a particular letter. For example, in the Windows' File Manager, 'Copy' can be selected by pressing 'C' whereas users selecting 'Create Directory' have to press the letter 'E'. Similarly, search is selected by the letter 'H' and select files by 'S' (see Figure 18.1). Fortunately, users have the choice of the drop-down menu cursor based approach in addition to selection by letter. Obviously the advantage of multiple selection methods within a system is that users may use their own preference.

Types of menu

There are four main types of menu:

1. *Pull-down menus*. These drop down from a single title bar at the top of the screen and a selection is made, either by clicking with the mouse or by typing in a code letter indicated by the choice required. These are typically found within WIMP interfaces. A variant on this

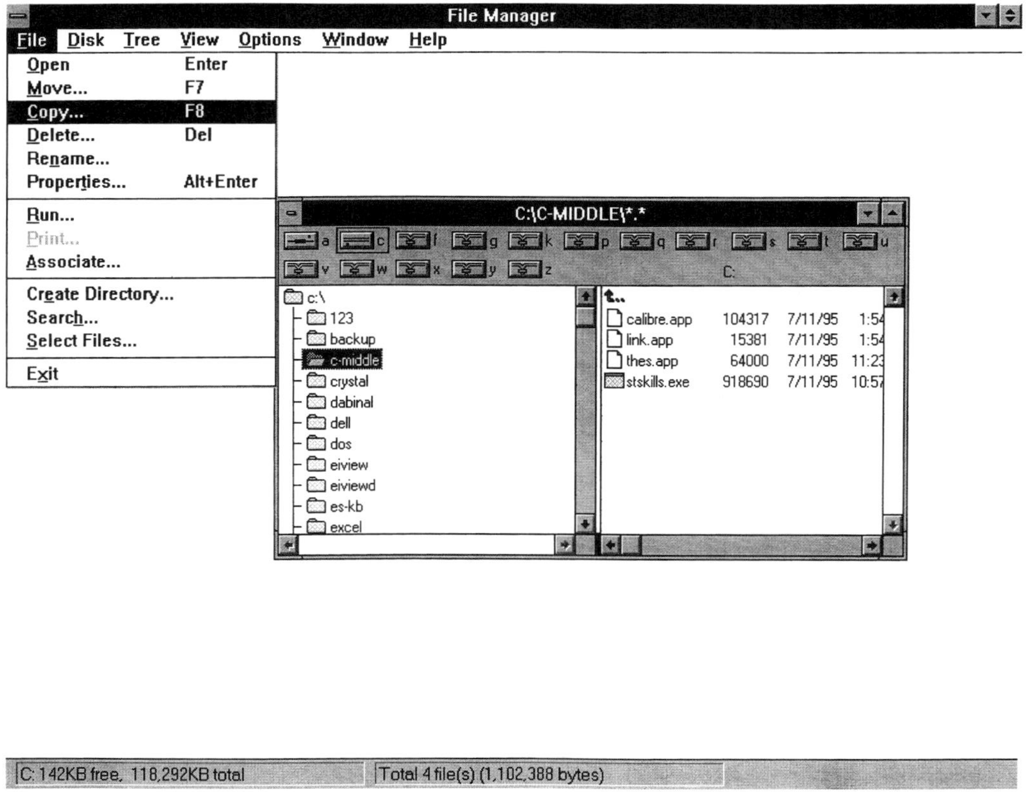

Figure 18.1 Cursor-based broad menu with optional selection using letters

and fixed menus is the tear-off menu where users have the option of creating a palette menu from a pull-down menu if they so wish.

2. *Pop-up menus*. These are also found within WIMP interfaces, and appear in a particular area of the screen when an on-screen icon is selected with the mouse. The menu is closed by clicking a close box within the menu or choosing a 'close' option.
3. *Fixed menus*. These menus stay in place on the screen and normally occupy just a small section of it, as in the case of palette menus (see Figure 18.2).
4. *Walk through or cascading menus*. The original and all subsequent menu choices are displayed on-screen until a choice is selected which activates a function (DTI, 1990).

Advantages and disadvantages of menu selection

The advantages of a menu style interaction can be summed up as follows:

- All the available choices at a given position in the system are displayed. This can be particularly helpful to the naive user who would not otherwise be aware of the options available. Hidden menus are sometimes available as a prompt to less frequent users in an apparently non-menu type interface. These may be available at the press of a function key, or there may be an option to continuously display or hide them.
- Choices and navigation are rapidly understood by novices, who would have major problems with the syntax and terminology of command languages and become frustrated when translating their queries into suitable commands (Shaw, 1991).
- Selection from a menu reduces the amount of keyboard input, which is helpful to those without typing ability, and quicker for non-expert typists.
- Training time for a menu-based system is generally shorter than for other interaction methods as recognition is used in the place of recall from memory.
- It is not necessary for the user to retain the complex syntax needed for command-type instructions as the menu selections are translated into the syntax needed by the system.
- Error handling is easily supported in the menu style of interaction (i.e. any non-prescribed input is an error).
- Decision-making is structured for naive users who do not know which options to choose.
- Exploration of the system is encouraged, provided that navigational facilities are made explicit and actions can be easily reversed.
- Faster route options can be provided for expert users. Some systems, for example, offer an alternative command system or enable users to

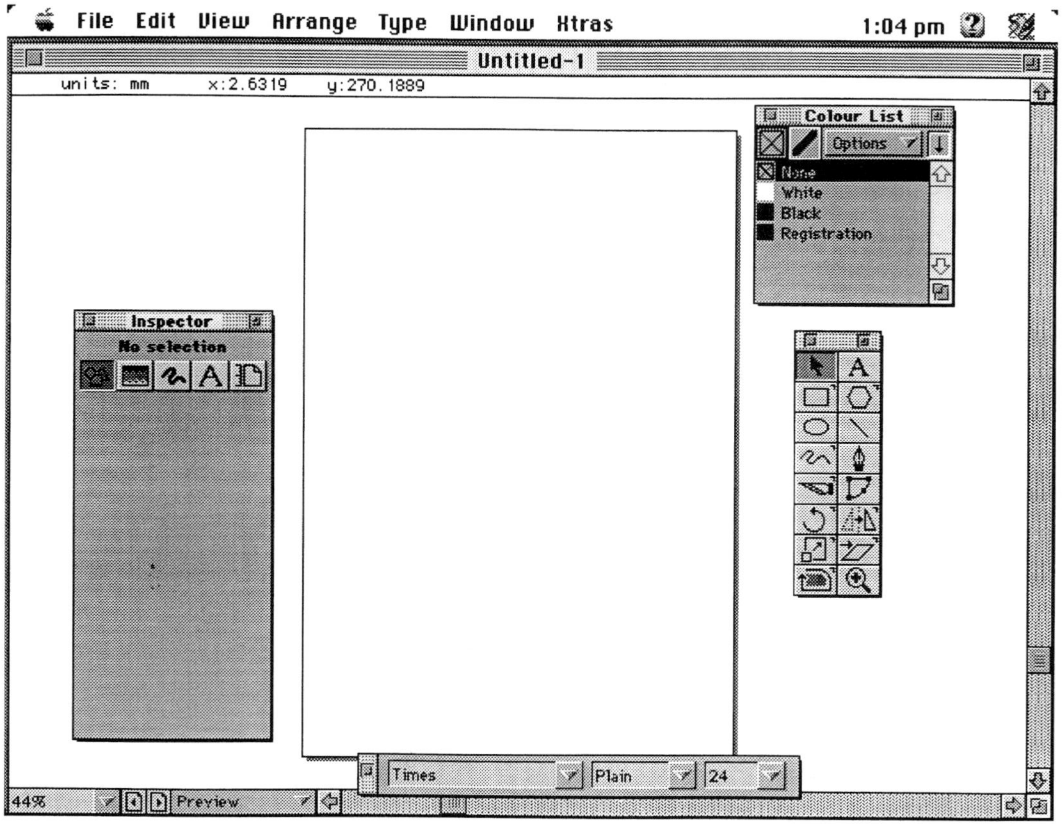

Figure 18.2 The use of palette menus on-screen

input a string of menu selections which take them directly to a function without displaying the menu. Alternatively, the system may offer a type-ahead facility, where the user types in all the known choices at the first menu screen, in advance of the display of subsequent screens and does not need to wait for each screen to be displayed in turn. Lastly, menu frames can be assigned a unique number or code which can be typed in at the main menu to go directly to the required menu (Shaw, 1991).

The disadvantages of a menu style of interaction are:

- There is a physical and cognitive limit to the number of choices that can be displayed at any one time.
- Frequent users may be slowed down and become frustrated by a menu system. The BRS After Dark service started as a menu-based system but within a year, due to frequent users demands, the interaction was changed to a mix of menus and commands (Spigai, 1987).
- A menu system is rigid and all paths must be prescribed when the system is built.
- Users with low spatial ability take twice as long as other users to complete tasks and become lost in an extensive menu system (Shaw, 1991; Chin *et al.*, 1988). It has been noted that it sometimes takes 'some minutes' for users to reach a required function in some menu systems (Henry, 1991). Menu screens should therefore include specific instructions on procedures to return to the previous or main menu screen when errors have been made.

The organization of menus

The organization of menu items has implications for how efficiently and effectively the user is able to interact with the system. Horizontal menus are harder to scan than vertical menus (Puttapithakporn, 1990). Factors affecting the organization of menus relate to the number of items displayed, the depth of menu nesting, the ordering and grouping of items, and the wording of choices and help given to the user for selection:

- *Number of items displayed in a menu.*
  The human capacity to remember unrelated items also has implications for the length of the menu. Long menus tend to confuse people, and it is recommended that the number of options in a menu should ideally be less than or equal to seven but never exceed nine.

  A small number of options is easier to handle but limits possible actions. To overcome this, Borgman *et al.* (1990) suggest that more frequently used options or option groups should be displayed initially,

together with a 'more options' option leading to further, but lesser used choices.

- *Depth of menu nesting.*
  Since the number of menu items which can and should be displayed on screen is limited, a menu may need to be 'nested' to various degrees in order to accommodate all available choices. This is commonly termed a hierarchical menu. The consequences of this are that many menus may have to be traversed before the required choice is found and activated. Full advantage should be therefore taken of including the maximum number of items in a menu, subject to the restrictions mentioned above.

  In order to cut down on the number of levels which must be accessed before the required item is found, more frequently used items should be placed at a shallower level (i.e. nearer the top of the hierarchical tree) than infrequently used items. If possible, the librarian should liaise with the software vendors to adapt systems to match local requirements and practices.
- *Ordering and grouping of items.*
  Three common bases for the ordering of menu items are frequency of use, logical sequence of operation and alphabetical ordering (Puttapithakporn, 1990). Alphabetic ordering is useful for new or infrequent users in public access systems, who tend to use scanning to select from all the alternatives offered (Card, 1982).

  Human decision-making processes mean that people tend to place much importance on the information collected early in a process (see Chapter 1). An alternative view, therefore, is that the most frequently used and most important items need to be placed towards the top of the menu or they are likely to be ignored. Hancock-Beaulieu (1992) suggests that, since subject searching is the most frequently employed method when using an OPAC, keyword search should be placed at the top of the options offered.

  With practice, the effects of menu organization disappear as users become accustomed to where they will find the appropriate item for the task in hand. However, menu ordering has a long-lasting effect on the user's mental model of the system, both in building the model and referring to it in subsequent use, and therefore, categorical menus are recommended if frequent use is expected (McDonald and Schvaneveldt, 1988).
- *Wording of choices and help given.*
  Users prefer menu labels to be short, concise and meaningful. One of the main sources of menu selection error is that users misunderstand the description of the menu item as contained in the displayed label. For this reason systems using many abbreviations and codes in labels are best avoided. Performance on menu selection is good when menu categories and labels are chosen by individuals similar to the eventual users (Shaw, 1991).

Ideally, menus should have on-screen or immediately available context-sensitive help to give the user more information on the choices available than can be given in the menu label (Norman and Chin, 1989). A menu label is good as an *aide-memoire*, but not very good at teaching users about the item.

There may be a need for alternative or subsidiary menus to be offered for professional use in some cases. For example, a professional search intermediary may need to search a database by ISBN, but for naive users this choice would be superfluous, and should therefore not be displayed.

Embedded menus/hypertext

In addition to the traditional types, embedded menus can be used, for example those used in hypertext systems. Hypertext is a 'structured browsing system' which is made up mostly of text, parts of which are designated as 'nodes' or 'hot keys'. These are usually indicated by highlighting or the use of a different colour for the text, and are, in essence, menu choices. By selecting a node of interest, the user can access more in-depth or related information associated with the text chosen. This is achieved by building links between words and other textual documents held in the system.

Hypertext provides a 'non-linear' form of navigation around information in comparison with the more linear nature of traditional hierarchical menus. The user chooses the order in which documents are examined or tasks are performed, and attempts are made to ensure that links reflect the processes of thought and memory of the human mind. This makes hypertext a powerful tool for information retrieval, because ideas are linked together associatively (Milne, 1994). It is also particularly useful when the users' questions cannot be anticipated by the organization of the text and for retrieving documents about a particular subject (Egan *et al.*, 1989).

Hypertext is being used increasingly to build different applications, including multimedia systems which illustrate and educate on particular subjects, for example, Penguin's CD-ROM version of Arthur Miller's play *The Crucible*. This system offers several access methods, including an interview with Arthur Miller, the text of the play, articles on the work, interviews with actors who have appeared in the play, and background material on the Puritan and McCarthy eras alluded to in the play (Oder, 1994). A prime example of a widespread use of hypertext can be found in the World Wide Web (WWW), a hypertext retrieval system for accessing files on the Internet, which can include sound, pictures and animation.

Milne (1994) suggests the following as useful examples of library applications of hypertext:

- *Information retrieval*. Emphasizing browsing via link traversal rather than specific goal-oriented searching.
- *Online databases*. Retrieving information according to user's need rather than the subject orientation of a specific database.
- *Citation indexing*.
- *OPACs*. Useful for examining a library collection by browsing.
- *Electronic books*. Overcoming the static, linear nature of paper, for example, electronic encyclopaedias enabling users easily to extract several different references to a particular subject from different sections of the book.
- *Computer-aided instruction*. Empowering a user to follow their own learning style.

One of the problems found with hypertext systems is that users can easily lose their way in the system, and clear and easy navigation must be provided. This includes good backtracking facilities, forward navigation, quit and exit, and the ability to keep track of one's position within the system. Another disadvantage is that nodes and links within hypertext systems tend to be predefined by the system designer, therefore confining the browsing abilities of the user to some extent.

NATURAL LANGUAGE

The ultimate goal of system builders is to create a system which will accept a natural language request from a user and provide an appropriate response. Much research is being carried out on natural language interaction, especially in the field of artificial intelligence, but with limited success. Vagueness, ambiguity and ungrammatical sentence construction have been found to be heavy burdens for natural language systems to deal with and to date, no true natural language systems exist. Most developed systems can only handle a subset of the user's vocabulary and are being used in very limited domains.

OKAPI is an example of a limited natural language system developed for naive bibliographic searchers. The system accepts a query typed in natural language terms. It then assigns a weight to each input term depending on the rarity of each term in the knowledge base, and output is then ranked with the records containing the rarer terms listed first (Hancock-Beaulieu, 1992).

The following problems have been found in developing natural language interfaces:

- The system appears to be more 'human' than it is, which can lead to unrealistic user expectations.

- Keyboard input places a heavy burden on users and leads to problems with spelling and typographical errors.
- The 'natural' way for a user to communicate with a natural language system would be through voice input. However, this has been found to lead to ambiguity in the system's understanding of like-sounding words (e.g. 'there' and 'their'), and of different accents, dialects and hesitancy.

Falzon (1990) argues that a user's behaviour is a consequence of the behaviour of the machine. If the machine's abilities were equal to those of a human, there would be little difference in the linguistic behaviour of the user. However, if the machine's behaviour differs, then the user tends to restrict the language used spontaneously. Humans have a natural tendency to restrict expression even when communicating with other humans.

SUMMARY

No single dialogue style can be pinpointed as the most suitable, rather a variety of dialogue styles is desirable to serve the differing purposes of task and user. There is an increasing number of systems which offer a mixed dialogue or Hybrid style. For example, the content of a database may be entered using a form-filling dialogue, and may be interrogated by an experienced user using a command-driven interface or a naive user on a menu-driven OPAC. The BLCMP Tallis system is another example of a hybrid system; both menus and form filling approaches are used. Systems which aim to cater for different levels of user or for changes in user ability are described as employing parallel styles.

CHAPTER

19

SCREEN DESIGN

Screens are used to capture or to display information. Galitz (1989), asking users what they wanted in a screen display, came up with the following rquirements:

- an orderly, clutter-free screen;
- an obvious indication of what is being shown and what should be done with it;
- expected information to be where it should be;
- a clear indication of what relates to what (headings, captions, data, options etc.);
- plain simple English;
- a simple way of finding out what is in the system and how to get it out;
- a clear indication of when an action could make a permanent change to data or system operation.

Shires and Olszak (1992) give a checklist of principles to consider when designing OPAC screens, but which may be applied to any interactive system. The screen display should:

- let the user feel in control of the interaction;
- reduce the load on the user's short-term (working) memory;
- be clear and consistent;
- require minimal action of the user;
- include only the necessary information;
- have all data relating to a single task on one screen;
- give information in a directly usable form;
- have information grouped logically;
- have a clear title for each screen;
- have labels for all elements on the screen;
- indicate what to do next, how to reverse steps, and how to exit;
- use a positive, professional tone;
- avoid condescension, accusation, humour and chitchat;
- use plain English and an ordinary level of language;

- address the user in the second person;
- use the active voice;
- use concrete rather than abstract terms;
- use short sentences;
- use abbreviations and acronyms with great caution.

The screen display has three main functions:

1. To display information.
2. To give the system status. It should always be clear to users where they are in the system.
3. To show the potential inputs (what information goes where or where the user can go next).

The design and layout of the screen will affect the ease and speed with which these functions are fulfilled and should be considered when evaluating software. The effectiveness of the screen design is determined by a combination of the following features:

- the *information content* of the screen, that is the data, prompts, messages and so on;
- the *formatting* of the information; the use of Windows, grouping, spacing and so on;
- the *coding* of the information; by the use of colour, for example.

INFORMATION CONTENT

CONSISTENCY

The information contained in the screen should be both consistent with other screens and with the task in hand. A menu system which lists an option as 'Search the catalogue' should display the same phrase as the heading for the catalogue search screen, rather than using a briefer but different term, such as 'Catalogue enquiry'. To the new or uncertain user, this practice can cause confusion even though to the initiated, the terms may appear to convey the same information.

ORDER

The frequency and the order in which certain operations must be performed should be mirrored as far as possible by the software interface. For example, the way in which information is arranged on the screens should concur, as far as possible, with the order in which this information will be used within the task and, preferably, should be compatible with any documents that are used (Sperandio, 1987). The screen fields for catalogue input, for example, should be in the same order as any hard-copy form which is used, or should be arranged to permit a logical and orderly input of data. That is not to say, however, that existing operations should not be changed, but rather that due consideration should be given to those which cannot.

AMOUNT

The amount of displayed information may depend on the user or on the task. Some systems try to combine a lot of information on a single screen. It is unlikely, however, that all the information will be required for a single task with the result that, whatever task is being performed, the user will have to scan a crowded screen to locate the relevant information, with subsequently reduced efficiency.

The display screen should hold only the information which is required for the task in hand. Additional information which may be required should be available in pull-down menus, rather than cluttering the screen. While endeavouring to restrict the amount of on-screen information, however, designers must be alert to the possibility of omitting items of required information. It is undesirable for related data to be spread over separate screens. If the page to be displayed is longer than the screen then the user needs to access the hidden information. This can be achieved by scrolling or paging. Long displays of multiple single-line records – for example, titles retrieved in retrieval systems – should show a line number on each record to enable the user to be able to maintain a sense of position within the display.

WHAT INFORMATION AND WHERE

When first encountering a screen of information, users from Western cultures begin scanning at the top left-hand corner of the screen. On screens of text, therefore, there have evolved certain conventions which instil a consistency between applications and enable the user to anticipate and understand what is being presented. Both Cole *et al.* (1985) and Galitz (1989) give various guidelines to be followed on what should appear in text screens and how they should be formatted. Briefly, these are as follows:

- A heading and instructions or prompts for dealing with the current screen of information should appear at the top of the screen. Further, the heading should be placed centrally and be distinguishable from the rest of the screen by the use of colour, a larger font, underlining, spacing and so on. Headings for different screens in a system should be consistent with each other in placement and appearance. The user input which caused the screen to be displayed should appear at the top right-hand corner of the screen unless it appears in the screen display. Otherwise a screen identifier or page number should be placed at the top right-hand corner of the screen. Many of these conventions can be seen on Windows screens, for example, title bars which act as headings, are placed centrally and are distinguished from the remainder of the screen by use of highlighting. They are also consistent with those title bars in other windows in placement and appearance.

- Instructions and prompts for leaving the screen should appear at the bottom of the screen.
- In command-driven systems, which are quite different from those using a graphical user interface, the bottom line of the display should be reserved for input of the commands by the user, with the line above it being used for error or status messages.
- All the information which is required for the task in hand should be available in the main body of the screen, including data, captions, section headings, completion aids and prompts. The various elements (headings, prompt lines etc.) should be clearly distinguishable by their size, colour or position. If the main task of the screen is to enable the user to select from a menu, the menu should be placed centrally, with the options in a vertical list. Menu options which are incidental to the main task of the screen (for example, in a word-processing application) should appear horizontally at the top of the screen if selected by using a mouse, or horizontally at the bottom of the screen if containing function key labels.
- Information which is not required for the task in hand should not appear on the screen.
- The information contained on the screen should be structured meaningfully.
- Only one task should be carried out on one screen.
- If a task requires more than one screen for completion, the user should not be required to memorize information from one screen to use on another.

FORMATTING

The layout of the data on the screen will have a considerable impact on the effective use of the data. Screens within an application should be consistent with each other. Users notice patterns in placement, highlighting, and so on, and expect any deviations from the general pattern to have a significant meaning, even when none was intended.

Users can become frustrated with inconsistencies to which they are unable to ascribe a meaning (Borgman *et al.*, 1989a). The following criteria may be helpful in determining what is or is not a good display.

PHYSICAL LEGIBILITY

It is important that the user should be able to perceive relevant data on the screen. Factors affecting the physical legibility of the display are the font, dot size, refresh rate and resolution, spacing, luminance and contrast. Also of importance to perception are the number of characters per row of text, and the number of rows of text per screen, the foreground and background colours used, the differentiation between headings, data and labels, between groups of unrelated data and between more important and less important data (Borgman, 1986a). Although these factors vary widely in the degree of control we can exercise over them, they should nevertheless be considered when purchasing new equipment.

CASE OF TEXT

Galitz (1989) recommends the use of lower case for data, with only the initial letter of each sentence capitalized. He quotes a 13 per cent advantage in reading speeds and even more for comprehension and proofreading over upper case text. Normal punctuation should be used.

TEXT AND SPACING

For text screens, there should be no more than 40–60 characters per line, or two 30–35 character columns with a five-character space separation between them. Paragraphs should be no more than four lines in length, and separated by a one line space. Text data should be left justified and there must be no hyphenation at the ends of lines of text. The ragged right margin helps the users to keep their place when reading and to progress easily to the next line of text. Fully justified text results in uneven inter-word spacing which is difficult to read on screen. Interline spacing within a section of text should be at least the depth of the character set, preferably more. Horizontal spacing between sections of the screen should be greater than this.

CAPTIONS

Captions identify data. They should be clear, easy to identify, and bear a consistent relationship to, but be distinguishable from, the data. For captions, the use of upper case helps to differentiate them from the data which results in faster search times. Captions should be brighter than the data fields which they describe.

Captions for single fields should be placed to the left of the data, and either left or right justified with a minimum one character space separating the caption and the text.

Examples of left and right justification of captions:

AUTHOR	: Smith, JR	AUTHOR	: Smith, JR
TITLE	: Travels in Singapore	TITLE	: Travels in Singapore
PUBLISHER	: Blacks	PUBLISHER	: Blacks

Captions for repeating data fields should be placed above the data.

Example of captions for repeating fields:

ACCESSION NUMBER	SURNAME	PUBLISHER
14326654	Smith, JR	Blacks
32547770	Brown, PT	Marshall
34874563	Canning, L	Alder

SIMPLICITY

Simplicity in a screen display results in ease of comprehension of the meaning of the layout pattern. Galitz (1984) quotes Tullis's work on mathematically calculating the simplicity of a screen layout and gives a simplified version for comparing two or more similar layouts as follows:

Add together: the number of fields on the display
and
the number of different positions across the screen which are the start points of one or more fields
and
the number of different rows down the screen which contain one or more fields.

The resulting figure gives a measure of comparative complexity for the screen. The screen with the lowest figure has the highest simplicity and lowest complexity and will be the easiest to read and comprehend.

DENSITY

The screen should have an orderly appearance and should not be overcrowded. A cluttered screen makes it difficult to locate specific information, causing an increase in search times and errors (Galitz, 1984). Galitz (1989), citing Dunsmore, reports on a system where separate items, which had been combined on the same display line to conserve space, were instead placed on separate lines, with the result that users of the new screens were 20 per cent more productive.

Reading from a VDU is slower and less accurate than reading from the equivalent printed version (Dillon, 1994) and readers perceive the screen to be more filled than it actually is. Various writers recommend that only 25–30 per cent of the screen space should be filled at any time. The remaining 'white space' aids the user to read the screen accurately and is as important as the 'filled' space. Time and errors in finding information increases with the density of the screen characters. A lower local density was found by a researcher cited in Galitz (1989) to create positive 'visually pleasing' feelings in users.

GROUPING

Grouping of information gives a structure to the display. Structured screens for alphanumeric information enhance performance for frequent users. This is because consistent placement of the same groupings develops spatial expectations which help the user to locate items more easily.

Information placed in each group must be semantically related to the rest of the group, and semantically distinct from information contained in other groups on the same display. This enables the user to 'chunk' all the data into logically distinct categories. Logical groupings are: by sequence of use; by function; by priority or importance; by frequency of use (Cole *et al.*, 1985); or alphabetical or chronological orders.

Groups can be distinguished by the use of separating graphic borders, lines, boxes, colour coding, highlighting, reverse video or column/line spaces. Where column spaces are used, these should not be as wide as the margins at the edge of the screen. Larger or more important groups should be placed nearer to the central vertical axis of the screen, with smaller or less important groups further away.

The partitioning of data should be consistent and while being clear, should not distract the operator. Large numbers of similar-sized partitions should be avoided. The displays in Figure 19.1 shows good grouping of data into logical sections: author; title; publishers; imprint; reference numbers; and notes.

Explicit labelling of separate data elements within groups is necessary to identify the data. Labels should use descriptive words familiar to the user group and consistent with similar systems. Hildreth (1985), however, found that variations in label names exist between different OPAC systems both in the choice and form of names used.

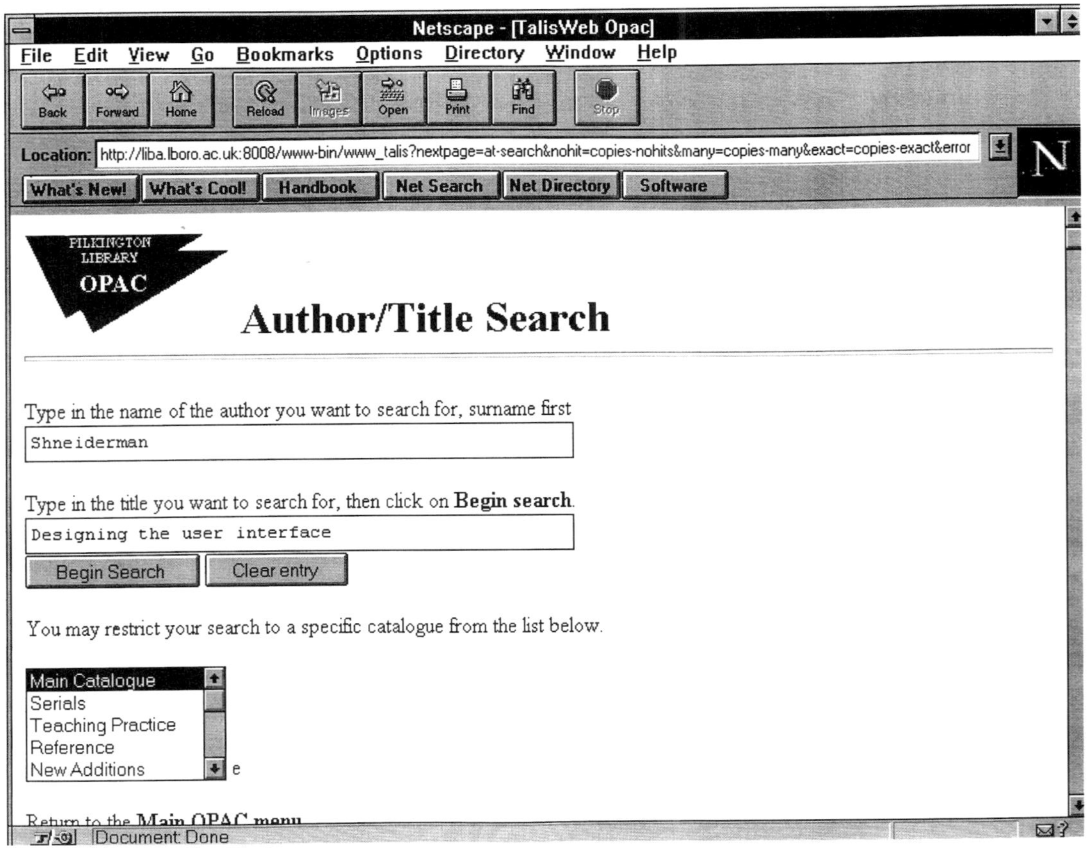

Figure 19.1 *Screen displays showing logical groupings*

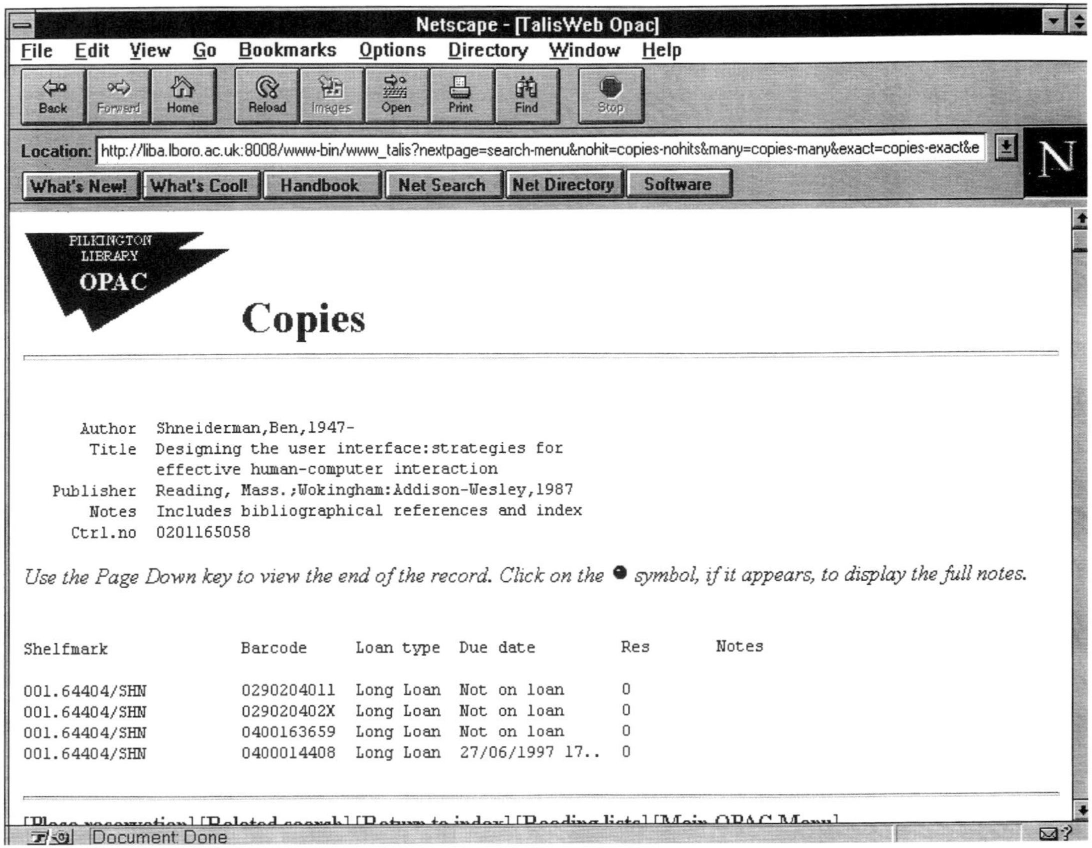

Figure 19.1 (cont.) *Screen displays showing logical groupings*

CODING

Screen information can be coded to attract the user's attention. For example, in a free text system, a word or phrase may be highlighted to indicate what caused the record to be retrieved, or a defaulting borrower may cause the issue system to bleep when the user next borrows or returns an item. The more conspicuous the coded target, the more likely it is to be noticed (Oborne, 1985).

CODING TECHNIQUES

A variety of coding techniques may be used, including alphanumeric codes, brightness, flashing, reverse video, position, shape, colour and size. Formatting techniques such as special characters, underlining, italicizing, and capitalization are also used. However, because of the way in which people respond to external stimuli (see Chapter 1), some codes may be more effective than others. A continuous beeping sound for example, is more likely to draw attention than a single beep (and will be more annoying!), and a combined visual and auditory stimulus will be more effective than either in isolation.

Many systems employ auditory signals to attract a user's attention. However, Davis and Shaw (1989), in comparison studies, found that these were disliked both by those working alongside others (it is difficult to ascertain from which machine the signal had come) and by those working alone. Fortunately, users can control the volume of the system sound, for example through the control panel of the Windows system, but this assumes that users are aware of this feature before annoyance occurs. In general, sound and flashing codes should be restricted to situations where the user must respond immediately.

Whatever coding techniques are used, it is important that there is consistency throughout the application, and that the coding is simple to understand.

Oborne (1985), reviewing the results of research on coding techniques quotes the following:

- Underlining has been found to be an effective coding device only if the user is not under stress because of time or other constraints. It may in fact make reading more difficult, because the size of the interline space is reduced by the underline and the lower profile of the character set is masked.
- Capitalizing is often used as a coding technique for captions, but Foster and Coles' (1977) work found that capitals (within text) may actually hinder reading performance. The use of bold is a better all-

round coding technique than capitalization.
- The use of colour (see below) is the most effective means of visual coding (Christ, 1975).

REDUNDANT AND MULTIPLE CODING

Redundant coding involves the use of more than one coding technique to highlight the same item to convey the same meaning. The use of more than one technique to code an item has been found to improve performance. For example, an item may be highlighted by flashing on and off and be coloured differently to non-highlighted items. It has also been found that redundant coding, capturing the attention of more than one of the senses (e.g. sight and sound) is particularly effective for warning situations. It must be remembered, however, that too many different stimuli will not be effective, because it is not possible for the operator to concentrate on several items simultaneously. The overuse of any coding bombards the user with items competing for attention which may result in confusion.

THE USE OF COLOUR

Due to the fact that colour monitors are becoming the norm, and the increase in use of iconic interfaces, the use of colour on screens is now common. Unfortunately, examples of the poor use of colour are also common. The use of colour may cause a person to notice differences in colour regardless of whether they have task-related meanings, or to group together, artificially, items of the same colour in a way which is unrelated to the task. When colour is used improperly, performance will show little improvement and may even suffer. Oborne (1995) points out that about 6 per cent of men and about 0.5 per cent of women have some form of colour vision deficiency; red-green and blue-green colour blindness being the most common. For this reason it is important to use more than one coding technique (see above).

Despite the drawbacks, colour can be used effectively in VDU displays in the following ways:

- *As a formatting aid*
  Colour can be used to differentiate between the foreground and the background display and to enhance monochrome displays by an appropriate combination of text or graphics colour and the background colour.

  The use of colour on conventional formats also makes it easier to understand the logical structure of the displayed data, for example

the use of colour to separate screen headings, captions and data where the existing discrimination is poor. It distinguishes the application being used from the information held in the application, but is more flexible than simple bold coding due to the number of colours available and the ability to combine with existing highlighting methods. However, some users only have access to monochrome terminals which interpret foreground/background colours in a way that is difficult to read (Harris, 1990).

Colours may be used to assist the reading of wide lines of text or data. This is achieved by using three row groupings of alternate colours to guide the eye when moving from one line to another (Galitz, 1989). The two colours used for the text, however, should not be too different in brightness, otherwise the darker colour will be overshadowed by the brighter one.

In the case of graphical user interfaces (GUIs), more icons or pictures will be found, giving more scope to the use of colour. It is used to enhance the meaning of icons, for example within File Manager the file icons may display blue lines as if they have writing on them, in order to signify that the files are text documents. Colour is also used to signify which windows are 'active', through the appropriate title bar, which helps users navigate their way round the screen. In addition, users are offered some opportunity to customize the colours used on the computer screen, which caters for individual aesthetic tastes, and which might be of benefit to those suffering from colour blindness.

- *As a visual code*

  Colour can help to show relationships between data items. The great advantage of colour coding is to show relationships and differences independently of other coding methods.

  Colour can be used as a coding device where the user must distinguish rapidly between several categories of data. Groups of related data may be tied together by using similar colours, and unrelated groups of information may be coloured differently to separate them. This enables the user quickly to locate the required category, before reading the actual detail, and helps the user to carry out tasks more quickly and easily.

  The use of colours to group or discriminate will depend on their position in the visible spectrum (see Table 19.1). Colours which are close in the spectrum should be used to link items together and those which are spread out along the spectrum should be used to distinguish items. Coding to represent the degree of similarity between items often utilizes the spectrum of colours, or degrees of colour saturation such as red for severe, and pink for less severe.

  The coding must be relevant to the task in hand. A display which

Table 19.1 The visible spectrum of colour (source: Galitz, 1984)

Colour	*Approx wavelength in millimicrons*
Red	700
Orange	600
Yellow	570
Yellow-green	535
Green	500
Blue-green	493
Blue	470
Violet	400

highlights short loan lending would be unhelpful or even misleading if the operator was searching for overdue issues. On the other hand, if the operator was trying to identify short loan issues, this coding would be relevant.

The emotive effects of colours in coding are important. Colour coding should be known and meaningful to the user and consistent with known conventions. This may differ in different cultures. In many countries red, for example, conveys the concept of a warning, of stopping, or of accounts which are in difficulties, while green signifies 'go', 'safe to proceed' or 'normal' for most people. The use of red for error messages and green for options to be taken would therefore be meaningful for end users. Additionally, the effectiveness of a colour in coding partially relies on whether the operator is looking for the particular colour at the time it is displayed.

- *To highlight items of information*
  Colour may be used to draw attention to important fields. Colour coding implies that the colour of an item or field conveys a definite meaning to the user. (One database system, for example, uses colour to draw attention to the terms within a field which have been indexed. It enables the user to attend selectively to the coded data. In order to attract attention, it is important that bright colours are used.) For example, hypertext systems commonly use coloured text to signify a node, that is, text which offers a link to related information. It draws the user's attention to designated text, and often changes colour when a node has been accessed, thus aiding in the navigation task.

WHICH ITEMS SHOULD BE COLOUR CODED?

Robertson (1989) gives guidelines for deciding which items on the screen display should be coloured:

- If the screen contains format types such as headers, captions, error messages, user input, these need to be colour coded.
- If different sets of fields are dealt with separately, these could be colour coded differently to allow selective attention to each set.
- If decisions are made based on the status of a field, the field could be colour coded according to the current status.
- If the user searches a record for fields of a particular type, the field could be identified by a specific colour to contrast with the rest of the fields.
- If certain fields are hardly ever used or options are unavailable at a particular point, and they cannot be removed, they should be coloured unobtrusively.
- If the sequence in which the user can access fields is constrained, for example in data entry, the next accessible field should be colour coded to guide the novice or infrequent user through the sequence.

The aesthetics of colour coordination should not be forgotten. Too many colours or the use of uncoordinated colours can make a screen confusing or unpleasant to look at.

SCREEN COLOUR

COLOUR AND LIGHTING CONDITIONS

The lighting conditions surrounding the VDU affect the ease with which screen colour is seen. The most easily visible colours under a range of lighting conditions are green and yellow. The human eyes perceive blue most easily in poor lighting and red in bright conditions. It should also be noted that if blue and red are displayed together, under normal lighting conditions, red will dominate. This particular combination therefore should be avoided. The human ability to focus on blue is very poor, and blue therefore should not be used for text colour, especially where the size of text is very small, except when it is wished to suppress a portion of text.

COLOUR COMBINATIONS

Colour combinations are important. Text screens are more easily read when there is a sharp contrast between the foreground (text) colour and the background colour. For example, black on white mirrors the printed page and forms the basis of most Windows-based text screens. However, some colour combinations are difficult to read, for example, blue on red or any colour except white on blue. The best combinations for text screens have been found to be:

Black on white Black on amber	These two give very fast reading times when used with left justified text.
White on black	
Amber on black	
Green on black	
White on blue	

THE NUMBER OF COLOURS

Galitz (1989) recommends that text screens should ideally display no more than four colours at one time. However, if comparative discrimination is essential then no more than six to seven colours should be displayed together. A greater number of colours on a text screen results in increased time needed to find, interpret and respond to a specific colour and increased confusion between colours (Cole *et al.*, 1985). No more than seven colours in total should be used in any one application using text screens. The general aim should be to use as few colours as possible for this type of application.

For graphics screens, however, a larger number of colours is often needed for complex pictures.

SUMMARY

Although colour should never be used in text systems purely for decorative purposes, its use can greatly facilitate functionality and provide an overall completeness to the interface design providing that:

- background colours are used skilfully;
- related elements are also related by colour;
- colours are used to indicate types of action (Norlin *et al.*, 1992).

The good use of white space, formatting, highlighting and messages are also effective as an aid to the user's understanding of the screen display.

CHAPTER 20

INTERACTION FEATURES AFFECTING USABILITY

The term 'user friendly' is often used to describe software which is easy to use, but the term is seldom clearly defined, and is often somewhat subjective. There is a distinction between software which is easy for a novice to use or is easy to learn, and software which is suitable for the experienced professional. The former usually offers choices or prompts, the latter relying more on commands issued by the user.

In addition to dialogue style and screen layout, there are features of the software interface which contribute to ease of use which can be viewed objectively. These are consistency, initiative, control, help, feedback, error handling, response time and recovery.

CONSISTENCY

A system is easier to learn and to use if it is consistent in its use of commands, its displays and its behaviour. Consistency of position of prompts and error messages and consistent use of commands and function keys will increase user expectancy and increase speed and accuracy of operation. One library system, for example, is inconsistent in its use of the RETURN key, sometimes requiring its use and at other times not. Consequently, errors are frequent and users become frustrated.

INITIATIVE

Initiative refers to who directs the interaction between the system and the user. It is sometimes known as the 'locus of control' (Marshall *et al.*, 1987; Shneiderman, 1992). Dialogues can be system initiated or user initiated. The amount of initiative taken by the software or expected from the user will vary. System-initiated dialogues offer limited choices to users and are suitable for novices, infrequent users or simple tasks. An example of software which takes the initiative is America Online, a general purpose information system with Internet access, where choices are offered by the software and selected by the user. User-initiated dialogues allow the user a free choice in order, choice and complexity of commands given, and are suitable for experienced professional users carrying out complex tasks. An example of software which leaves the initiative to the user is that used by online hosts, where there is a wide range of commands which can be issued by the user at any time and in any order.

Flexible systems offer a number of ways of carrying out tasks which vary in the amount of initiative needed by the user, and can accommodate various user types and changes in user types (Cole *et al.*, 1985). It is recommended that systems display such flexibility by building in experience levels to be chosen by the user, where the expected user population is heterogeneous.

The degree of initiative offered to a user also depends on the task in hand. In data entry tasks, possible input is limited to what the system expects in a particular field. Order and type of entry are circumscribed by the system and the user has little or no choice in controlling the interaction. In command-driven online information retrieval systems, however, the user has complete control of the interaction and can specify the order and complexity of search commands according to knowledge and requirements.

CONTROL

The degree of control which users have over the interaction will also affect their feelings towards the software package. The users should always be in control of the interaction and able to proceed at their own pace, backtrack or leave the system as desired. A task which is paced by the machine or which in any other way makes the user feel the machine is in control, can create stress (see Chapter 6 and Chapter 22).

Different dialogue styles affect the degree of control which the user experiences. The success of Direct Manipulation interfaces such as Windows is due to the fact that the user has the feeling of being in control. The users receive the impression that they are actually manipulating the objects/data/structures rather than interacting with a computer system. The interface itself becomes transparent. So-called 'conversational' interfaces, on the other hand, rely on the understanding and use of a representational language to interact with the system. Failure to use the language correctly means that the user loses control of the interaction (Hutchins *et al.*, 1985).

HELP

All users of computer systems, whether professional users, novices or in between, need help in using the system. Help can be provided in several ways:

- by initial formal training in the use of the system, by online tutorials and training aids such as brochures or leaflets;
- by telephone help desks provided by the system developers;
- by online help systems which can be called up when required;
- by hard copy reference manuals which can be referred to when needed;
- by feedback from the system such as meaningful error messages or context specific help.

TRAINING

Cherry and Clinton (1989) carried out research into user satisfaction with OPACs and the methods of learning the system. It was found that satisfaction with the system was linked directly to the principal means of learning. Those who quoted ‘Friends’ as being the principal learning method showed the highest satisfaction, although this was ignored in the study as the number involved was too low to be meaningful. Of the remaining methods, brochures and printed aids learning produced the highest eventual satisfaction, followed by formal training, self-teaching (trial and error), online help, and training by library staff, in that order.

The following conclusions were reached by the research:

- Formal instruction results in high user satisfaction, but is high in cost, does not satisfy the requirements of independent learners and is not always available when required by end users.
- New users of an OPAC system bring with them experience of other similar systems and of computer systems in general, which should be taken into account when designing both the instruction programme and the interface.
- Some users wish to train on their own. An online training programme should be provided for those who wish to learn in this way.
- Brochures are an effective, low cost, highly satisfying method of training large numbers of OPAC users.
- Library staff should be available to give help when needed, that is trouble shooting, but should not get involved in training individuals.

Formal training

Formal training in the use of a system may be undertaken under the guidance of instructors from the software developers, in-house training centre or department, or by attendance at a higher educational establishment prior to using the system. Formal training is usually restricted to members of staff.

Human trainers can react to individual problems and changing situations more effectively than on-line tutorials and are more able to explain semantic rather than syntactic difficulties. Users who are trained are more likely to understand what system functions do and how to use them.

Online tutorials

Online tutorials are interactive environments in which the system itself instructs the user. They are popular as a means of helping novices to get started with a system. However, the training offered is often in the form of procedural instructions rather than conceptual training, and is often unrelated to the actual tasks the user wishes to perform with the system. It has been found that users often abandon the tutorial in favour of learning by doing, which can result in the user only learning a few of the functions available (DTI, 1990).

Brochures and leaflets

Brochures and low-cost leaflets are effective in providing initial training and continuing support for large numbers of novice or infrequent end-users of simple systems such as OPACs.

TROUBLE SHOOTING

Trouble shooting help in the form of a readily available expert or telephone help desk to solve problems occurring after training, will solve problems not anticipated in formal training, online tutorials or training materials.

ONLINE HELP SYSTEMS

Online help systems provide in-depth explanations of functions and syntax which cannot be accommodated on the task screen. They also provide ready assistance at the terminal so that the user does not have to leave it to seek help elsewhere. Online help systems are unsuitable as a means of initial training, but are useful as a continuing support system for the user. Access to information is usually through menus and can sometimes be tailored to specific user groups by, for example, providing brief reminders only for expert users, or more extensive explanations to novices.

Research shows that online help is one of the most common but least successful features of online search systems (Norlin, 1992). Difficulties experienced by users of help systems can lead to their not being used. Shaw (1991) gives the following reasons for non-use:

- difficulty in finding information;
- failure to find relevant information;
- difficulties in switching to 'help' and back to the task;
- complexity of the help system;
- quality and layout of the help information.

Borgman (1986a) stresses the importance of allowing the user temporarily to leave the task at mid-point, use the 'help' and then quickly return to the task in hand.

Where the help system is in the form of a full online manual, the user can be overwhelmed by the amount of information and the difficulty finding a required item. The problem of unmanageable amounts of help information is alleviated by the use of context-sensitive help. This is restricted information which is displayed, according to the user's position in the system when the 'help' is called. It is in effect a subset of all the 'help' which is available.

Context-sensitive help should be brief and to the point, to solve the problem in hand. Well-chosen examples should be used to convey the necessary procedure or syntax, which the user can then mimic to achieve the desired result. Directions to more in-depth information on a given or related topic can be included, or the user referred to the manual. To be effective, the help screen should be just as well designed and easy to read and understand as the task screen.

REFERENCE MANUALS

Reference manuals can be particularly useful for explaining how a system works and for giving any in-depth information which would be inappropriate for an online help system. The advantage of hard copy reference manuals is that the user already carries a mental model of a reference manual which can be used to locate required information more easily than with an online system. On the other hand, there are difficulties in ensuring that manuals are always readily available when needed, especially for networked and public access systems, and they occupy desk space.

FEEDBACK

Feedback is the system's communication to the user and gives information on the current state of the system. Users rely on feedback from the computer screen to evaluate the success of actions taken (Grossman, 1991). Hence, every action taken by the user should produce an immediate, identifiable and understandable reaction on the screen. If the system is able to respond instantaneously, for example when a 'next screen' command is given, feedback to the user is the appearance of the required screen. However, if there is a delay of more than a few seconds between a user action and the screen reaction, then an appropriate feedback message should be displayed. For example, the message 'Searching ......' informs the user that a search system is responding to a search request, but the search is taking some time. An additional feedback feature, used in some systems, is a cursor which alters shape as the search or function is processed, for example, the hour-glass cursor found within Microsoft Windows. This can be particularly useful because the user can tell when a system has crashed and is not just searching.

Feedback messages must be appropriate to the user and strike the right balance between terseness, which can be off-putting, and chattiness, which is annoying to frequent users. Jargon and the specialized use of words should be avoided in end-user systems. Maddox (1989) advises that potential end users should be consulted on questionable terms to identify jargon and simplify the language used. Additionally, the terms used should be consistent to avoid confusion that would arise if more than one word was applied to the same concept. Galitz (1984) advises that screen messages should be brief, simple and clear, and should not be authoritarian, threatening, anthropomorphic or patronizing,

Feedback can appear in several forms. These are:

- welcome screens with instructions;
- status messages;
- prompting messages;
- task performance messages;
- confirmatory messages;
- error messages.

WELCOME SCREENS WITH INSTRUCTIONS

Welcome screens, encountered when first entering a system, reassure the user that the correct application has been selected for the task in hand. They introduce the system and give navigational options. Ideally, welcome screens should be uncluttered and the information contained should utilize no more than 15–30 per cent of the screen space available (Shires and Olszak, 1992).

STATUS MESSAGES

Status information relating to the preceding and current screens should appear at the top of the screen display. Information relating to the possible choices for following screens should appear at the bottom of the display (Marshall *et al.*, 1987).

PROMPTING MESSAGES

Prompting messages ask for input from the user and may act as a reminder to the user as to the type of input required, for example, typed input or a menu choice, and how to carry out the input such as 'Type in your search term'. Prompts can suggest a default input which can be activated by pressing the <Enter> key, or overtyping the default if other input is required (Shires and Olszak, 1992). In Windows, for example, a default is highlighted by the selection box being outlined in black. Pressing 'enter' initiates the default, although other options can be selected.

TASK PERFORMANCE MESSAGES

Nothing is more irritating than a system which does not inform the user of progress. Task performance messages inform the users on the current activity which the system is carrying out. If a file is being accessed, or an item is being updated, it is helpful if the system displays a message to that effect, especially if there is a delay of more than 10–15 seconds (Galitz, 1984). Task performance messages reassure the user that actions taken have been accepted and are being acted upon by the system. Task performance messages are also required when the system is performing a fully automated procedure in which the user takes no part, for example, 'Building search ....'.

Task performance messages may be dynamic, for example '35 per cent records searched'. There may also be additional messages to confirm completion of the task, for example, '45 titles found' (Shires and Olszak, 1992).

CONFIRMATORY MESSAGES

Confirmatory messages are provided by a system as part of its error handling capabilities. In essence confirmatory messages ask for positive confirmation from the user when so-called 'data-fatal' commands have been given. 'Delete File' command from the user, for example, should be followed by a message requesting confirmation of the action. The system default should be to cancel the command so that the user cannot simply press the <Enter> key to activate the command. Confirmatory information when a menu selection is made is provided by the highlighting or video-inversing of the chosen option so that user can double-check the selection before activating the command.

ERROR MESSAGES

Error messages are displayed when the system is unable to carry out a function, or to accept input from a user. It is important that the system should anticipate all likely errors and provide an appropriate message for every contingency. In order to avoid anxiety, error messages should be accepting, understandable and positive in tone. The message should be brief and specific, be phrased in user-centred terminology, and contain neither cryptic nor coded information which is not universally known to the user population (Bernstein, 1989). Messages should be consistent in grammatical form, terminology and abbreviations, visual format and placement on screen (Shneiderman, 1992). Buxton and Trenner (1987) in their study of error messages in information retrieval systems found that even those systems which claimed to be user friendly tended to respond with hostile error messages. They argue that error messages should not use 'hostile' features such as cryptic codes, cryptic vocabulary (e.g. 'Storpstg ovfl'!) or dominant commands (such as 'LEFT HAND TRUNCATION IS NOT ALLOWED. REENTER THE QUERY CORRECTLY') or emotive vocabulary. The tone and content of error messages, they believe, can significantly alter users' behaviour, sometimes heightening their anxiety. They cite a paper by Shneiderman which concludes that 'improving messages can play a significant role in improving user performance and attitude'.

Error messages should suggest specific correction methods to assist the user, and Shires and Olszak (1992) suggest that different levels of error messages should be employed so that if the user is unable to correct the error following the initial feedback, more in-depth information is displayed on subsequent messages.

ERROR HANDLING

How a system handles errors is often the key to how acceptable a system is to end users. However experienced the user, some mistakes are bound to occur, and these will vary in type and frequency.

Selectors of software should look for systems which:

- Provide meaningful feedback when things go wrong. This has been discussed above.
- Allow the users to recover from error situations easily (see 'Recovery' below).
- Prevent serious errors being made. Methods of preventing errors include validation of input, reducing command names to single memorable characters and the use of menus. The detection of invalid input should be followed by feedback information enabling the user to re-evaluate input.

Borgman (1988), in referring to the use of information retrieval systems, divides errors into two main types: those which are due to the mechanical aspects of searching and those which are conceptual in nature (i.e. are due to a failure to understand search logic). Mechanical errors may be of two kinds: logical errors, such as an incorrect command sequence, and typing or spelling errors. Reporting a study by Dickson (1984), Borgman claims that 15 per cent of the errors in title searches could be attributed to typing or spelling mistakes, the remaining errors being conceptual in nature. Such a high percentage of errors, caused by an inadequate understanding of the system, points to a need to establish good training and support procedures. As experience grows and training is undertaken, the ratio will undoubtedly change, more of the errors being due to typing or spelling inaccuracies, and the overall number of errors should fall.

RECOVERY

All system users, especially novices and infrequent users, will at some stage wish to back-track on their actions because a wrong selection has been made from a menu or an incorrect command has been given. The system should therefore provide simple methods to 'undo' the last action(s) taken. All systems should provide editing facilities to enable the user to amend the input with minimum effort.

Another form of recovery, is the ability to quit from any point in the system. This is particularly important for those users who like to learn by exploring the system and who may then get 'stuck' or for users who simply wish to start again from the beginning. The WWW enables users to navigate back screen by screen, for example.

RESPONSE TIME

The system response time is the time taken for the system to react to a user command. The speed of response of the system should be compatible with the task to be performed, but this will not be dictated solely by the software. A task requiring continuous data entry, for example from catalogue cards, requires a rapid response between records, whereas when inputting from a book or journal a longer delay may be acceptable as time will be taken in handling the items.

If response time is too fast, the operator may begin to feel paced by the computer which is stressful (see Chapter 22). During data input, particularly, a fast response may encourage intensive, repetitive work, with its attendant problems.

A slow response time is likely to worry novice users and frustrate experts. Therefore, if the system response shows a perceptible delay, which is necessary in the case of a particularly long process or calculation, then an appropriate screen message should be displayed. When choosing a system, buyers should be particularly aware of obvious delays in response times which are not accompanied by an explanatory message (Norlin *et al.*, 1992).

A variable speed of response can also cause frustration and stress. This is more likely to occur when using mainframe computers or networks where heavy usage slows down the system.

RULES, GUIDELINES AND STANDARDS

Whichever method is used evaluators must be aware of the available rules, guidelines and standards relating to the software interface. Some general guidelines have been discussed in the previous chapters, other sources of information are given below.

RULES

Originally computer systems were built according to the designers' own view and knowledge of the system, and often with the user interface tacked on as an afterthought. There were no conventions to guide the designer and little research to investigate what worked and what was unacceptable.

As systems grew in size, teams of designers became the norm, each working on a different part of the system, and design teams set in-house rules and standards to achieve compatibility between the various parts of the individual system. This was extended to give a 'house style' across many products, enabling procedures and skills learned on one system to be transferred to others. The use of in-house rules and standards meant that teams had to spend less time on agreeing individual standards for each piece of software written, and that general-purpose programme modules could be used in many applications, thus speeding up development time.

In 1992, Shneiderman proposed that design teams should agree standards for:

- *Screen-layout issues*:
  - menu selection, form filling, and dialog-box formats;
  - wording of prompts, feedback, and error messages;
  - data entry and display formats or items and lists;
  - terminology, abbreviations, capitalization, and justification;
  - character set, faces, fonts, icons, markers, line thickness, and white space;
  - use of colour, highlighting, blinking, and inverse video.
- *Input and output devices*:
  - keyboard, display, cursor control, and pointing devices;
  - audible sounds, touch input, and other special devices;
  - screen layout and use of multiple windows;
  - response times and display rates.
- *Action sequences:*
  - direct-manipulation clicking, dragging, and gestures;
  - command syntax, semantics, and sequences;
  - programmed function keys;
  - recovery procedures.
- *Training:*
  - online help and tutorials;
  - training and reference materials.

The aim of this was to reduce the frustration and anxiety encountered by end-users. He also advocated the prototyping of systems under development, and the use of acceptance tests to ensure the usability of the end-product and the production of well-designed and satisfying systems.

GUIDELINES

Guidelines have been published on the Internet, in professional, trade and academic journals, in handbooks and in organizational literature. Although lists of guidelines are intended to improve user interfaces, by their very nature they tend to be phrased in general terms, to enable their application to a variety of situations. This lack of specificity can make them difficult to apply to any particular interface. However, they may form the basis for the development of rules for specific applications.

General guidelines on GUI design, for example, are offered by Weinschenk and Yeo (1995). These cover controls, data fields, dialog boxes and windows, menus, message, colour, icons and toolbars, metaphors, fonts, coding, graphing, navigation and interaction and online help. Galitz's *Handbook of Screen Format Design* (1989), on the other hand, specifically addresses at the appearance of the text screen display with the intention of producing clear, meaningful and easy-to-use displays. He produced extensive and precise rules governing what information should be displayed, where it should be displayed and how it should be formatted. He also gives extensive advice on the use of colour in display screen formatting. Galitz's guidelines, although rather dated, are still pertinent to the display and evaluation of bibliographic software and text databases.

Guidelines also exist for OPACs. Shires and Olszak (1992), for example, give clear and concise guidelines for the design of OPAC screens by librarians and give check lists for the physical screen, general principles, menus, commands, enquiry screens, messages and organizational considerations in the design process.

Despite these and many other efforts, designers tend to make little use of external sources of information. This is because information is often fragmentary, contradictory and overwhelming in nature. It is often difficult to find needed information. Guidelines are best used as sources of information for the physical aspects of the display and presentation rather than for the interaction aspects of the interface.

However, the universal acceptance and use of computerized systems and equipment in office work, and the identification of problems associated with these phenomena, found by research in ergonomics and cognitive psychology, has led to legislation in the form of standards covering the design of hardware, software and the work environment and organization of VDU work and to proposals for industry standards.

STANDARDS

Standards seek to promote safety, efficiency, quality, usability and consistency of use across products of the same type. In addressing the benefits of standards, the International Organization for Standardization (ISO) state that standards aim to achieve:

- mutual understanding;
- health, safety and the protection of the environment;
- interface, interoperability and interchangeability;
- fitness for purpose.

There are a number of bodies which are responsible for making standards but only the key standard bodies relevant to the software interface are discussed here. The ISO, mentioned above, is developing standards for display screen equipment which will apply to countries world wide. These will be adopted wherever possible by the European Standardization Organization (CEN), the body responsible for developing standards applicable to the European Union and European Free Trade Area member states. The British Standards Institution (BSI) is the independent national body responsible for preparing British Standards and representing the UK views on standards in Europe and internationally.

The current standards cover:

- the design of the dialogue between the user and the interface software;
- the quality and consistency of the components of the user interface;
- user guidance and documentation.

While these standards are important in promoting common interfaces and consistency it is important to remember that the adoption of standards does not automatically produce usable designs. As Hicks and Essinger (1991) point out, visual design, cognitive psychology, ergonomics and instruction design skills must also be applied.

CD-CINC

The growth of the use of CD-ROM technology in libraries has resulted in many different packages, each with its own user interface and set of terminology. Users are faced with mastering an ever-increasing assortment of software interfaces and command sets, and the resulting problems are beginning to impair further development of the technology (SIGCAT, 1992). The CD-ROM Consistent Interface Committee (CD-CINC), formed by SIGCAT, the Special Interest Group on CD-ROM Applications and Technology, investigated what could be done to encourage some consistency of the user interface in terms of functionality and terminology, throughout the industry.

The resulting report, submitted for consideration as an industry-wide standard and concerned with the conceptual, rather than the physical, interface recommends the use of good design principles. It identifies 13 basic functions needed for smooth running at three levels, top level, operational level, and navigational level. Each of the functions is given a definition, list of related terms, description and remarks in the report. Guidelines for developing installation procedures are also included. Top level functions are identified as Help, Browse, Index, Search, Display, Print, Download, Restart, Change, Quit. Operational level functions are Execute, Break, Escape. Navigational functions refer to movement through the data and 10 types of movement are identified.

Wherever the functions are equal, CD-CINC uses the same terms as those in the standard for the Common Command Language (CCL), which is used for interactive online information retrieval. PC software environments have also influenced the choice of terms for function names.

SUMMARY

An appropriate software interface is a combination of functionality, dialogue style, screen design together with a satisfying and satisfactory interaction between the user and the system. Human characteristics of the users have a large part to play in the interaction and should have been taken into account when the interface was designed. When choosing an automated system for a library, librarians need to be aware of these factors and the rules, guidelines and standards in order to evaluate the options available.

CHAPTER 21

SOFTWARE EVALUATION

The introduction of a new automated library system has many implications for a library's operations, staff and users and therefore all available options must be evaluated carefully before a system is chosen. The wrong choice of software will have a deleterious impact on budgets, efficiency, motivation and ultimate usage.

The functionality of a computer system alone does not guarantee its suitability for any given situation and group of users. The terms 'usability' and 'user friendly' are often applied to computer systems and refer to the characteristics of the system which enable the operator to interact with the system effectively and easily. These characteristics are manifested in the software interface, dialogue and the additional features discussed in the previous chapters.

Shackel (1991) gives a model for usability and related concepts to be used when evaluating a computer system, where utility (i.e. does it do the job you want) + usability (i.e. can it be used by your staff) + likeability (i.e. is it enjoyable to use) are balanced against cost and consequences to arrive at a decision on acceptability.

THE TYPES OF LIBRARY SYSTEMS

Library systems may be commercially developed turnkey systems, or systems developed by a consultant or in-house by an individual library, for its own use, or by a consortium of libraries as a cooperative system (Msuya, 1990). For most library applications, Rowley (1990) recommends that pre-written, commercially available software is purchased. The advantages of this approach are as follows:

- A tried and tested package should be much easier to implement.
- Less time and effort are spent on systems analysis, communications, liaison and project management.
- A ready-made package which is not tailor-made encourages the adoption of standard practices and discourages an individual library's idiosyncrasies.

Rowley also classifies three types of systems likely to be encountered in a library environment:

1. Text retrieval/text information management software.
2. Library housekeeping/library management software.
3. Database management systems (DBMS).

EVALUATION METHODS

Evaluations of systems software fall into three general types, each requiring a different approach. These are prototyping, re-evaluation of existing software and the selection of a new commercially available system.

PROTOTYPING

This is a process of designing and evaluating the software interface of a system under development and involves the construction of an inexpensive mock interface showing all the features of the software interface, for presentation to the client, which would be the library manager and/or senior staff. The client is able to try out and comment on the interface. Any necessary adjustments to the interface or the system requirements can then be made before the software is actually written. Prototyping therefore reduces the chances of surprises to end users and enables designers to obtain a complete specification, by revealing misunderstandings between developers and users and by showing up errors in the requirements specification (Hartshorn and Hix, 1989).

Prototyping varies from pen and paper sketches, through to automated prototype construction programmes. It provides the means to involve users in the design process and enable iterative refinement to the interface early in the design stage. Other methods of prototyping include the use of storyboards, flipcharts and animated video. Programming languages, graphics tools, hypermedia systems and data manipulation languages have also been used in the production of inexpensive simulations (DTI, 1990). The experimental 'Wizard of Oz' method (Green and Wei-Haas, 1985) even had the developer's terminal linked to the unsuspecting user's terminal, with the developer keying in the 'system' responses to the user's input. In fact, prototyping covers any method which quickly and inexpensively simulates the interface, to enable the interaction to be tried out and evaluated before the system is built.

Software interface design can also make use of user interface management systems (UIMS) which are intended to fulfil, in interface design and management, the role occupied by computer-aided software engineering (CASE) tools in software development and database management systems (DBMS) in database management. The objectives of UIMS are to improve productivity and shorten development time and cost, to enable development of the interface without programming and to improve interface quality (Rantanen, 1991). The speed and ease with which interfaces can be constructed make UIMS very suitable for prototype production. The premises on which the design of UIMS is based are:

- *Function and dialogue of the application under development can be separated.* This means that several interfaces can be built for one application, which is important where the range of user expertise is wide. Additionally, many applications can share a single intelligent interface, which has implications for consistency across applications and the transfer of skills from one application to another (Hartshorn and Hix, 1989).
- *Rapid prototyping is the preferred method of interface design.* The importance of user involvement in design and user-oriented systems points to this as being the logical way of producing the system interface for user evaluation.
- *Applications programmers are not the best people to design the interface.* It has been recognized that human factors specialists and clients (in our case library managers) have an important part to play in interface design and evaluation, and the UIMS helps to define the interface as a part of the system demanding its own experts.
- *Tools are needed for maintainability and portability.* The ability to adapt interfaces to suit changing needs without the need to rewrite the entire application and the use of 'reusable' interfaces both prolongs the usability of systems and cuts down on development costs.

Other methods of evaluating the software interface include predictive tools which assess a typical user's performance with a given interface (Hammond *et al.*, 1987), and programme tools which evaluate screen layout according to established principles. These tools assess aspects of the layout including overall and local density, number and size of groups, number of items per line and layout complexity, consistency between screens, predicted ease of use and search time (Williams and Leaf, 1986).

Stages in the development of software and the interface are given later in the chapter.

RE-EVALUATION OF EXISTING SOFTWARE

This procedure should be carried out periodically by libraries with existing automated systems to ensure that installed systems are still meeting the library's requirements, and to identify any areas which may need improvement. Any organization's needs will evolve and change over time and the systems which support it will inevitably have to be changed.

There are many methods recommended by various writers for carrying out re-evaluations, such as measurements of performance accuracy interviews with users, written and online surveys, examination of 'Help' requests, observation or video recording of users, verbal protocols, where users 'talk through' their actions, and the use of computer-monitored data (Rice and Borgman, 1983; Christie and Gardiner, 1990). Feedback from the re-evaluation is used as a basis for a new system's requirements or to upgrade an existing system.

The use of computer-monitoring for evaluation of the software interface is a valuable tool which will probably become more important as a supplement to other monitoring methods in the future. It measures quantitative rather than qualitative information in that it automatically logs the type, content and time of each information exchange, that is behaviour, but not the user's impressions or attitudes towards the system. There are both advantages and disadvantages to its use.

The advantages can be summarized as follows:

- Collection of data is unobtrusive and elicits no response bias from the users.
- Information can be collected from more users than a sample survey.
- It records actual use rather than reported use of the system.
- It can be used to record data over a continuous time span or at discrete intervals and therefore be used to uncover trends.
- The same data can be analysed from different viewpoints and may be used for various research projects.

Disadvantages of using computer-monitored data are not insignificant and include the following:

- Computer-monitoring has implications from an ethical point of view and steps must be taken to ensure that there is no invasion of privacy for individual users.
- Computer-monitoring rapidly creates vast sets of data which have to be managed and analysed. The relevant expertise and time must be available for these activities.
- There is only a very limited correlation between computer-collected and user-reported data. The discrepancy between the two could be inaccuracies on either side in any given situation.

SELECTION OF A NEW COMMERCIALLY AVAILABLE SYSTEM

The selection of a new or replacement computer system in a library must be viewed as a major project, the main stages of which should be as follows:

Need for a new system?

Automation is a major investment in time, money and people and should only be undertaken if there is no alternative method of meeting the organization's needs. The system may be needed in order to increase efficiency, to eliminate or reduce clerical routines, to replace an existing system or to offer services which only automation can make possible.

Pachent (1996, p. 215) when discussing factors in choosing the replacement system for Suffolk Libraries and Heritage said that the automation needed to meet one or more of the following criteria:

- to carry out an operation or service more efficiently;
- to carry out an operation or service not possible using manual systems;
- to provide management information about operations and services in order to inform decision making.

Project team

Ideally the project team should involve a broad range of staff. In the case of Aston's project to find a third generation replacement system, the team consisted of the Acquisitions Manager, the Document Supply Manager, the Bibliographic Records Manager, the Head of Systems and an Information Specialist interested in OPAC design (Matthews, 1995). The project team responsible for choosing a system for Suffolk Libraries and Heritage consisted of sixteen people from all grades of staff (Pachent, 1996). Although the whole group only met together about six times, smaller subgroups met frequently and were responsible for detailed decisions.

Project plan

A project plan needs to be developed to establish tasks, timescales, responsibilities and deliverables. Some projects have found it helpful to use project management techniques, Lewis (1995), for example, describes the use of PRINCE (PRojects IN Controlled Environments) project management methodology in choosing a new library system at the University of Wales, Bangor. PRINCE is often used within government departments but can be applied to any size project. The objective of PRINCE methodology is to make sure the deliverables (in this case the selection and implementation of the right library system) are produced at the right time at the right cost (Bently, 1993). This is achieved through careful planning by a project manager, who answers to a Project Board. PRINCE plans require all products and activities to be listed, sequenced and assigned to individuals who have responsibility for their completion. Lewis (1995, p. 239) concludes that 'the use of project management techniques is an invaluable aid to achieving a successful project'. Matthews (1995), however, is rather less enthusiastic about the use of project management software. He describes the use of the Macproject software package at Aston University and advises unfamiliar users to leave plenty of time for adequate training. Ultimately, he believed that the use of simple critical path techniques would have worked just as well for the selection of the system.

Potential replacement systems

Details about potential replacement systems can be obtained through scanning the professional literature, talking to colleagues in other institutions, visiting exhibitions, such as the Library Resources Exhibition, attending conferences and seminars, reading literature freely distributed by system suppliers and consulting published directories and handbooks. Marshall Breeding's handbook (1994), for instance, evaluates 20 PC-based integrated library systems for functionality and includes background company information and numbers of systems sold. Blunden-Ellis and Graham (1994) may also be a useful source. They present the findings of a UK market survey of library automation suppliers. However, since technology is changing rapidly it is advisable to check the current 'state of play' directly with suppliers. Background information found in this phase will provide useful input into the operation requirement (OR) document.

OR document

The purpose of an OR document is to outline the minimum and desired functionality requirements for the new computerized library system. They are used as guidelines for suppliers who may wish to submit a tender. A number of libraries have made their ORs available on BUBL (1996) and these provide a very useful starting point. The involvement of library staff may also be useful at this stage. Pachent (1996) describes how all permanent library staff (over 300) were sent questionnaires asking them to say what they wanted from a new system. The functionality requested covered circulation, inter-library loans, serials, catalogue, OPAC, acquisitions, community information, management information, electronic mail and dial-in facilities. The views of approximately 250 library users were also sought when choosing the system for Suffolk Libraries and Heritage. Their views were found to be consistent (Pachent, 1996, p. 224):

- library users were keen to have a new system;
- many users were knowledgeable about the existing system and had constructive ideas for a new system;
- many users wanted new services, not just a more attractive version of the existing public terminals;
- many users wanted access to networked CD-ROMs, to remote databases, to images, and, to a lesser extent, to self-issue terminals and to the Internet;
- a significant number of users wanted magnified screens;
- many users wanted self-service printers, and scanners which work;
- the users wanted a sufficient number of public terminals.

In addition to the functional requirements the OR document should include introductory information about the library and a specification of the technological requirements. Heseltine (1993) suggested that an automated information system should feature:

- industry standard hardware such as using PCs and a server running the UNIX operating system;
- industry standard communications (such as X.25, Z39.50, TCP/IP);
- industry standard UNIX-based relational database management systems.

Added to this list might be client/server architecture or specific requirements to connect to existing equipment.

The tendering process

The European Union's tendering process must be adhered to if the system is likely to cost above a certain amount (£158 018 in January 1996). Details of this process are given in a pamphlet (Harvey, 1993).

Shortlisting

Once the tenders have been received it is often necessary to eliminate some systems from further evaluation. Usually, institutions only have time to undertake detailed evaluations on four or fewer systems. Each tender document must be examined carefully to make sure that all the essential requirements can be met by that particular system. If it does not comply then it can be dropped from the list.

Final evaluations

This is generally the most critical part of the project and, assuming that all systems meet functionality requirements, is concerned with evaluating 'usability' and 'likeability' aspects. Generally, the starting point for this phase is demonstrations of the shortlisted systems by the suppliers (see LITC Report No. 2, 1992). It must be remembered, however, that demonstrators often use small data collections which may affect response times and are experienced in hiding any weak aspects of the system. For this reason it is advisable to make visits to other organizations already using the systems. This can provide considerable insight into the true usability and functionality of the system. Discussions with existing customers or user groups can give valuable feedback on the producer's performance, potential problems and the producer's willingness to amend and improve the system in line with customer requirements. Many suppliers will also install their demonstration system on site for one or two weeks. This allows staff to explore and evaluate the system more thoroughly.

Evaluations can be formal or informal, but should be structured and recorded in detail for later analysis, and involve the carrying out of real tasks by evaluators representing the types of user groups who will be using the system. The benefits of carrying out realistic tasks are given by Johnson *et al.* (1989) as follows:

- Real tasks are the most effective way of exploring the functionality of the system.
- Evaluators can see the interface as part of the overall system rather than an artificial series of screens and actions.
- Many problems, difficulties and aspects of usability are only exposed when carrying out real tasks.
- Observation of evaluators carrying out tasks and recording certain aspects of task performance is extremely valuable in identifying areas of difficulty for end users.

Matthews (1995) describes how Aston University tested the circulation components of systems with 'typical counter scenarios' involving renewing a book, checking for a reservation, discharging the book, displaying fines due, placing a reservation, and finally issuing a book. He recommends tests of this type to find out the true strengths and weaknesses of a system. Their evaluation also involved getting feedback from library customers using each of the system's OPACs.

During the evaluations it is particularly important to assess all aspects of the interface. The use of checklists is a good way of doing this. Checklists of interface features can be devised locally or taken from the published literature (e.g. Hildreth, 1985; LITC Report No. 2, 1992; Ravden, 1989) and adapted for local use.

Whatever evaluation methods are used, rigorous recording of results should ensure that candidate systems can gradually be eliminated until a system is chosen which most closely reflects the organization's needs and the users' requirements.

CUSTOM SOFTWARE, OR WHAT TO DO IF YOU CAN'T FIND A SUITABLE SYSTEM!

There are occasions when commercially available software is either not appropriate or unavailable for a particular application. In such situations an organization may decide to contract for the design and development of a custom-made system or to develop a system in-house.

THE SYSTEMS AND SOFTWARE DEVELOPMENT CYCLE

The process of creating applications software for a custom-made system involves several well-defined steps. Knowledge of the steps to be taken and how the development of the software interface fits in with the development of the whole system is essential for organizations contemplating a custom software. Some of the steps are taken by the members of the organization, some by the software house, consultant or in-house developer, while others are carried out jointly.

Bernstein (1989) gives the stages in developing applications software as follows:

1. *Assess the present system*
   As previously mentioned, this is an evaluation process which should be carried out by all organizations periodically, whether current systems are manual or automated. During the evaluation, problems will be uncovered. Options to solve these problems are then investigated and evaluated. For example, Yee (1991) listed 16 problems he found when he evaluated OPACs. If automation is found to be the method of resolving the problems, then the next stage can be undertaken.
2. *Assemble the team*
   The development team will consist typically of a project manager, systems analysts, human–computer interface specialists and software engineers/designers/programmers. Ideally, user representatives from the management of the organization should also be involved. The user representatives can ensure that the design meets with organizational requirements and be responsible for liaison between the design team and the end users.
3. *Write a brief specification*
   This will include organizational requirements regarding functionality, capabilities, reliability, speed, accuracy, benchmark tests and the minimum operating environment.
4. *Conduct a task analysis*
   Systems analysts will need to investigate the goals and sub-goals of the present system, information flows in and out of the system,

procedures followed (the what, why, where, and when), and problems to be overcome. The analysis may involve observation, interviews, protocol analysis and document analysis and will take place in the workplace involving end users of the system.

5. *Conduct a user requirements analysis*
   Human factors specialists will need to identify the potential system users, their domain knowledge, skills, literacy level, computer experience and how they will use the system. The user requirements analysis may uncover user behaviour in carrying out tasks which suggests enhancements to a system. For example, the results from a user study (Belkin *et al.*, 1990) resulted in the following suggestions to improve the functionality of an OPAC:
   - a table of contents or index for each bibliographic record to assist browsing activities;
   - a locator map and/or directions on screen to give the physical location of a retrieved item;
   - a record of an individual user's past borrowing record to enable the same or similar items to be located;
   - subject searching by genre to enable the location of similar items.

   A statement of user requirements is produced at the end of the analysis which is used as input to the interface design.
6. *Design the interface*
   In this stage the most appropriate interface style or styles for the target user population are chosen and screens designed.
7. *Prototype and test with users*
   One or more mock-ups of the system interface designs are constructed and these are then tested by actual end users of the system rather than the organization's management. Feedback from users is used to clarify system and user requirements.
8. *Revise the prototype*
   This is needed to incorporate omissions and errors discovered by user testing. Where a choice of interfaces is offered, the most favoured is developed further. User testing and revision of the prototype is an iterative process which continues until the users are satisfied with the result. Maddox (1989) notes that end-user involvement in the design stages almost invariably ensures the ultimate acceptance of a new system.
9. *Write functional specification document*
   Detailed notes and diagrams on how the software will look and operate are prepared as a result of the final prototype. This is a complete blueprint of the system under development which is then agreed with the client organization and used as a basis for preparing the programme code.
10. *Write programme code*
    It is recognized that the writing of the programme code should

occupy only about 20–30 per cent of the whole project time (Maddox, 1989) and result in an acceptable system providing that the previous stages have been rigorously applied.

11. *Test the system*
    This is a process of testing for average, boundary and exceptional value ranges, and ensuring that the system is bug-free, crash and fool-proof. The process is usually carried out by quality assurance staff, but at least must be carried out by someone other than the designers.
12. *Carry out acceptance tests*
    Previously specified benchmark tests are carried out by the user to verify that the originally agreed specification has been fulfilled.
13. *Introduce the system*
    The new system may be introduced to the organization in various ways. Incremental development involves the development and introduction of parts of the system, one at a time, as part of an overall project, so that the new system gradually replaces the existing system. Parallel running means that the new system is introduced and runs alongside the existing system for a period of time to ensure that there is a fall-back position. The so-called 'Big Bang' implementation closes down the use of the existing system and commences use of the new system immediately.
14. *Evaluate the system*
    Unanticipated problems and suggestions from users for future improvements are uncovered during the periodic evaluation of an implemented system which are then used for input into the next version.

SUMMARY

The acquisition of software systems to aid a library's operations is an important investment in resources of time, people and money. Thorough evaluation is necessary, whether the system is bought off the peg, developed in-house or commissioned. In this chapter we have discussed various methods of undertaking the selection and evaluation of software. We have also discussed the process of software development which is needed when contemplating custom-made systems. Once a system is acquired, periodic re-evaluation is needed to ensure that the library's current needs are being met.

PART

VII

JOB DESIGN

CHAPTER 22

JOB DESIGN

INTRODUCTION

Even though software may have an excellent user interface, interaction with it makes up only part of a person's job. To optimize staff well-being and productivity, attention needs to be paid to the design of the job a person is employed to undertake. The introduction of a new library system is a unique opportunity to reappraise work activities and conditions.

A well-designed job is one factor influencing staff motivation, and well-motivated employees have the potential to contribute beyond the narrow specifications of their jobs. However, motivating staff is a complex and sensitive issue. There are many variables determining an individual's sense of job satisfaction. A worker's own attitudes play a part, as do the prevailing management style, the nature of the work group, and the nature of the task. An organization cannot afford to disregard motivational theories or poor morale (which in turn leads to haphazard performance and general indifference) will result. Since the whole concept of technology revolves around changing the way in which tasks are done, it is not sufficient to superimpose one system on another: attention should be paid to the impact on job design. Sadly, all too often little or no thought is given to job design but rather jobs consist of those tasks which the computer cannot handle with little thought to their coherence as a job.

Information technology will impact in different ways according to the individual functions and the structure of the overall information service. Some tasks or even whole jobs will disappear, others will be created, merged or expanded: responsibilities will alter. Technology's effects may differ according to whether staff are professional or non-professional and its impact can be construed as either positive or negative according to the individual. No hard and fast rules can be made, but some general principles and guidelines are offered below.

JOB DESIGN AND JOB SATISFACTION

Job design is important from the twin standpoints of the organization and the individual. Failure on the part of management to devise satisfying and complete roles for their employees can create intolerable levels of strain and stress, which in turn can lead to a high rate of absenteeism and labour turnover. Informing staff that their jobs are structured in order to improve efficiency may imply that worker satisfaction is not a high priority on the part of management and may not in itself guarantee productivity; it is better actively to involve employees in the quest for increased effectiveness and thus promote team effort. The principles set out here are primarily concerned with employees' welfare, which may lead to increased productivity but may need to be combined with other incentives.

PRINCIPLES OF GOOD JOB DESIGN

There is a vast literature on the topic of what constitutes a 'good' job and the many factors which influence people's criteria once their basic requirements of security, reasonable working conditions and adequate pay have been met. Starting from this documentation and other personal studies, Sell (1986) – of the Work Research Unit – has compiled a list of principles on which to base sound job design: variety; autonomy; feedback; social contact; identity; responsibility; achievement; opportunities to learn and develop; to which should be added the need for adequate rest breaks (required by the European Union Directive on VDU workers), and some areas to watch for: role under/overload; role conflict; and role ambiguity. These are discussed in more detail below. Other factors which affect worker motivation according to Herzberg (Bridger, 1995) are the company policy and administration, supervision, relationship with supervisor, work conditions, salary, and security.

Variety

Routine jobs are likely to induce feelings of fatigue. Ideally, the tasks that the worker performs should provide variety of pace, method, location and skill. Such variety can help to minimize fatigue and promote interest and motivation, but should be under the control of the worker where possible. Variety can also help the worker acquire a number of skills. Promises are frequently made that technology will eliminate dull and repetitive tasks, only for them to be replaced by equally unexciting tasks. It is ideal if the time saved by the new computer system can be used to enrich the job with an interesting non-VDU task, such as dealing with customer enquiries. This can also be one way of providing productive rest breaks (see below).

Autonomy

Staff should have some degree of control over their jobs, in particular how and in what order they perform individual tasks to achieve an agreed goal. Management style and software packages both vary in the flexibility they afford individual workers.

Feedback

Employees need to know how well they are performing on a regular basis. Feedback can be obtained from co-workers, management or directly from the work itself. Ideally, it is desirable for employees to be able to ascertain their performance level without requiring feedback from management – although this is also important. A computer system should be able to provide feedback in the form of statistics, although it should be recognized that bare statistics do not tell the whole story. It is also important that if such statistics are transmitted to management it is with the knowledge (and preferably also the consent) of the staff involved.

Social contact

This, for most people, is a desirable part of any job. While it is appreciated that time must not be wasted by unproductive chatter, it must also be recognized that the opportunity to see and converse with others can be beneficial to staff morale. Indeed, isolation from fellow employees can lead to stress, poorer performance, lower satisfaction and even obstructive behaviour (Zuboff, 1988). Workstation layout therefore should enable social contact while minimizing unnecessary interruptions.

One area within a library where automation has an obvious impact on social contact is the issue counter. The time taken to issue or discharge a book using the Browne issue system used to provide an opportunity for social interaction with the user, whereas the speed of throughput with most computer systems reduces contact with readers. Increasingly now, too, borrowers are able to check out their own books and to make reservations without interacting with library staff. This may result in a need for a reduced number of staff or affect the nature of the work, depending on the situation in individual libraries.

The implications of VDU usage are, however, far wider than the immediate library context. Electronic communication has revolutionary potential for society as the pervasion of the Internet has shown – Parker (1996) anticipates that every citizen will have Internet access in the twenty-first century. Group teleconferences, data and text transmission, and electronic mail all theoretically allow work to continue independently of geographical location. People are no longer restricted by the physical transfer of material and many academic libraries are linked to campus networks. This will have a major impact on social contact as it becomes increasingly adopted by libraries. Borrowers may become quite remote and most contact may be via electronic means. It may even be possible for more information work to be done at home, although this is unlikely to happen on a large scale for some time. The need for social contact should be considered when such changes are made.

Identity

Tasks should fit together to form a coherent job. Incorporating preparatory or sequential tasks can result in more coherent work as well as ensuring variety and challenge for the worker. At the issue counter, for example, it may be possible for the worker to be able to deal with all of a reader's requests rather than having to refer them to someone else to alter their record or to make a reservation. Similarly, stock preparation and cataloguing could be grouped into a more coherent and satisfying job than having the functions divided into a number of different processes performed by different workers for increased efficiency.

On the other hand, incorporating all aspects of a job may include some less fulfilling tasks. For example, on-line searching, itself a professional task, demands a number of record-keeping activities which, although clerical in nature, fall to the search professional (German *et al.*, 1987). Some professionals enjoy clerical routines from time to time as a change from more intense professional work. The exact balance of tasks can be decided in each situation by management and employee.

Responsibility/skill

The automation of tasks can affect the level of responsibility held by the worker and also the skill required to do their job, although technology itself neither automatically degrades nor improves a job (Iacono and Kling, 1987).

An example of a task being de-skilled is when, using a cataloguing cooperative or utility, the employee may spend most of the time copy cataloguing instead of creating original records. In such situations, perhaps a different level of employee could be used to perform the tasks. What could be a lower level of responsibility for the professional cataloguer may be appropriate for a senior library assistant and it is not uncommon in larger libraries for a different level of employee to copy catalogue while the professional cataloguer deals with complex items and with the supervision of the clerical staff. It is worth noting that an operator whose job has been de-skilled will experience more stress than one who has always performed the less skilled task (Cakir *et al.*, 1980).

Responsibility can also be increased. A redesigned job with a coherent set of tasks tends to increase the level of responsibility over a job which consists of a series of disparate tasks performed by different people and requiring a supervisor to keep an overview. Interestingly, Johnson (1996) found that supervision as a task had itself declined in the years between 1975 and 1990. She surmised this to be as the result of automation. Thus increased responsibility and job satisfaction for one level of worker may result in reduced responsibility and possibly satisfaction for their supervisors. Increased responsibility or a demand for increased skills can enhance the job satisfaction of those involved or add an unwelcome burden of responsibility, depending on the employees involved. Johnson (1996) found a general de-skilling among the group of subjects she studied and yet found ironically that entry requirements for positions had increased, despite the static or reduced task skills requirements.

Achievement

It is important that people go home at the end of the day feeling that their effort has been worthwhile. Achievement is closely related to feedback, but also relates to the library's objectives and the employee's goals. A clear understanding of the library's purposes and the staff member's contribution to achieving them can help. It is unhelpful if the two sets of goals conflict and discussion between staff and management to clarify their respective positions can be useful at times. Feedback on a task can help the worker to assess the quantity and quality of their work, but only an assessment of the importance of the overall goals being achieved will contribute to a real sense of the worker's value.

Opportunities to learn and develop

The level of responsibility and skill has been referred to above. However, not only do people like to develop in their work, but also their jobs change over time and in the light of automation (Johnson, 1996). The introduction of a new system may mean that while old skills are discarded, new ones are learned.

Managers often underestimate the abilities of their staff and fail to offer them the chance to develop existing and new skills. This may be out of indifference or ignorance. Ideally, however, workers should see a future for themselves within the organization and their potential should be developed where possible. Regular software updates mean that the learning process is never ending, but it should be at such a pace that staff feel stimulated rather than overburdened by the changes. The reaction to the opportunity to learn will vary from one employee to another.

Role under/overload

Similarly, individuals vary with regard to the amount of work that they feel able to achieve in a day. Everyone has an optimal rate of working, and it is desirable that people should be allowed to regulate their own work. As a rule, people achieve more and work to a higher standard if they are allowed to choose their own pace. If the pace of work is too fast for the operator, they may feel that they are constantly under-achieving, and emotions of poor self-esteem and frustration may be experienced. Alternatively, they may try to increase their own pace of work. It has been found that if the system response time is consistently less than half a second, the operator's keying rate tends to rise above normal, with the result that muscular tension, stress and fatigue also rise, causing the operator to feel 'washed out' afterwards (National Electronics Council, 1983).

If the system response is too slow the operator may feel that they are over-achieving and become frustrated that they are not in control of the pace of the work. Generally speaking, a task which takes the same overall time to perform, but during which the operator is active, is more satisfying than a second one in which there are periods of operator inactivity waiting for the machine to process an item. This may be observed with many automated issue systems. The rapid issue and discharge of items which reduce queues at the counter, can also create periods of enforced idleness. These problems may be exacerbated by technical faults or systems failures. It can also be true of so-called 'user-friendly' interfaces, which may become irritating when greater skills are acquired. The option to 'bounce' through menus is highly desirable (Pearce, 1986), giving operators the impression that they are deciding the pace at which work should proceed. Rather worringly, Kahn and Cooper (1986) note that work underload has been linked with ill-health.

Role overload, too, can create problems. As staff costs increase, so it becomes necessary to think of automating the system rather than increasing the number of employees in order to maintain the existing level of library service. Undoubtedly the computer takes much of the workload, but can bring with it optimistic assessments of increased productivity. It also brings its own workload, needing to be fed data, maintained and backed up. Initially, too, extra work is required to train staff in the use of the system. In consequence, some people find that their workload has increased rather than decreased. Yang and Carayon (1995) found that in cases where worker overload was unavoidable, stress could be minimized with supportive supervision.

Role conflict

Where there are doubts as to the priorities in a job there is a risk of stress. The implementation of an automated system often creates more work for those in charge. Conflicts can arise between, for example, the need to create a usable staff manual for the system (an important, but not usually urgent task) and the pressures of the other tasks the job holder continues to perform (often urgent but possibly less important in the long term).

Role ambiguity

Another cause of stress arises when there is doubt about what support will be available for decisions taken in the absence of a superior. Junior staff, for example, may be unsure how far to 'bend' the library rules if at all, or whether inflexibility will engender poor public relations. Lovecy (1984) comments that automated systems are more likely to pick up on infringements such as over-borrowing which may make the library personnel seem stricter than before, creating its own conflicts.

The need for adequate rest breaks

The Display Screen Equipment Regulations (HSE, 1992d) requires that VDU operators have periodic breaks or changes of activity to reduce their workload at the display screen. The break should allow a change of work for the hands, the arms and the eyes as well as a change of posture, and should ideally be taken away from the screen. The break should be included before the onset of fatigue rather than as a recovery afterwards; the timing of the break being more critical than its length. A two-minute break every 15 minutes is better than a 15-minute break every two hours. Ideally, the job should be designed to include non-VDU tasks into the daily work schedule at intervals and this has been found to be more effective in relieving visual fatigue than formal rest breaks. However, it is also possible to incorporate exercise routines which include stretching, blinking and focusing the eyes on distant objects.

The need for the breaks is threefold. First, an operator performing a visually demanding task will rapidly tire (see Chapter 5) and time spent working away from a VDU will enable the eyes to re-focus and will allow the eye muscles to rest. It is not only the eyes which will tire during concentrated VDU work. Chapter 3 outlines the postural problems which may result from prolonged work spells (among other causes). Figure 22.1 indicates graphically the percentage of employees (the number studied is not given) reporting selected adverse health symptoms. It can be seen that as the number of hours worked on a VDU increases so, too, does the percentage of operators reporting problems, with a noticeable increase when more than four hours per day are worked. Åborg *et al.* (1993) came to a similar conclusion, but found that women reported a larger number of symptoms than men. This implies that the total time spent on a VDU will affect the comfort of the operator and it is suggested that a total of no more than four hours per day should be spent on intensive VDU work. The chart in Figure 22.2 shows an increase in reported adverse health symptoms as the length of time worked on a VDU without a rest break is extended. These charts merely give a guide, however, as no details are available as to the workstation layout and the type of task being undertaken.

Mentally, too, operators will tire if they are forced to work on one

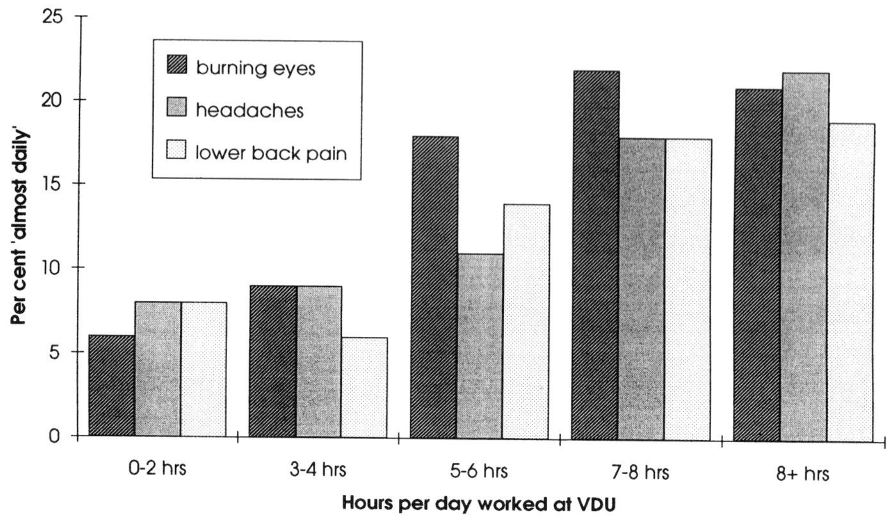

Figure 22.1 *Percentage of workers reporting selected health symptoms 'almost daily' by hours per day worked on VDUs. Source: Canadian Labour Congress as reproduced by the London Hazards Centre, 1984.*

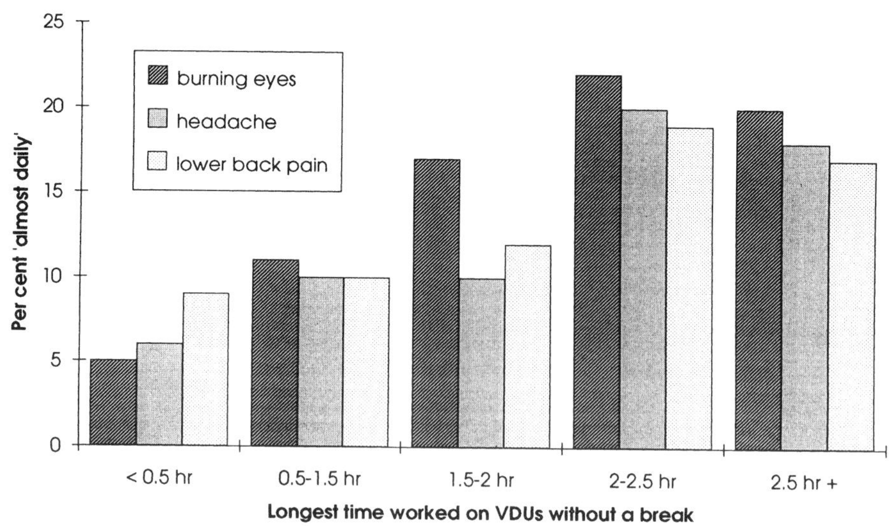

Figure 22.2 *Percentage of workers reporting selected health symptoms 'almost daily' by longest time worked on VDUs without a break. Source: Canadian Labour Congress as reproduced by the London Hazards Centre, 1984.*

task for a long period. Attention will depend on an operator's motivation (see Chapter 1). If attention waivers, operator performance will diminish. Adequate rest breaks are essential for all VDU operators, but the length and type of interlude needed will depend on the operator and the type of task performed. As a general rule, work which is physically, mentally or visually demanding requires more frequent rest periods than less intensive work. Henning *et al.* (1993) found that exercise with frequent short rest breaks increased both operator well-being and productivity.

TECHNIQUES FOR IMPROVING JOB SATISFACTION

There are a number of techniques which can be used to minimize dissatisfaction and improve the quality of working life where tasks are intrinsically dull. These have been described by Argyle (1972) as job enlargement, job rotation and job enrichment, and are most applicable where staff have least control over their own assignments (although it must be acknowledged that, in the broader sphere, team staffing structures in public libraries are also a form of 'enrichment').

JOB ENLARGEMENT

Employees' jobs are made to contain more tasks, and hence are enlarged. For example, the serials check-in clerk may be given, in addition, the task of assembling items for binding or the interlibrary loans clerk be asked to collate statistics about the service. Increasing the breadth of a worker's role can also encourage them to take greater responsibility (see 'Responsibility/skill' above).

JOB ROTATION

Staff can be moved regularly from one job to another, and bring about an increase in job satisfaction, especially if the work tends to be dull and routine. Job rotation has the added bonus of giving the employee experience of several aspects of library work and an overview of the whole system. This is particularly important for library trainees. However, critics have rejected it on the grounds that each task is likely to be as boring as the last. It is also difficult to determine the frequency of rotation. If staff are transferred from section to section or department to department too often they may experience a sense of disorientation or that they have only just mastered one set of routines when they are once again confronted with the unfamiliar. Job rotation can therefore, if not monitored closely, lead to dissatisfaction and additional stress. If well-implemented, job rotation can add variety and increased social contact to an otherwise routine job.

JOB ENRICHMENT

Job enrichment requires tasks to be organized so that the employee has some degree of responsibility. They may become a specialist in inter-library loans or the short loan collection and have a certain amount of authority. Organizing staff into small teams in which they work together towards set targets can also give meaning and direction to their labour. At the very least, enrichment can consist of giving workers responsibility for their own quality control (Bridger, 1995).

Caputo (1991) argues that not all jobs can be sufficiently enriched and suggests that job modification is the next best alternative. Job modification involves redesigning the tasks within the job within the limitations imposed by the needs of the organization to match an individual's needs.

It must also be said that uninteresting work may be unavoidable for practical reasons. Holroyd (1986) therefore recommends the techniques of job enrichment and semi-autonomous work groups. These are superior to the strategies of rotation and enlargement, he declares, because they involve an analytical approach to employees' work. The others are in some measure an admission of failure to create sufficient variety from within the job itself. An element of challenge is also a recognized criterion of good job design. Building this feature into basic library work which involves automated routines is very difficult. One way, however, is to emphasize the importance of the employee's work role to the overall organization. Counter assistants are ambassadors for the whole library service and should be acknowledged as such. Library assistant posts do not offer many chances for progression and personal growth. Management consequently has an even greater responsibility to ensure that other 'satisfiers' are met. There must be room for human interaction within the framework of the job. This can partly compensate for boring or repetitive work and is especially vital in a machine-dependent environment. Realistically, not all jobs will ever fulfil all principles of good job design, but, with thought, the quality of working life can be improved.

Introducing a new computer system provides such an opportunity. It is a chance to capitalize on what people find interesting and satisfying at work and to review total organizational efficiency at the same time. Sell (1986) sees it as a unique event because, with the introduction of a new system, employees' 'psychological contracts' with their jobs will inevitably have been broken and they will therefore be more receptive to change. He recognizes just how crucial is the management of change in this instance.

To descend on staff and suggest that they have 'bad' jobs which need altering will not only be a blow to their own self-respect but will undoubtedly damage personal relations, perhaps irrevocably.

Any change must therefore be presented to staff in a positive manner. A consideration of job design is necessary in order to maintain morale. Unfortunately, this again is likely to be overlooked when time is short. In the long term such an attitude can prove to be very expensive. Involvement is crucial to success in this phase of the implementation process. The job redesign exercise should have a joint commitment from both management and unions. Damodaran *et al.* (1980) note that workers' participation in the redesign of their jobs is vital too, for they often place different values on aspects of their tasks which might not seem important from a cursory examination of their duties. 'Routine clerical procedures' and 'helping the public' may be descriptions of the same set of tasks, manager and job holder interpreting the situation differently according to their emotional involvement. When employees are actively engaged in work role analysis, the factors that motivate them will emerge: new roles can be designed accordingly to enhance their own views. Increased efficiency is not in itself a sufficient motivator.

Acknowledging that job redesign is a necessary complement to technology is not enough. The introduction of a new system provides the opportunity for an innovative approach but must also build on the past. This is particularly true where work groups are concerned. One of the reasons that people go to work is for their social contact with others. Great dissatisfaction can result if the composition of any group is disrupted through change. Productivity levels may even drop. Finer (1982) stresses that existing relationships should be respected wherever possible. It is not advisable to reorganize tasks so that staff who dislike each other should have to work closely together. Wherever feasible, Finer continues, physical alterations to the library caused by computerization should take into account established 'territorial rights'. Compromises may be necessary but should be reached after amicable discussion.

It has already been intimated that the jobs of professional librarians are not (as a general rule) so adversely affected by automation as that of their junior colleagues. This is largely because the global function of their jobs incorporates more varied tasks and they can also exercise greater control over when and how to do things. The hypothesis is supported by a case study by Waters (1986) of the effects of new technology on employees at the University of Tasmania. A positive picture emerges from his (self-admittedly) limited survey: the minority of dissenting opinions coming from the non-professional echelon. Where problems have occurred at senior level it is usually the result of de-skilling, and job redesign is an obligatory solution.

Management frequently has a carefully developed scheme for the introduction of advanced new technology systems but no similarly imaginative plan for the design of future jobs and work organization. Corbin (1985) outlines a methodical approach:

- Prepare new job descriptions for each staff member whose duties and responsibilities will change once the new system is installed.
- Discuss the new duties and responsibilities with the staff, stressing their continued importance to the library and the system.
- Orient each staff member in general to automation (if needed) and the new system.
- Train each staff member for his or her specific new duties and responsibilities.

THE IMPACT OF AUTOMATION ON STAFF STRUCTURES

New computer systems give opportunities to redesign not only individual jobs, but also the structure of the whole organization (Dyer *et al.*, 1993). Some jobs may no longer be significant, others may be merged. As job design changes take place, it is worth considering a total restructuring, in consultation with the staff.

Situations differ from library to library. In one, clerical staff may be able to answer routine queries because the information is available on the system, freeing the professionals for more in-depth work. In another, the ready access to information by the public may pose its own threats to professionals, making them less inclined to allow non-professionals to undertake some of their duties. Departmental boundaries may also be blurred as technical services burdens are eased by new computer systems and greater emphasis is placed instead on training the public in using the information systems.

Promotion prospects, too, may change as greater emphasis is placed on technological skills, encouraging those with such an aptitude to progress. Although it is anticipated that this will be relatively minor, attributes such as interpersonal and managerial skills still weighing heavily.

Changes to staffing structures will probably be gradual, new posts evolving rather than being created. Although gradual, it will follow and keep pace with technological change.

PART

VIII

PLANNING AUTOMATED SYSTEMS

CHAPTER 23

TECHNOLOGICAL CHANGE

There has been an unprecedented escalation of technological developments in the last few decades and it seems unlikely that this trend will alter in the foreseeable future: technological change, it seems, is here to stay. Responsiveness to the demands of an organization's clientele and its employees also involves being adaptable. Rooks (1988) observes that transformation is mandatory if an organization is to survive in a world of developing technology. In a volatile environment, skilful management is all important.

In order to respond to user and organizational demands and to keep pace with technological developments, library systems must be upgraded. This may be moving from a manual system to a computerized system, sometimes called 'first time around automation', or from one automated system to another – 'second time around automation' (Sykes, 1991). During the changeover, it is very easy to become obsessed with the technical aspects of information technology, yet 'no matter how well a project to introduce new technology is managed, it is likely to be a failure if the organization as a whole is not prepared to accept the change' (Klobas, 1990, p. 344).

An understanding of the complex and varied attitudes that people have towards change and technological transformation in particular will help its management. The adjustment will affect the organization's decision-makers, the library staff and the library users. This chapter deals with people's reaction to technological change and suggests steps for managing the process.

REACTIONS TO TECHNOLOGICAL CHANGE

Clarke (1995, p. 194) defines change 'as external ... [and] when something starts or stops, and takes place at a particular point in time'. She distinguishes this, which she regards as inevitable, from transition, an internal process that people go through to accommodate change. The process of transition has three phases:

1. *The ending*. The more you have identified with the past the more painful and stressful it is to let go.
2. *The neutral zone*. Here people have let go of the old and face the new. This phase is marked by a lot of chaos, uncertainty and confusion.
3. *The new beginning*. The final phase is when people have accepted and are now committed to the change.

Clarke (1995) argues that the transition period can be by-passed, but people who do this find it too difficult and painful to let go of the past and are often lacking in areas of personal growth, future direction and commitment.

People may react positively or negatively to change and take longer or shorter periods of time to go through the process of transition. Any alteration in circumstances is almost invariably accompanied by an element of stress – even when people have a positive attitude towards the change. However, it appears that when the change involves technology, a greater strength of reaction occurs. The spectrum of responses will vary from mild apprehension mingled with excitement (which, in itself, can be stimulating) to fear. Other authors confirm this view. It would seem, according to Fine (1982), that there are few changes in our social lives (other than war or other calamities) which seem to produce the same intensity of reaction that the coming of technology has aroused. Turner (1981) suggests that the most remarkable feature about computers is their ability to provoke an emotional response from human beings and Burton (1987, p. 211), warns that: 'the introduction of IT-based systems generates more conflict and stress than any other type of change'.

ADVERSE REACTIONS

The stress to which Burton refers may manifest itself as opposition to the new system. Döckel (1992), in a similar way to Clark above, likens the response to a grieving reaction whose phases include shock, fear, anger and depression, culminating in final acceptance or rejection of the new situation, depending on the individual. Some may complain loudly or predict failure, others may be aggressive or, at best, unhelpful, while still others attempt to ignore the situation.

Resistance can show in gossip, slow down, or even refusal to learn a new skill (Döckel, 1992). This may be expressed by an individual or a group. However, even if the problem manifests as a group problem, the manager needs to tackle it as a series of individual problems. Talking to the people concerned, listening to what they say and taking it on board is crucial. In talking, it may be helpful to remember that reaction against the transfer to the new system may be for one or more reasons. These usually come down to one or more of a number of fears, the most common of which are the following.

Fear of the unknown

Uncertainty or fear of the unknown is a strong influence against change, particularly for people experiencing 'first time around automation'. Change is threatening to most individuals because no matter how unattractive the present may be, they have largely adapted to it. White (1985) notes that most individuals not only bend to meet the requirements of their job, but that to some extent they also adapt the demands of their job to meet their own preferences – with or without their supervisor's knowledge and agreement. The perceived virtue of any long-standing system thus lies in its familiarity. New technology represents change: it means giving up familiar ways and facing the unknown; it is insistent and demands effort; it requires the acceptance of new ideas and concepts and it disrupts well established behavioural patterns, habits and securities. As Drake (1993, p. 50) states: 'it means giving up comfortable ways of doing things, and established ways of relating to colleagues, customers and the parent institution'. Often the most vehement opponents of change are, not surprisingly, those who have been in their particular job the longest. They have the most invested in the status quo and have fixed ideas on what is necessary, whether or not this coincides with any broad aims of the organization.

The introduction of computer-based systems for the first time, or even replacing an outdated system, can thus be a stress-producing experience for the individual employee within the organization. There may be a great desire to hang on to the old way of doing things, which can last for years. It can be avoided by taking an active interest in staffs personal and professional development, although it may take a lot of time and effort (Storey, 1992).

Individuals will not only be concerned about what a new computer system can do for their library but also about what effect it might have on them. Resistance to change is not exclusive to junior or middle-grade staff. Rooks (1988) observes that change is often resisted by administrators – those who have the authority to accept or reject any proposal – which she calls 'resistance with veto power'. Again, the resistance may stem from uncertainty or fear of the unknown. The argument runs that, while there are acknowledged faults and inefficiencies in the present system, they are at least a known quantity. An overworked supervisor may only be able to envisage the problems of any new scheme and fail to see beyond them to the subsequent advantages.

In tackling fear of the unknown, one of the key elements is to communicate hard facts and give ample opportunity for people to gain experience of the new system, so that it begins to become known. (Methods of doing this are given in 'Planning', see below.)

Fear of obsolescence and failure

Learning a new system inevitably results in a short-term reduction in our efficiency, and may call our self-esteem into question (Fine, 1982). There is a feeling that the knowledge and judgement acquired gradually over a number of years may become obsolete with the introduction of new systems and procedures. Experience suddenly seems to count for very little. Hand-in-hand with this fear of obsolescence goes the fear of failure, of an inability to adjust and do the new work. Status, privilege and authority are all threatened by lack of proficiency causing enormous apprehension and potential resistance to change.

An example of this is cited by Clayton (1992): a traditional reference librarian may consider that his own professionalism is being undermined by the presence of a relatively junior member of staff with experience of online searching. Rooks (1988) articulates fears that the re-distribution of information as a result of new automation can lead to loss of ownership. Staff who no longer find themselves in charge of critical information may experience a sense of loss or a diminishing of self-worth or value to the organization.

Bichteler (1986) recognizes the particular problems faced by managers who may have felt pressurized to purchase and use new or more powerful computers, yet who were simultaneously suffering from similar anxieties. They may fear that many of their decision-making and supervisory functions will be taken over by sophisticated software, causing them to feel less important to the organization. One of the points raised by Mann (1984), in his survey of public libraries, is that many middle and senior managers received their training before the advent of computers in libraries and their lack of awareness may make them feel vulnerable to replacement by younger librarians who are more knowledgeable about computers.

Librarians may feel that their own importance diminishes in comparison to that of the system designers and programmers. Experienced staff may be largely ignorant in this particular sphere and consequently feel downgraded. Zoltan and Chapanis (1982) call this a fear of 'technocratic elitism'. Fear of obsolescence is perhaps greatest in middle-aged workers or those near retirement. They may feel either that retraining is not worthwhile or that management holds that opinion.

Fear of economic insecurity

Employees may feel that the introduction of a new computer system is a step towards eliminating their jobs, if not immediately, at some point in the future. Economic insecurity is a prime reason for responding in a hostile fashion to new concepts. This fear is sometimes coupled with that of failure, the employee worrying that an inability to perform a task may lead to redundancy.

Automation has undeniably brought redundancies in some industries and therefore anxieties about job losses will exist. The extent of this fear will largely depend on the economic climate of the individual organization and Lovecy (1984) admits that in academic libraries, where senior staff have contractual job security, it will mainly concern clerical grades. In his survey of 54 libraries, Johnson (1991) found that nearly half (48.2 per cent) of the respondents reported a decrease in the number of clerical positions as the result of 'first time around automation', although 35.1 per cent reported no change in the number of positions at this level. Few libraries reported decreases in middle or top management positions. The outcome of 'first time around automation' on staff seems to depend very much upon the individual library, but there tends to be greater fear among non-professional staff. Shaw (1986) found that 42 per cent of respondents to her survey feared being laid off in the first year, but this contrasts with Winstead's (1994) findings that job losses were anticipated by only a minority of staff (18 per cent).

It should not be assumed that automation and staff reductions go hand in hand. Staff reduction can be avoided by re-allocating staff to other positions. Of the libraries in Johnson's study, 74.1 per cent of them retrained and moved their staff to other functions within the library instead of making them redundant (Johnson, 1991). In fact, only 9.3 per cent reported that they had terminated employees and over half the libraries reclassified some staff members to higher level positions.

As redundancy is certainly not inevitable, it is wise to dispel unfounded fears at an early stage, especially before rumour can take hold. Honesty is important, however, so if redundancies are anticipated, it is best not to deny the fact; honest, open communication with all concerned usually works better in the long run.

Fear of changed relationships

The importance of social relationships in the workplace as well as in our private lives has been established by Argyle (1972). They are particularly important during a time of change (Goulding, 1996, p. 83). People may resist change because it seems likely to alter long-established work groups in some way. Individual members of a closely knit group may be carried along by the feelings of their peers in their attitude towards change. They may not be antagonistic but will not wish to hold out against an apparent consensus of opinion.

A library research project was conducted by Fine (1982) under the sponsorship of the US Department of Education, to look at why people respond to technology in such complex ways. Library administrators, practising public librarians, library school staff and postgraduate students drawn from a national sample were interviewed. The conclusion was that the single most important fear that people have about new technology is that interpersonal relationships will suffer. This overrode all other fears. Employees may be worried about who they will be working with, and who will be their supervisors, and whether their opportunities for interaction with other people will decrease.

This change is particularly important when redeployment is considered. While it may be considered preferable to redundancy, a transfer carries its own fears since the individual often has to adapt to a new set of colleagues as well as new tasks.

Fear of changed working practices

Even if a person is not redeployed, a change in the way they work is likely as a result of the automation of their own, or an interrelated, task. Staff may wonder whether their 'new' jobs will be as interesting and carry as much responsibility as before the change, or if the functions of one department will be merged with those of another. Lovecy (1984) comments that some staff may be concerned about the possibility of an increased workload. Many will be more concerned about their ability to master the new skills necessary for the job. New systems tend to streamline complex tasks, but may also make things routine as a result. The change to work patterns must be recognized and good job design should ensure creative, satisfying work for employees. If staff are to be transferred to other work, the ideal is to aim for the new posts being more interesting, challenging and less routine than their previous post.

Not only will working practices change, but the level at which people are working will also change. More and more paraprofessional staff are becoming involved in hitherto professional areas (Goulding, 1996, p. 6). This may be seen as a challenge – welcome or otherwise – by the paraprofessionals and a threat by the professionals. Goulding (1996) argues that frequently this overlap of work areas leads to tension between professional and paraprofessional rather than cooperation, because they are regarded differently by the employers.

One area in which non-professional and professional staff are differentiated is in the area of pay. Jones (1989) found that 55 per cent of her study felt there was an increase in responsibility with no increase in pay. A suitable pay rise can help (Cleeve, 1995). Johnson (1991, pp. 74–5) found that 83.3 per cent of libraries reported that lower level positions, with poor quality of working life, had been upgraded and carried greater autonomy and accountability following automation. He also found that new skills were required from all levels of employee, in particular analytical skills – using the raw data from the computer and interpreting the vast amounts of information that are now available. He argues that this can, in time, lead to an improved quality of working life.

In addition to changing tasks, Biddiscome (1996) observes that some roles will change. For example, he believes that the role of the reference librarian will change as the Internet becomes more common as an information-seeking tool among the general population. The reference desk, he says, is becoming more of a help desk to advise library users on the information systems they themselves are searching.

Fear of computers

Most workers are by now used to computers in offices, at home and in society. However, while the machine itself is no longer feared by the majority of employees, other factors relating to working with computers have become more common. Winstead (1994) found that there was an increase between 1987 and 1993 in the people who feared the health effects of computer-related work. The main worries about operating VDUs as identified by Cross and Bawden (1987) are: radiation hazards; visual problems; musculoskeletal problems; tenosynovitis – work-related upper limb disorders; facial dermatitis; psychological and psychosomatic problems, all of which are discussed in detail in Part II.

Some people, however, still fear the computer itself. Lovecy (1984) identifies several common concerns: lack of a printed record; unwittingly wiping out a set of data; corrupting data; and damaging equipment. Some people never learn to trust the machine and wish to check and double-check its operations at all stages. Sykes (1991) confirms that nearly all the non-professional staff interviewed had been apprehensive about the prospect of automation. While several library assistants talked of sleepless nights before the system went live, mention was also made of two older members of staff who left shortly after the computer arrived, largely because they could not overcome their fear of it. The fears centred around inadvertently harming the computer or computer records. An investigation by Daniels (1995) into the effects on non-professional staff of the implementation of computer-based library systems in college libraries reveals a similar picture. The main cause of stress, voiced by the interviewees, was the fear of the system breaking down and not knowing what to do in the event of a crash.

Computers also carry their own mystique. The complicated equipment, jargon and very cost can all be off-putting to the individual lacking self-confidence. Mistaken impressions can also exist about the level of expert knowledge required to operate computers. Competence to operate the new system so as not to destroy data was, in fact, the principal fear identified by Sykes (1991) in his study of automation and non-professional staff at the former Polytechnic of the South Bank (now South Bank University). People need to be informed that perfection is not expected and mistakes are allowed. Training, given at the earliest opportunity, should help to alleviate some of these fears (Chapter 24).

Fear of increased management surveillance

There are also concerns that the introduction of new systems may offer an opportunity for increased management supervision and surveillance. Errors and inefficiencies become more apparent in automated systems and output is more easily measured. This facility may be seen by some employees as a form of 'Big Brother' tyranny. Readers, too, may resent personal details being held on computer and worry about the use that is made of them.

The Trades Union Congress (1996) reports the use of a system in the Employment Service which records the time each officer (employee) logs on and off the system; the number of decisions each individual makes; the rates at which decisions are made and whether or not individuals are meeting their output targets. The information is used by management to assess the performance of staff and ultimately their levels of pay. Not surprisingly staff are uncomfortable with this, believing the use of technology to be intrusive and stressful. It is helpful if people are informed honestly about what information is held and how it will be used.

POSITIVE REACTIONS

Conversely, the positive response to automation is not without its problems. Fine (1982) describes how an 'overboard positive' reaction can be as damaging as a negative one. Members of staff who throw all their energy and commitment into a new scheme and totally abandon the old can create difficulties for management. There is usually a lengthy time lapse between deciding on a replacement system and implementing it and, during this period, service has to be maintained. Those who have mentally discarded the old system will not operate it as effectively.

Even though not all staff positively welcome the new system, most staff will welcome the introduction of a system that will improve on the existing situation. Expectations will include the anticipation of a faster throughput of work with less effort and improved job satisfaction. Winstead (1994) found 98 and 99 per cent of staff surveyed in favour. Jones (1989, p. 450), however, warns that positive reactions may be a 'cover up for fear or confusion'.

PEOPLE AFFECTED BY TECHNOLOGICAL CHANGE

THE DECISION MAKERS

The organization that funds the library will be made up of people who will respond to automation in their own way. Usually, with this group, efficiency and cost will be deciding factors, but they may hold back for other reasons. As they hold the purse strings, one of the first tasks is to convince them that any proposed change to the system is worthwhile. They may resist such a change, however, not for economic reasons but because they themselves are familiar with the old system and unaware of its limitations. Or, they may be over-enthusiastic with the anticipation that the new system will deliver better and faster than the old and expect benefits sooner than is practicable. The danger then is that when expectations are not fulfilled within the expected time scale, negative reactions can occur.

It is wise to get to know who the key players in this group are and to try to understand how they might react to a new system before beginning to ask for funds. From then on, good communication at all stages is important.

LIBRARY USERS

Library clients, will also vary in their response. Some will be enthusiastic to see new features and will even push the library to move ahead. Others will be resistant to losing what they have come to know and trust.

STAFF

A change of system will most directly affect the staff. It is therefore essential that library managers support staff and give them the time to go through the various phases. In particular, as said earlier, employees need time to 'grieve' for the old way of working.

In most libraries, staff can be 'divided' into two main groups: professional and non-professional. The distinctions in terms of the work undertaken and the status held by each group varies from library to library. It would appear that the non-professional staff are most affected by a change of library system. Sykes (1991) claims that 67 per cent of the UK library workforce is made up of non-professional staff. Not only do they outnumber professionals, but they also tend to hold a more prominent position in most libraries – on the issue counter, for example, as a result of which, their attitude and competence will affect the users' perceptions of the system. For these reasons much attention should be paid to the training and reassurance of non-professional staff. The reassurance is important because, as we have seen in 'Adverse reactions' above (p. 298), non-professional staff are the most likely group to have their numbers reduced by the introduction of a new system.

UNIONS

The trade union movement has always sought to protect employees from the adverse effects of new technologies, library automation being no exception. The widespread use of computers in the early 1980s encouraged trade unions to look at the situation more closely. Various guidelines for negotiating 'new technology agreements' with management were established. Jones and Jordan (1987), for example, set out a comprehensive list of points on which agreement between management and unions should reach before the implementation of a new computerized system:

- An assurance that jobs will not be lost.
- Agreement by management to retrain staff to use the new equipment.
- Divulgence by management of future plans for introducing new technology.
- Permission by management for trade union representatives to attend new technology courses so that they will be adequately informed on the implications of introducing new technology.
- Where redeployment or alternative work becomes necessary, existing staff are to be protected, and joint management/union discussions held on suitable alternative work.
- Where new technology gives certain staff new responsibilities and demands new skills, this should lead to discussions on re-grading or additional pay.
- The level of job satisfaction should be safeguarded, and tendencies towards de-skilling should be monitored.
- Negotiators should insist on certain standards of health and safety for staff working with VDUs, and monitor the effects on staff conditions in the workplace.

The Trades Union Congress (1996) admits that trades unions now rarely have a say over whether information technology is introduced and state that the purpose of new technology agreements should be aimed at influencing how it is introduced. They point to the need for trade unions to have policies in the following areas:

- consultation with staff;
- training and re-training of staff;
- health and safety;
- job security;
- worker equality (to ensure no section of the work force is discriminated against).

Union representatives should be involved from the beginning and should be free to decide their level of involvement. Failure to consult can result in setbacks. Dean (1987), County Librarian of Durham, reports a delay of almost two years while agreement was reached with trade unions on the expansion of their automation programme. Unnecessary bad feelings can be generated when consultation does not take place: unions can even prevent systems from being operated. Early involvement of staff and unions is the key but as Clayton and Batt (1992) point out, much will depend on the prevailing local atmosphere in the field of industrial relations.

Consultation with the unions can bring with it the benefit of the considerable help and advice some unions are able to offer, especially in the areas of health and safety and job design. Many unions issue guidance on the Display Screen Equipment Regulations (Health and Safety Executive, 1992d), leaflets or information packs which can be given as additional information to staff, but it may be advisable to check for any bias and to supplement the leaflet with additional information if need be (see Useful WWW Sites and Addresses).

STEPS IN THE MANAGEMENT OF CHANGE

Jones and Jordan (1987, p. 8) comment: 'automation makes it necessary for librarians at all levels to be more conscious managers, if their staff and users are to accept changes positively'. Olsgaard (1989) has indicated that 85 per cent of all failures during the implementation of automation can be attributed to people problems. Being aware of the various stages in the change process may help to minimize such problems. The steps include:

- preparation;
- choosing a leader;
- setting up a planning group;
- participative management;
- setting aims and objectives;
- planning;
- communication;
- timing;
- changeover procedure;
- anticipation;
- feedback;
- evaluation.

These steps are described below.

PREPARATION

It is important to have a clear vision of the system to be introduced and be able to anticipate its likely impact. All staff should be familiar with the visions, goals and objectives of the library – why, what and how automation should take place (Bichteler, 1987). It is important to communicate to all groups of people at this early stage (see also 'Content' in Chapter 24).

Cleeve (1995) argues that successful change can be assisted by communicating a new model of the organization for the future to help staff see their proposed place in the new scheme of things. She argues that a shared vision can be fired initially by enthusiasm for what is good about the library already; that the vision is shared by, rather than imposed on, the staff is important. For staff to welcome change they will need to feel some level of dissatisfaction with the status quo; communicating the limitations of the existing system and the potential of the new system may help, but it is better if ideas can be drawn from the staff themselves. While commitment by staff and users is important to a successful changeover, it will not come without adequate communication of the idea (Klobas, 1990).

CHOOSING A LEADER

A project must have a leader to control and to guide the changeover. Klobas (1990) suggests the leader could be a consultant, the chief librarian, a senior member of staff or even a member of the general staff who 'champions' the change. Cleeve (1995, p. 41) proposes that an independent consultant may '[help] staff, particularly those of long standing, to give up some of their professional blinkers with dignity'. However, it is probably best if the leader is someone from within the organization, although outside consultants could be used from time to time. As a person, the leader should be well respected within the organization, have sufficient authority to be able to communicate with decision makers and all levels of staff and should command the respect and loyalty of both staff and management. In terms of skills, the project head should have good technical understanding of the library systems, for which special training may be needed but yet be able to maintain a broad overview of the development and exhibit a detailed and sympathetic understanding of human concerns. Understanding the financial implications is also important. The assistance of a planning group may be helpful too.

SETTING UP A PLANNING GROUP

A planning group or automation working party should comprise representatives of all levels and sections and exist for the express purpose of planning and implementing the new automated system. All staff should be represented – non-professional as well as professional. In the case of 'second-time around automation', non-professional involvement is highly recommended because they have a great deal of practical experience of the old system, and will be among its shrewdest critics (Sykes, 1991). Baddeley (1988) suggests a two-way benefit from this approach; not only will all views be represented, but if chosen with care, the representatives can act as 'missionaries' to their peers. Ideally, the members of the working party should reflect a spectrum of views from the enthusiast to the 'Luddite' and a mix of personality types (Döckel, 1992). It is useful to include: people who can plan and initiate, people with an eye for detail, people who complete tasks and 'carers' who will nurture the people affected by the change (Cleeve, 1995). Different interests are therefore represented and benefit may be derived from the exchange of ideas within the group. In small libraries, the project team may consist only of the librarian and one other staff member, but a considered approach towards involving everyone is still necessary. A cooperative working relationship is highly desirable.

Individuals or sub-committees can concentrate on particular aspects of the automation plan, such as the requirements of different functions, health and safety aspects, furniture and so on. The frequency with which the planning group meets will vary according to the situation in different libraries. Allen (1991) had a steering committee which met once or twice a week for two and a half years.

PARTICIPATIVE MANAGEMENT

The most cooperative working relationship is participative management, which involves a wider group than the system working party. The subject has received much attention in the literature, and it has generally been found that job satisfaction is greater when people are allowed to contribute ideas (Marchant, 1982). Dakshinamurti's survey (1985), to study the reaction of library personnel towards automation, found that the most desirable characteristic in a manager is a participatory management style. Decisions that are reached through participative, rather than hierarchical, management are of a higher quality and have greater group acceptability, particularly when the composition of the group is representative of different areas of expertise and points of view (Sidgreaves, 1985). West (1987) refers to the 'emotional investment' that people have in decisions to which they have contributed and this, in turn, encourages them to share the results of discussions with their colleagues.

Where true participative management occurs, decisions are left to those with the most relevant first-hand experience, a principle endorsed by James (1987) in his account of automation at Kingston Polytechnic (now

Kingston University, UK). The participation of non-professional staff was actively sought on decisions relating to the system as a whole and particularly on those that affected a task which they performed; the rationale being that the junior staff would have greater knowledge of the individual clerical routines. James argues that such involvement is effective rather than merely democratic.

Lucas (1974) has summarized the benefits of staff participation in the management process as follows:

- participation can be ego enhancing;
- participation can be intrinsically satisfying and challenging;
- participation results in more commitment to the change;
- participants are more knowledgeable about the innovation;
- participants can observe better solutions to change problems;
- participation means that users retain some control over a system that will affect them.

The concept of participative management, however, does not have universal and unequivocal acceptance. Some hold the view that the involvement of all staff is unnecessary because the management team contains adequate expertise. Marchant (1982) acknowledges that staff participation consumes valuable working time. Often, too, funds may have to be spent within a restricted time period, rendering user involvement impractical within the time allowed.

It cannot always be assumed, either, that staff will want to participate. Sykes (1991) found that junior staff at the Polytechnic of the South Bank (now South Bank University, UK) did not want to be included in the debate about the system's design. They had insufficient knowledge about automation to make constructive suggestions and, while their understanding could have been improved through visits and lectures, this was not thought to be economically viable. Where a 'second or third time around' computer system is installed, however, junior staff could prove useful as shrewd critics of the existing system.

Staff involvement in the planning stage will therefore be affected by various factors, such as the time available, the size of the library and the willingness of staff to become involved. While direct participation of all staff and other interested parties is not always feasible, it is important that some form of two-way discussion is established to keep staff informed about the progress of the project and to enable them to air their ideas, their hopes and their fears.

SETTING AIMS AND OBJECTIVES

The team, under the guidance of the leader should be responsible for goal setting and planning. Goals or aims are the main purpose for which the system is being selected. For example, to reduce queues at the issue desk. The methods which might be employed, such as using a self-issue system or introducing a laser bar-code reader should be subordinate to this.

Clearly stated objectives are important. Döckel (1992) suggests asking the questions 'who?', 'what?', 'with what result?' and 'when?' and also recommends that any possible negative outcomes are considered at an early stage. For example the threat of redundancy can be addressed by setting the objective that higher productivity must be achieved with existing staff and job satisfaction can be listed as a precise system objective.

It is also advisable to consider other aspects of the library when automating, so that outdated objectives are not assumed along with the new library system. Introducing a new system gives an opportunity to consider changes elsewhere, bearing in mind that staff may not relish too many changes at once – another reason why the planning team may be helpful.

PLANNING

Harley (1995) argues that unforeseen change in library organization can cause disruption to staff, and to minimize such disruptions, the change should be planned. In devising a plan for the change, all aspects of the change, from both the technical and the human factors perspectives should be considered. This will include: what parts of the system are to be implemented first, and in what order; the likely impact these sub-systems will have on both staff and clients and how to deal with it; what might go wrong and what contingency plan to adopt. Those with a human factors perspective are discussed in more detail below.

Human impact

The success of a system does not depend merely on whether it is technologically sound, but also on staff and user reaction. Planning for automation therefore involves planning for the people and their interaction with the computer. This means taking into account the human factor in the choice of system: the hardware; the software interface and any consequent furniture requirements, which have already been discussed. It also implies preparing staff and other interested parties for the changes that computerization will bring. It is therefore necessary to consider the communication requirements: who needs to be kept informed of the project and its progress; when they need to be told; what they should be told (and why) and how the information should be imparted. Some idea of the time scale should also be available. Rarely is a single promotional technique enough; different audiences will naturally have different information needs and each should be considered on its own merits. This stage in the implementation process is vital. Keeping all those with a vested interest in a project informed tends to generate goodwill in itself and valuable reciprocal advice can be received.

Planning for setbacks

Thorough planning and involvement of all those affected by automation should go a long way towards making the actual installation of the system relatively problem-free. However, it is almost inevitable that unforeseen setbacks will occur. Even the most organized library has to rely on outside agencies for the delivery of equipment, and delays in its arrival frequently occur. Problems are further compounded when the equipment is found to be faulty or incomplete. Testing of hardware and software can be a difficult time for the library and is often a lengthy business. This is the point at which even the best timetables tend to go awry.

Procedures must exist for informing staff and users of the reasons for delay, especially if it inconveniences them in any way. An account of what is being done to remedy the situation should also be given and a revised time-scale suggested. Information of this kind cannot dispel doubts about the future operational reliability of the system, but it can at least help to keep the issue alive. Damage to public relations can be minimized by broadcasting the facts.

Planning for disruption

Any computer development is bound to cause a number of disruptions to the normal running of the library. These include discussions with library staff with respect to the system specification, the physical changes which need to be made to the workplace, the installation of new furniture and equipment, the staff training programme, and the testing of the system (Damodaran *et al.*, 1980).

Staff should be aware of the possibility of disruption before it occurs and patience needs to be exercised all round. Productivity is likely to decrease over this period and due allowance must be made. A number of practical points need to be considered.

Time should be allowed for staff to adapt to the new working environment and, as alluded to earlier, to mourn for the old. Nadler (1980, p. 6) describes it thus:

> People need to mourn for the old system or familiar way of doing things. This is frequently manifested in the emergence of stories or myths about the 'good old days' even when these days weren't so good. The process of dealing with loss takes time, and those managing change should take this into account!

Adequate alternative workplaces need to be made available when physical alterations are made to enable staff to continue their work with minimal inconvenience. Health and safety factors should be considered. Boxes should not be left lying around where they are likely to constitute a hazard, either through blocking passageways and exits or protruding where people are likely to trip over them. The same care should also be taken of discarded furniture.

Installation can be a hectic and fraught period for all concerned. Fine (1986) stresses that people's dependency needs are greatly intensified during change. They seek support, feedback and direction from one another and from those in charge. Communication, which takes on an even more vital emphasis at this time, is less easy to achieve. Gratton (1983, p. 23) makes a perceptive remark on the situation in Derbyshire County Libraries:

> From the outset it had been the intention to keep staff fully informed as ideas and activities progressed. I have however to acknowledge a degree of failure . . . [because] 'At the time of greatest need for communication there is the greatest shortage of time to communicate.' When things are happening everyone appears, not surprisingly, to be wholly involved in making it happen.

COMMUNICATION

The various groups of people affected by the change will need to be kept in touch with progress. At different stages in the changeover, different groups will need to be informed of different aspects of the process. How they are to be informed is also important.

Decision makers

Before consulting with decision makers, it is necessary to be aware of the key people. Once that has been ascertained, well thought-out and detailed communication can help build support for the project over and above the purely financial. Influential friends and allies can be won who may be able to exercise their authority at critical stages later on.

The initial information usually takes the form of explaining the current and likely future benefits, in order to justify the expenditure on the system (see 'The decision makers', p. 304) and the cost of the attendant disruption. This should be presented in report form, but may be backed up by verbal and visual presentations. Care should be taken to avoid over-enthusiastic estimates or encourage unrealistic expectations which may lead to subsequent staff or book fund cuts. Possible extensions to the service should also be noted.

Clayton and Batt (1992) suggest that a good way to 'win friends' is to arrange a demonstration of the potential new system (see also

'Shortlisting', Chapter 21, p. 276). It can be incorporated in a cheese and wine evening. Nevertheless, she warns, it must be well organized in order to serve its purpose; invariably something goes wrong when the occasion is important. The operator should appear confident and be able to meet all contingencies, otherwise the demonstration may have a negative effect. Demonstrators must convey their enthusiasm for the system but must not forget that they are trying to convince users, rather than librarians, of its advantages. The talk should be pitched at an appropriate level. One method of personalizing demonstrations is to include names of participants in a borrower database: a simple but successful ploy even at senior level.

Once implementation is under way, management meetings and bulletins can be used to inform officials of the progress of the development and anticipated completion dates. At later stages, it is important to inform them of staff and system performance. Two communication methods that Klobas (1990) has found successful are: regular management meetings to provide a summary of new developments (a verbal summary, backed up by a written review works well); and sending an irregular 'Technology update' to the manager to whom the librarian reports 'for information'. A balance needs to be struck between keeping decision makers well informed and overloading them with information.

Staff

An efficient automated system will only come about if the staff who operate it are happy to work with it. Negative reactions to technological change have already been discussed (see 'Adverse reactions' above, p. 298) and it has been indicated that the underlying cause of all of these reactions is fear. Uncertainties and fear are further compounded by rumours and speculations which spread in the absence of hard information. Thus it is argued that effective communication at all stages is one of the best ways of conquering misplaced fears. However, it is insufficient to assume that the existing methods of communication in an institution are adequate. Quinlan (1987) argues that no organization should consider embarking on major change without first closely examining the effectiveness of its communication systems. Handy (1981, p. 353) warns that 'if there is one general law of communication it is that we never communicate as effectively as we think we do'.

The information passed on to staff should address expressed and unexpressed fears, most of which are possible to anticipate. For example, reassurance should be given about personal matters such as job security and job content. Complete honesty and openness are crucial. Reeves (1984) states that, in the automation of Essex County Libraries any reductions in staffing levels were dealt with by natural wastage. This is probably the least distressing solution for all those involved, but exactly what is meant by the term should be carefully explained. Staff should not think that pressure will be exerted upon them to leave. In the exceptional event of redeployment or redundancy, Lovecy (1984) advocates involving the organization's personnel department.

Other specific fears should be anticipated and tackled – for example, health and safety, and the use of work monitoring. Lovecy (1984) recommends that the manager should admit if the latter facility exists and explain under what circumstances (if any) he would use it. Confirmation should be made that adequate training will be given (assuming that it is).

System maintenance and failure can create a lot of stress and staff need to be warned that problems are likely to occur, especially in the early days. Atkinson (1984) advises that when we plan for change we have to accept a certain error rate. No change will be perfect and where computers are involved there will be a certain amount of downtime, a certain amount of failure. It is unlikely too that the full potential of any system will be realized in the first instance. Acknowledging early on that there may be unexpected setbacks helps prepare the ground for when they do occur. Often there can be problems getting through to someone who can help and, even then, there may be considerable delay before the fault is found and rectified. In one library it took a year to get the system running because of an incompatibility between the network and the system, neither supplier being willing to accept the responsibility.

As automation should bring benefits to the library, these should be outlined – especially how the new computerized system will help staff in their work. When the potential advantages are appreciated, there is more commitment to the change. White (1985) maintains that such a perspective helps to prevent any feeling of lament for the old way of doing things. Staff are also in a better position to deal with readers' queries if they themselves are up-to-date with developments.

Although the positive aspects of automation should certainly outweigh the negative, staff expectations must not be raised unrealistically. Promises that computerization will make the tasks of library assistants more interesting and rewarding may prove to be rather hollow. Computerization is not a panacea for all ills; there are still boring jobs to be done. If unbalanced claims are made, then staff will be disappointed and may become disenchanted with the whole scheme. Daniels (1995) argues from a number of case studies that a positive attitude can make all the difference to the success of a new system.

A variety of communications methods are recommended for different purposes:

Written communications

News bulletins are a popular way of disseminating information, especially when large numbers of staff are involved. They are useful in dealing with both general topics and the more specific. It may also be useful to issue general bulletins to staff not directly involved with change. At a later stage a manual for the system adapted to the local situation can be useful, but it is time-consuming to produce.

Quinlan (1987) has remarked on the problems of achieving a happy medium between satisfying people's information needs and inundating them with so much paper that the exercise becomes meaningless. Sykes (1991) argues that 'little and often' is probably the best approach. Written communications should constantly be reviewed and improved, and the opinion of the recipients sought on the usefulness of their content. Another drawback to this type of communication is that it is essentially passive. Recipients do not have the chance to react to the information in two-way discussion; consequently it should only be used as a support to other methods, not as a substitute.

Lectures

Lectures are also, for the most part, a passive form of communication, but are useful for transferring information to a number of people simultaneously and for explaining complex issues. They are particularly useful for training (see Chapter 24).

Visits

Travelling to see other automated libraries provides staff with the opportunity to see a working system in operation (see also 'Final evaluations', Chapter 21). At the initial stages of the process, especially with 'first time around automation', any automated library system in a similar library can be useful. At a later stage, it is probably more helpful if the same system can be seen, preferably in a library which has been using the system for a while. It is best if as many as possible of the staff can go, but appreciate that this can create problems with staffing levels back at the library. Visits help the visiting staff to assess the effectiveness of the new system and learn how staff who use it feel about it once they are familiar with it. It can be especially helpful, therefore, if visiting staff can talk to their opposite numbers in the site library and assess the impact it will have on their own jobs. There can be surprisingly beneficial effects: some people find that seeing the system in operation reduces their feeling of alienation towards machines.

Lovecy (1984) warns that visits should be carefully planned, the library selected being one in which there is a positive attitude towards automation. Employees will get more from a visit if they know the type of things to ask. At the end of the visit, it is good to have time to debrief and for the visitor to report back to others who were unable to go.

Consultation

When a change is introduced without the 'ownership' of all concerned it can give rise to employees feeling a lack of control over their environment. This in turn can lead to anger which can then be transferred to people, or even things (such as the computer). Staff may, for example, begin to withhold communication or try to make things difficult for others (Clarke, 1995). To counteract this, Cleeve (1995) argues that regular consultation of staff on all levels and on an individual basis helps towards acceptance of the new. The more that staff and users are aware their feelings are being taken into account, the greater will be their willingness to go along with the plan. Winstead (1994, p. 19) also found that good communication was essential. Respondents to her survey said that 'being patient, accepting suggestions from staff, allowing time for staff to learn, being realistic about expectations, and being sensitive to staff needs' were essential qualities of management. In addition, Quinn (1980) recommends widespread consultation, arguing that one benefit is for the management in that consulting those with knowledge can help provide information.

Departmental meetings

Staff meetings are a useful method of communication and can be used to address all topics, but can also tend to be one-way unless management is open to comments and suggestions. Good listening skills are an important tool here.

Staff fears are an important area to be addressed, but this will only be successful if those who are in charge are seen to be sympathetic. Many of the staff fears relating to re-grading of jobs and de-skilling of tasks are perfectly legitimate and it is important to acknowledge this. There can be a danger with being too sensitive to the person who shouts the loudest and so a balance must be struck. However, resistant responses are far from being totally negative. They serve a purpose in causing management to re-examine proposals and act as a natural check on too avid an acceptance of innovation for its own sake. Reassurance of the value of any dissenting staff member and their role at this stage can be crucial. Sykes (1991, p. 35) argues that reassurances should be 'specific, concrete, realistic, and carry conviction'.

Meetings over and above the usual departmental staff meeting may be useful. For example, Astle (1995) used a day-long staff retreat at the planning stage. All staff were expected to attend. The seminar, which was facilitated by a consultant, was seen as invaluable. Such a day can include small group sessions arranged so that like-minded individuals are together, enabling any techno-phobic managers to express doubts or ask simple questions without feeling that they are 'losing face' with junior staff.

General training sessions on the new system which allow opportunity space for questions can also help (see Chapter 24).

Library users

A central argument for introducing computerized systems into a library is the improvement that it will bring to reader (as opposed to technical) services. The readers should not be left to discover these benefits for themselves, rather they should be informed through a well-constructed publicity campaign, which in turn will help to increase their acceptance of the system. Library users will need to be told why automation is necessary, how it will affect services and when it will take place, although it may be wise to be a little vague about the last until dates are certain. The benefits, such as the savings in time and improved access to the stock, should be spelled out, as well as warnings of the short-term problems of staff training and getting the system running. The possibility of down-time should also be mentioned, thus preparing users for the shaky start that so often accompanies a new system. Assurances should be given that there are adequate back-up procedures in this eventuality. Users will also need to be informed of any requirements the system will make of them, such as the need to bring their ticket with them in order to borrow books.

While the benefits need to be outlined, the dangers of oversell must be avoided. An honest, but positive, tone is recommended. For example, there is nothing wrong with boasting that an online catalogue will provide an up-to-date record of stock, providing that the drawbacks of splitting the catalogue (where it is necessary to do so) are likewise mentioned. Too little information can also be dangerous – for example, when readers are told that the library is going to introduce a self-issuing system but given no idea of the time scale involved, they may make aggravating demands on the counter staff.

Progress reports are essential in any situation but particularly so when there is direct inconvenience to the reader. Special graphic charts may be used to illustrate how much of the book stock conversion has been completed and to give an indication of how much longer disruption can be expected. One of the simplest and most effective methods of presenting this core information is to produce a library leaflet outlining the automation plans.

Herring and Mackenzie (1986) report that this was the principal method of communication used in Aberdeen City Libraries before automation, where separate leaflets dealt with bar-code labels, overdues and reader registration. Leaflets need to be attractive and easily understood, so care should be taken to avoid the use of technical wording and abbreviations (many readers in rural public libraries, for example, still may not know the meaning of 'VDU') or else explain them; otherwise they may be self-defeating.

Alternatively, any newsletter which is regularly produced should include a section on automation. The language used in the bulletin should also be simple and straightforward. Leaflets can be supplemented by displays, posters, bookmarks, audio-visual presentations and electronic mail, depending on the library's situation. The public library may take the opportunity of any coverage in the local press to remind people of its existing services, or an academic or special library, the organization's bulletin.

Equipment whose demands on the user are more complex, such as CD-ROM or end-user online systems may imply user training in addition to information (see Chapter 24).

TIMING

The project manager or system working party will need to consider timing. In particular, they need to think about when to commence the project, in what order and over what time to carry it out, and when to evaluate it.

If a computerized system is being introduced for the first time, both the library and parent organization need to be prepared for introduction of new technology. This entails having the financial and physical resources as well as appropriate human resources.

Having decided to forge ahead, the next major timetable hurdle is when to perform the changeover. Often people choose times such as summer holidays or parts of semesters or terms when the library is least busy for the changeover (Hallmark and Garcia, 1992), but the vendor will have a say in the timing as well.

It is important to remember to keep change gradual and not to alter everything at once. It can be unsettling if too many things are changed at the same time. While the change process may take longer if done gradually, it is likely that more people will be carried with you. Clarke (1995), however, suggests that people are able to handle vast amounts of change if it makes sense and seems to be necessary. It is helpful, she says, if changes are packaged and as much as possible introduced at one time. Otherwise, she says, there is a continual feeling of uncertainty over what is happening next. A library manager may know best what local staff would prefer. As said earlier, consultation and good communication at all stages can assist.

Allowing time for change may be important to employees' health. Clarke (1995) argues that we have only so much 'adaptive energy' on which to draw to cope with changes in our lives. If it is all used up very quickly it is possible that people will experience physical illness, which is not only bad for the employees, it adversely affects the staffing needs of the library.

CHANGEOVER PROCEDURE

There are several different approaches to the changeover from one system to another which can be employed: some cause fewer problems for staff than others. Tedd (1993) identifies four methods: complete changeover; phased approach; parallel running; and pilot operation.

Complete changeover

On a predetermined date, one system is completely replaced by another. This has two main advantages. It is cheap (the library does not have to meet the cost of two systems at the same time), it can be quickly effected, and the benefits rapidly appreciated. On the negative side, it is only really practicable if the system is simple and not too many departments are involved. Also, it offers no 'safety net' if anything goes wrong, and it does not give employees time to adjust. A complete changeover must be supported by excellent back-up facilities.

Phased approach

The system is divided into sections and each is installed as a separate unit. This enables problems with one module to be resolved before another is introduced. A controlled, gradual method, the phased approach allows operators time to adapt and to learn, and also reduces the likelihood of a complete system's failure. Nevertheless, it can make the implementation process rather a protracted one and the full benefits of the system cannot be felt until the changeover is complete.

Parallel running

The old and new systems run side by side until the new one is performing satisfactorily. The advantages of parallel running are that the library service is uninterrupted and, in an emergency, it is possible to revert to the old system. However, it is an expensive solution and staff may become confused and frustrated over how to use each system. Lack of space, for example, may also inhibit the simultaneous operation of two different VDUs or of a manual system and VDU input.

Pilot operation

The system is tried out in one small branch before being installed in the main library. Difficulties can be sorted out with less embarrassment in a 'guinea pig' branch, and the prestige value of being the first to acquire a system may also offset some of the accompanying traumas. Public relations, too, are easier to safeguard on a smaller scale. This is a useful approach for a multi-site environment. Several factors will determine which changeover method is preferred, including the particular characteristics of the system to be installed, the type of library, and any time constraints.

ANTICIPATION

In any situation it is far easier to prevent fear and adverse reactions than it is to remove them once they have become established. The library manager should therefore anticipate responses to the introduction of new systems and wherever possible take appropriate action to reduce stress. One of the responses will be that changed behaviour takes time and a good manager will allow for this.

In a situation where an existing system is being replaced, it may be useful to identify the person who installed the previous system and give them credit for their actions in the past before moving on to extol the virtues of the new system that is to replace theirs (Klobas, 1990).

Extra work is likely to occur during the changeover – for example, in dealing with readers who will need help in dealing with the system and in issuing publicity about the automation programme. Daniels (1995) points out that an improved system can itself give rise to work that was not hitherto considered possible, such as tracing books between branch sites and also that the level of user expectation – and therefore demand – can increase. There is also back-up and maintenance of the computer and the data. Simultaneously, staff will need to be trained and so fewer of them will be available to carry out the excess work. The manager will need to decide between working his existing staff harder or hiring extra staff to cope with the temporary increase in workload. Consulting staff about this may be helpful and the manager needs to consider what employee reaction would be to redundancies following a period of excessive work.

Another area that needs forethought is the scheduling of the automation plan. Timing often goes wrong according to the survey by Hallmark and Garcia (1992). They found that more than half the libraries in their study were unable to keep to their proposed schedules for a whole variety of reasons. They advise people to build in sufficient time for data transfer and be sceptical about timings from the vendor. They warn librarians to be prepared to hold them to their promises.

FEEDBACK

After the change has occurred, it is important to give staff some indication of how they are performing in their new roles and to give them the opportunity to feed back their responses to the new system and their place within it. Goulding (1996, p. 39) argues that staff need to know not only how they are performing but how they might improve, what is expected of them and what they can expect in return. This is especially true if restructuring of roles has taken place and staff are performing different functions than before.

The need to set up a structured or informal vehicle for feedback from the users is often overlooked, problems being dealt with locally as they occur. However, it is desirable that all staff should be able to learn from the discoveries made by their colleagues.

Comments from all staff should be given careful consideration, not just those from staff who use the system regularly or organize its use by others. Lovecy (1984) points out that the reaction of an employee unfamiliar with the system, or a new staff member, may give some indication of how an untrained library user might perceive the system.

Evaluation should not cease with the implementation of a new system because it is unlikely to remain static. However, time-consuming methods of evaluation are impractical in the long term. Good two-way communication is best. The person in charge of the computer will have information to communicate to staff, not just on the faults of the system, but on new ways of using it. Details about major changes may be best passed to users through a formal training programme (Damodaran *et al.*, 1980). Smaller changes do not warrant the same approach, although users should know of the change and its implications before it takes place and they should have the opportunity to ask questions. Accompanying documentation should also be available.

EVALUATION

Whichever changeover procedure is adopted there will be some relief when the system is finally up and running, but it must be recognized that this is not an end in itself. Time must be made to appraise the project's success or otherwise. It is perhaps best to do this after a year or so, when much of the heat should have disappeared from the situation and constructive comments can be made. Future plans should also be outlined at this time. At this stage, a more formal method of evaluation may be required. Evaluation needs to be carried out with sensitivity and tact. Interviewing the staff and users and holding formal meetings to discuss the system are two methods that may be used. The nature and the size of the library will obviously have an impact on the most appropriate method chosen.

A number of things need to be evaluated, such as:

- the implementation and how it went, especially the ability to keep to the time of the implementation schedule;
- the extent to which the new technology meets the goals and objectives set for it;
- job satisfaction with the new roles.

Each of these sections can be broken down into smaller units as befits the library.

Part of the evaluation may take the form of a testing period. Immediately after installation, there is a trial period during which 'bugs' in the system can be removed and various adjustments made. Difficulties in using the system or errors should be reported and logged. This is an essential part of the overall plan. Quinlan (1987, p. 36) comments that 'the trial period should not be shortened to offset delays in earlier parts of the project'. Above all, the process of communication should continue: the systems librarian will want to make adjustments to the computer program and users will have their own ideas to contribute, especially as their confidence in using the system grows.

On an informal level, evaluation can be achieved by watching operators at work. Clayton (1987) admonishes that techniques for measuring systems' performance may hint at 'Big Brother' tactics. When the time taken to perform tasks and productivity are under scrutiny, staff must be aware of the reasons for it or they will misconstrue management's aims. If tasks are taking an inordinate amount of time this is probably due to as yet unidentified faults in the system. The mutual benefit to all in isolating these faults should be stressed. The best way of avoiding suspicion is to involve the staff in the evaluation exercise. Sensible suggestions must be followed up whenever possible.

A less tangible, though equally important, gauge of the system's performance is employees' morale, which in turn reflects their job satisfaction. Staff themselves may not be aware of the causes of their anxiety. Signs to look out for are: any sudden, unexplained increase in absenteeism or poor time-keeping; irritability; headaches and increasing complaints from and between staff.

The reaction of the library's clientele is also significant. Complaints and praise alike can be valuable indicators, though here there is an even greater communication problem – unless people feel very strongly they are unlikely to voice their opinion effectively. Comments books are one way of establishing views, though these are open to abuse. Marchant (1982) states that as a library expands the librarian's direct contact with its users decreases. Library staff should therefore play an important part in reporting reader feedback. This assertion brings the discussion full circle: staff are unlikely to relay readers' comments unless they already have confidence that their own remarks are given serious attention.

SUMMARY

Library automation is a complex and demanding management task. Turner (1981) emphasizes that there must be a mutual sense of purpose and a mutual sense of direction between all involved for the project to be a complete success. He gives four important pointers for creating a positive attitude among staff which can be equally applied to all users. They should become: *confident*, *informed*, *aware* and *trained*.

Cleeve (1995) summarizes the management tactics during the change as:

- being persistent;
- being patient;
- maximizing the positive aspects and minimizing the negative;
- concentrating on influencing decisions in an effective way;
- encouraging, praising and thanking;
- if a few staff do decide to leave for other jobs, not taking this personally.

CHAPTER 24

TRAINING

Education and training can be thought of as an extension to the communication process: they are a means of informing people in more detail of how changes will affect them and of teaching them to cope with these changes. They are, in turn, a logical way of overcoming fears and misconceptions. Thorough training is of great importance. To a large extent the effectiveness of a system will depend on those who use it, and adequate training is needed in order to exploit the system's capabilities. Well-trained staff are generally confident in handling the computer, a confidence which transmits itself to the users. Conversely, poor or insufficient instruction can cause stress, poor morale and low productivity (Welsch, 1990). Automated library systems are very expensive, so too is quality training, but there seems little point in spending large sums on the one without a commensurable outlay on the other. Unfortunately, all too often librarians place more importance on acquiring hardware and software than on training in its use (Hepworth, 1990).

THE STAFF TRAINING PROGRAMME

The training needs of staff must be carefully analysed well before the installation stage. Again managers must ensure that their organization does not merely pay lip service to the idea. A comment by the Library and Information Technology Centre (1987, p. 86) is worth bearing in mind:

> The scale and success of training appears to vary enormously and there is often a considerable gap between the stated aims and objectives of particular programmers and the actual implementation.

The staff training programme will vary according to time and funds available, the abilities of staff, the scale and type of computerization (whether up-grading or installing a new system), and the size of the library service. Some of the common features which can be identified are analysed below.

TRAINEES

The manager's first task is to identify all the staff who will need some form of training. There appears to be a general consensus in the literature that this should have as broad a definition as possible. In addition to everyday users, trainees should include those who may use the system occasionally as well as those who work in the library but do not have to operate computers. However, it should be noted that the same type of training will not be appropriate for all. If staff are not involved in the training sessions, they may feel ostracized, which can have a detrimental effect on morale. Bichteler (1986, p. 126) expresses it thus:

> The [people] who don't use the computer feel socially isolated since they can't contribute their 'war stories' . . . Psychologically, it would be valuable to train everyone.

Problems may also occur when staff change jobs within the library system and with readers. Readers seldom discriminate between library personnel, and anyone working in the library is likely to be asked for help in using an OPAC, for example.

Particular attention should be paid to training senior and middle-grade personnel, as they may have missed out on opportunities for IT training and education earlier in their lives, although this is becoming less common. Training senior staff first makes managerial sense, since they are then in a position to give direction and support to the project. Such a policy will help to prevent the feelings of inferiority that may arise if all grades are trained together. The sense of inferiority may be on either side, since junior staff often learn more quickly than their older colleagues: damaged pride should be avoided.

Where time is short, however, top priority must be given to those who have to use the computer systems in their daily work. The needs of part-time and evening workers should not be overlooked, and special contingencies should be made to ensure that employees do not miss out altogether on training through illness or leave.

TRAINERS

The training provided by systems' suppliers varies in extent and quality with many libraries experiencing problems (Hallmark and Garcia, 1992). Typical problems reported include a lack of understanding on the trainers' part of the individual library's needs, insufficient knowledge of the new system, inadequate preparation for training, and poor documentation and training materials. To minimize these types of problems training aspects must be thoroughly investigated when negotiating the contract. Connor (1992) points to the need to negotiate the total number of training days given, the number of days spent on each module (if appropriate), the number of people to be trained and the timing of the programme. The 'when and where' of training should also be considered and preferably be decided by the individual library, and not by the commercial organization, but in reality, a compromise may have to be sought. System operator training is usually limited to a few staff, but it is important that, even in a small library a couple of people receive this type of guidance to ensure continuity if one person leaves or is away for any reason. Some contractual arrangements with the vendor provide for general training of staff on site. Although this is generally regarded as the best option it may have its own drawbacks especially if the trainer is unfamiliar with library procedures and practices. Sykes (1991) reported that this was the case at South Bank Polytechnic, England (now South Bank University) which meant that the representative emphasized the system and hardware, rather than the application and purposes.

Staff may feel happier learning from people whom they know and who will have some idea of their abilities. If the trainers are themselves unsure and hesitant, however, they will only perpetuate unease with the system. Trainers must be competent at their task and able to communicate well. When training is provided in-house, managers must allow ample time for trainers to familiarize themselves with procedures, otherwise it will be a false economy. Epple *et al.* (1992, p. 88) advise that trainers should attend sessions offered by the vendor to learn the system. They should then adapt vendor training materials to reflect local procedures and operations, and should pre-test the training programme on a small group of staff. The value of advice gained from other clients of the vendor should also not be forgotten (Hallmark and Garcia, 1992).

LOCATION

Where the training takes place will largely depend on who is responsible for training. There can be logistical difficulties involved in travelling to a firm's headquarters for a demonstration. Connor (1992) advocates that training should take place in the library and on its own equipment wherever possible. There is a sense of security to be gained from the known environment. The practicality of this will very much depend on the premises available. Above all, staff should be removed from their normal duties in order to give their undivided attention to the training.

TIME SCALE

Different aspects of training can take place at different times. General orientation cannot be started too early. According to Epple *et al.*, (1992, p. 89), more specific training should be scheduled as close as possible to the date when the new system will be implemented, so that use of the new system will be readily remembered. The logic of this is frequently defied in practice. However, training that is left too late can be rushed (if not forgone) in the implementation process. Examples abound. Bagley and Oyston (1982, p. 51) remark on the situation at Hatfield Polytechnic in England (now University of Hertfordshire): 'Plans for a comprehensive training programme for library staff were seriously affected by the unavailability of the system until several days before the beginning of the new academic year.' Staff may be apprehensive enough at this juncture without worrying whether they are actually to receive any instruction.

It is difficult and unwise to predict the exact time required to train individuals. Dyson (1986) says that 30–36 hours proved sufficient at Westminster (which included 18 hours of practice and revision), but there are too many variables to give even a rough estimate. What must be stressed is that sufficient time should be allowed and physical and mental overload avoided. Learning anything new is exhausting. Concentration can only be maintained for short periods. Lovecy (1984, p. 168) comments: 'People take quite a while to familiarize themselves with a system, and even longer to convince themselves that they know it.'

This is another reason why training should not be a last minute affair. Sufficient time should also be left to change training methods if they prove inadequate or unsuccessful. Burton and Petrie (1986) maintain (perhaps rather idealistically) that the changeover date should be amended if training has not been completed on schedule.

CONTENT

The same level of training and education is not appropriate for all staff within the organization. System staff, managers, operators and library users will require different skills and their training will need to be adjusted accordingly. This is discussed below, but the special needs of library users are considered at the end of this chapter.

All staff

Ideally library staff should be involved in the library's decision-making processes but, if this is not possible, they should be at least kept informed about any proposed changes. In particular, they need to be told:

- why the library system is being automated or upgraded;
- which vendor is supplying the system and why;
- what hardware and software is to be used;
- where the hardware is to be sited;
- any environmental changes that will be necessary;

- how the system will affect jobs and organizational structure;
- who is to be trained and where and when.

System staff

System staff will need to know in detail the operation of the system and may need comprehensive training depending on their competencies. System-specific training is usually provided by the vendor. At a minimum, staff will need to know how to:

- assemble, operate, test and maintain equipment;
- install software;
- load data possibly from internal as well as external sources;
- handle software utilities;
- run file backups and restore operations;
- maintain logs of operations;
- manage networks;
- troubleshoot and solve problems;
- handle emergencies.

Library managers

Key management staff and middle managers who will be directly responsible for supervising and training staff to use the system will need to know about:

- the overall system capabilities and limitations;
- the use of functional components;
- how to operate the equipment;
- how to train other staff;
- how to handle emergencies;
- how to troubleshoot simple problems.

Generally this detailed level of training is given by the vendors, although, in special cases, it may be given by the system staff.

Operators

The operators of the system will need to know a variety of things including:

- how to use the equipment;
- how to use the system;
- how to identify problems;
- how to use the documentation;

- how to use online help facilities;
- what to do when things go wrong;
- what to do in emergencies;
- how the system will affect their job;
- what responsibility they will have;
- health and safety implications;
- what future changes are likely.

Determining just how much background information, if any, to give staff on how computers work is a contentious issue. On the one hand, it is argued that such knowledge will help to increase their understanding of why certain things happen (or fail to happen), while on the other hand, it is seen as an unnecessary burden likely to cause even more stress. Some understanding of what computers can and cannot do may, however, prevent the situation where staff blame faults on the computer. Computer 'errors' are far more likely to be human errors.

A fear, identified in Chapter 23, is that of data corruption. An explanation to employees of inbuilt checks in the system itself can remove these unfounded worries and eliminate the tendency to repeat operations. This is important not only from the point of view of system efficiency and staff confidence but from users' trust as well. Readers must not be suspicious of the computer's ability to produce correct overdue notices, for example, or they will question the validity of any transaction.

Part of the training sessions must be devoted to familiarizing people with unknown equipment. Knowledge should not be assumed, even that of the layout of a new keyboard. The various hardware and software components should be identified and some explanation made of what they can do. Operators should at least know how to turn their terminals on and off and how to adjust and clean them. This not only reduces the likelihood of any damage to the equipment through misuse but again gives staff a sense of personal control. Commings (1991, p. 43) advises training staff to troubleshoot their own equipment:

> Not only does local troubleshooting develop back-up staff for those times when the system administrator or designated computer person is unavailable, it also saves branch staff time when they can solve their own problems locally rather than waiting for someone to come and do it for them.

Commings adds that a manual offering a checklist of typical problems, such as loose cables or peripherals, and the logical steps needed to solve them should be developed.

Another matter, potentially more sensitive, is how much to teach staff about the overall system and its capabilities. It may be worth spending time providing staff with information on library functions other than those with which they are involved. Begg (1991, p. 115) describes how staff at Boston College were taught not only how to perform their own tasks, but how the part of the system that they use fits into the system as a whole:

> We have seen the benefits of the more integrated approach not only in how staff work with the NOTIS system, but also in how they now have a greater understanding of how the library works overall, as well as an increased appreciation of the work of all library departments.

This is also supported by Connor (1992) who suggests that training staff beyond their immediate job requirements may help them to feel more involved in the whole process, and increase their awareness of the new system.

Health and safety issues likewise must not be bypassed. This is particularly important in light of the recent legislation on health and safety issues when working with VDUs (HSE, 1992d). This requires employers 'to increase the user's competence to use workstation equipment safely and reduce the risk to their or anyone else's health' (p. 24). To do this the guidelines recommend six interrelated aspects of training (p. 26):

1. The user's role in correct and timely detection and recognition of hazards and risks. This should cover both the absence of a desirable feature (chair comfort) and the presence of an undesirable one (screen reflections and glare), together with information on health risks and how problems may be manifested.
2. A simple explanation of the causes of risk and the mechanisms by which harm may be brought about, for example poor posture leading to static loading on the musculoskeletal system and eventual fatigue and pain.
3. User initiated actions and procedures which will bring risks under control and to acceptable levels. Training should cover the following:
   - the desirability of comfortable posture and the importance of postural change;
   - the use of adjustment mechanisms on equipment, particularly furniture, so that stress and fatigue can be minimized;
   - the use and arrangement of workstation components to facilitate good posture, prevent over-reaching and avoid glare and reflections on the screen;
   - the need for regular cleaning (or inspection) of screens and other equipment for maintenance;
   - the need to take advantage of breaks and changes of activity.
4. Organizational arrangements by which symptoms or problems with the workstation can be communicated to management.
5. Information on the regulations, particularly those concerned with eyesight and rest pauses.
6. The user's contribution to assessments.

A reminder about the Data Protection Act may also be useful. Staff should know what details about readers, if any, they are allowed to disclose to others. Procedures should be taught on how to cope with systems' failures and breakdowns. In this eventuality, the employee must know how to ensure that service can be maintained and the most appropriate contact to inform of the problem. They must also be aware of how to catch up once things are running normally again. Set formulae will lessen any sense of importance in the face of machine malfunction.

METHOD

How the training programme is run will be dictated by time, available resources, and the functions to be taught. In large organizations, a structured approach is necessary in order to make quite sure that everyone receives instruction. A variety of techniques are available. Using a combination of several is advised since different methods are more suitable for different aspects of information and different activities help stimulate learners' interest as well as facilitate the transfer of learning (Cheung, 1993). Matthews (1980) also recommends that a series of training sessions should take place rather than just one; knowledge gained from one session can then be reinforced and consolidated at the next. Training can be carried out through:

- lectures;
- classroom sessions;
- audio-visual presentations;
- demonstrations/visits;
- computer-assisted learning (CAL).

The list is by no means comprehensive but whichever methods are chosen it is important that the trainee has an opportunity to ask questions and clear up any queries they may have. Practical experience should form part of the module wherever possible, and some kind of assessment of the course and of the trainee's progress made.

Lectures

Lectures are a useful means of disseminating general information to a wide audience. The drawback is an administrative one – scheduling them at a convenient time when a worthwhile body of staff can attend. If the library already opens late one day a week for a regular training slot this could be used.

Classroom sessions

One of the best ways of giving intensive training is through a classroom type situation – although care should be taken to maintain a relaxed, informal atmosphere. Staff can receive individual attention and the 'hands on' experience which is so vital to give real meaning to the exercise. There should be a one-to-one ratio of people to terminals and sufficient trainers. If there is not enough equipment or trainers the value of the sessions is limited. Epple *et al.* (1992, p. 89) recommend two trainers at each training session:

> This allows more control over the pacing of training: a second trainer can help all trainees keep up with the instructor and can give individual attention to staff who are having problems without interrupting the flow of the overall training session. A second trainer can also provide immediate feedback concerning trainees' comprehension of the material being covered, allowing the session to be more effective.

Changing from one automated system to another carries its own peculiar problems: Lefebvre (1985) notes that staff then have to 'unlearn' techniques. Sessions should be broken up, detailed instruction interspersed with group discussion, role play and other activities.

Audio-visual presentations

Audio-visual presentations are an effective method of teaching but again can only serve as an introduction. However, they give no opportunity for practice. They can be useful for subjects like health and safety for which films, videos and tape slides are available (see Useful Addresses).

Demonstration/visits

Demonstrations or visits to a vendor or another library allow many staff to see the system. Sykes' account (1986), however, highlights the problems. Interviewees complained that they had only been allowed brief access to the terminals, and the large size of the groups deterred nervous people from trying out the system. They did not want to make mistakes in front of others. Leaving staff alone to 'play' with the computer afterwards can, however, be very beneficial.

Computer-assisted learning

Computer-assisted learning is another possibility. Munrush (1990, pp. 50–1) identifies its advantages as follows:

- trainees may learn at their own pace;

- trainees incur no travel expense as training may take place at their workplace;
- once acquired, the same package may be used by many trainees;
- trainees learn on the equipment they will be using daily;
- content delivery is consistent among all trainees.

This technique permits people to proceed at their own pace: mistakes are witnessed only by the machine and the individual. The drawbacks are, however, self-evident in the title. Those who are nervous about automation will not feel at home with this medium. In addition, it may be difficult or costly to obtain appropriate software, although many of the well-known microcomputer software packages are now supported by their own online tutorial which takes the users systematically through the main features of the system.

SUPPORT MECHANISMS

Simulation techniques and training sessions can only go part-way towards preparing staff for 'going live' with a new system. The real problems and difficulties cannot be envisaged. Gratton (1983) makes this point with some feeling. Practice cannot be given on the scale of transactions that staff will be dealing with, the speed with which they will have to respond, or the pressure from readers. Mechanical failures and breakdowns can only be discussed in the abstract. He says that there is much learning to be done on the day. There are two major sources of support to help meet short-term user needs, human and documentary.

HUMAN SUPPORT

During the transition period, ease of access to human support is crucial. When problems develop, people tend instinctively to turn to others for guidance and advice. Local experts, supervisors, and computer experts are all much in demand.

A primary source of support is an employee's fellow workers: staff generally ask each other about queries concerning the system and its operation as they do on other routine matters. Questions can often be resolved by one member of the workgroup, an unofficial leader or a 'local expert'. This person does not necessarily have a higher technical knowledge than the other members of staff but has absorbed more information overall and taken a greater interest in the implementation of the system. A humane local expert can be indispensable to the organization. One of their greatest assets is accessibility; they tend to be around when things go wrong and can also answer enquirers on their own level. Prompts can be given on everyday procedures such as how to log on and off and the meaning of certain messages. Although the position of local expert has no formal standing, managers should recognize the valuable role they fill and give concrete acknowledgement wherever possible. The goodwill of a local expert is worth cultivating.

When local experts cannot provide information or are themselves in doubt, staff look to their immediate supervisors or managers for help. Consequently, managers must know more about the system than their employees, which confirms the need for training senior staff first. For the sake of their own esteem, managers must be able to provide solutions at least part of the time. They must be approachable in themselves so that staff will confront them with their problems.

In the event of major systems' breakdown, computer personnel have to be consulted. The amount of computer expertise readily available will vary greatly between libraries and depend on the systems that have been implemented. Where computer personnel are not intrinsic to the library, certain dilemmas may present themselves. A central difficulty is communication. Librarian and technician may appear to speak in different languages. Should the librarian have to seek advice over the telephone they may have problems making the crisis understood. Similarly, if this obstacle is overcome, they may very well find the diagnosis incomprehensible. This particular dilemma may disappear over a period of time as each becomes used to the idiosyncrasies of the other, but only if regular contact between the same people is maintained.

Librarians, experienced in their own field and used to resolving things for themselves, may also hesitate to contact systems' experts over every detail. If an identical fault has occurred before, they may feel that they should know the procedure to rectify it. This sort of unease should be avoided and a good relationship established. Library staff must feel able to call in outside help whenever it is needed and, if brought in early, computer personnel may find it easier to deal with the problem. Coaching can be given as the fault is corrected: there is little profit to be gained from a 'magical' resolution with no attendant explanation.

Although there may be a predetermined sequence of people to consult in the event of a computer fault, Gratton (1983, p. 19) argues that staff will still call upon those they know or someone who has proved helpful in the past: this is human nature. He also notes that the urge to call on others or to declare an emergency fades as familiarity with the system grows. Huff Arneson (1993, p. 117) notes the value of cultivating relations between library staff and computer staff. She describes how staff at one university were given tours of the computer room, enabling them to meet and become acquainted with systems staff who would also be working on the implementation of the system.

Human support is of great value to the smooth running of a system and the promotion of a healthy atmosphere within the library. But, by its very nature, human support can be erratic and unreliable: employees can leave organizations, or be absent on business and through illness. Fine (1986) has remarked that when staff are experiencing the greatest stress of change it is not unusual for personnel resources to be over extended and withdrawn. There is a need, therefore, for accompanying documentary support.

DOCUMENTARY SUPPORT

Damodaran *et al.* (1980, p. 124) draws up a list of five documents that are required to accompany any item of computer equipment:

- installation instructions;
- maintenance instructions;
- technical specification of the equipment;
- users' guide to operating the equipment;
- users' guide to operating the system.

The first three documents are of limited interest after installation has been completed. However, they should all be retained in a safe place: where systems have been adapted to meet local requirements and are therefore unique, the documentation is extremely valuable. The most essential piece of documentation from the staff point of view is the operator guide – this is frequently supplemented (if not replaced) by the organization's own manual.

Documentary support can prove invaluable in many instances. Even extensive training cannot prepare staff for all the situations they are bound to meet and they are unlikely to retain the finer details in any case. Since libraries are generally open long hours and staffed by people who only have intermittent contact with the system, a written instruction can provide a vital support link for them. Staff turnover, either that within departments or that of new employees joining the organization, is also inevitable. It is rarely cost effective to hold major training sessions for them and documentation can be used as an aid to learning. Documentation can also serve as a tool to further self-development and as a reminder of procedures. Tangible instructions can be reassuring when human support is unavailable.

In order for the documentation to fulfil these multivarious purposes, however, much effort has to be put into the preparation of the literature. Handing over a manual provided by the vendor may induce panic and an unnecessary sense of insecurity. Commings (1991, p. 43) suggests that a member of staff who is knowledgeable about both library procedures and the automated system, and who has good communication skills, should add a local commentary to the vendor's documentation to make it easier for other staff to understand. A more satisfactory alternative is to have a staff manual written specifically for the library.

The effective staff manual should satisfy several basic criteria, simplicity and clarity in presentation being high on the list. Burton and Petrie (1986) advise that only one person should actually write the manual, although many can be involved in draft copies and suggestions. This will give uniformity to the text and so facilitate use. Information should be concise and a suitably easy indexing system employed. There is little point in providing documentary support if its very appearance deters use. Where the system is complex there is usually a need for several different manuals to deal with the various circulation, acquisitions and cataloguing modules. One compendious volume might be off-putting, although an authoritative copy should be kept. The manual will need to be updated, in any case, and attention should be paid to its physical format to make sure this can be done readily: a simple ring-binder file could be used. Included or in addition to the manual should be what Saunders and Lee Kwon (1991) call 'cheat sheets' which succinctly list the steps necessary to perform each function. These are particularly useful for staff working under pressure at the issue desk.

Manuals should be available to accompany initial training sessions and staff guided in their application. They will then have an idea of where to look when problems occur. Maximum use should be made of the documentation at these sessions. Staff who come to the manuals 'cold' when they need assistance are far less likely to derive much help from them. The meaning of new terms and phrases can also be covered in systems' documentation.

Detailed operational procedures should be set out in manuals and minute, logical instructions given. Examples should be relevant to the particular library: manuals can thus be used to establish library policy. Lovecy (1984) says that when the computer offers a choice of responses there should be written advice on how to reach decisions. He gives the common example of a reader wishing to exceed his borrower limit. A further advantage of this is to ensure the library actually *does* have a policy on which line to take. Clayton (1987) remarks that it can promote consistency of practice. This helps to prevent the situation where readers feel victimized if they receive conflicting responses from different assistants. It is particularly important that standards are maintained between separate branches of a single library.

Just as the system changes, the manual should change too. The manual produced before implementation is far from being the definitive version, especially when it is based upon theoretical problems. A revised edition should reflect the concerns of those who have had to use the system. Instruction may, for example, be necessary on how to hold a light pen, or a better way of doing things may become apparent in practice. Burton and Petrie (1986) stress the importance of feedback mechanisms to make sure that these amendments are incorporated. An out-of-date manual is virtually useless and can be very misleading. When system changes are made there is a tendency among suppliers to neglect a similar update of the manual. Where a library relies on documentation from a supplier this could be a serious problem.

Computer-based documentation is also commonplace. Essentially similar to paper-based documentation, it is available online and has all the advantages of ease of update associated with electronic publishing. Such documentation must, however, be easy to access and not interrupt normal work. The use of Windows facilitates this. Where it is used purely as a learning medium there must be enough terminals to cater for this. Most computer systems now have built-in 'help' facilities. These can guide the user when they are unsure what to do next or can give them an explanation of abbreviations or codes being used.

It has been shown that training and documentation have a vital link but Baddeley (1988, p. 79) warns that they: 'are not substitutes for each other. A good training course does not eliminate the need for well-prepared documentation, and reprinted course notes are not a user guide.'

Over a period of time, staff will become less reliant on manuals, but they do have a continuous purpose and should not be abandoned. They can be used for reference and in the training of new staff.

FURTHER TRAINING

The effectiveness of any training programme needs to be assessed and the need for further training identified. One method of assessment is an anonymous questionnaire, completed at the end of the training session. This, however, only judges the relevance of the training session to attenders at that time. Other needs may become apparent as operators become more proficient with the system. The staff of one library expressed regret that they had consumed the total number of days' training offered by the supplier shortly after implementation. In retrospect, they felt they would have benefited by waiting until they had become more familiar with the software. Additional training sessions are therefore necessary 'to ensure that staff maintain a good working knowledge and are aware of any changes resulting from updates' (Daniels, 1995, p. 4). Epple *et al.* (1992) also recommend using library newsletters for keeping staff updated or as a forum for questions about the system.

Refresher courses should be arranged for all staff from time to time. This is important in academic libraries, particularly where staff employed during term time only have long breaks away from the system. It is particularly vital to remind staff of what to do in the event of a system's failure: the fact that a computer is reliable will mean that this eventuality is met infrequently. Comprehensive records should be kept of who has been trained in what and when.

New staff have particular needs, especially junior members joining just before a new system is installed. Everything will be strange, not only the people and place but possibly library procedures as well. The newcomer is likely to have missed out on some of the general orientation lectures. It is unfortunately also true that other employees may not be as cooperative as normal if they are experiencing their own anxieties.

New staff joining after the implementation of a new system are often trained by their peers. Graham (1987) can see disadvantages in this. Although the library assistant may be extremely good at their job, that does not automatically qualify them for communicating their skill to others. Indeed, proficiency may in itself be a barrier: the trainer may be unable to envisage someone else's problems or remember their own hesitant start. Technicalities can be glossed over and the trainer's own particular method of performing transactions (rather than the standard approach) perpetuated. New employees may also miss out on instruction about adjustable furniture and terminals, which may have become second nature to more established staff.

TRAINING THE USER

The ultimate beneficiaries of a new automated library system should be the library's users yet they are often forgotten in the rush to get a system operational. Training the user is partly realized through contact with the staff, and partly through good publicity. It is to be hoped that well-trained employees communicate their positive feelings for the system to the readers. Timely publicity and information also prepares them for automation or system changes. Publicity about the new system should be given well in advance of the installation date. Successful techniques used in the past include posters, newsletters, bookmarks, handouts, public meetings and the use of media such as local newspapers, radio or television. Library users need to be kept informed of the changes that are likely to affect them and, in particular, should be told about:

- when the new system is going 'live';
- what benefits it will bring to them;
- what changes are going to occur;
- what they need to do, if anything, about their registration.

One way of directly involving the public is that reported by Clarke (1985) at Hillingdon. When enrolling new members into the library, both VDU operator and user sit down together at the terminal. The user can then see what details are being put onto the computer and can correct them as necessary. This helps to accustom the user to the machine, is beneficial for public relations, and helps to reduce any feelings of anxiety the user may have towards 'secret data' being kept about them. When dealing with any reader query it is useful for VDUs to be mounted on swivel stands or angled so that assistant and user can view the screen alike without either straining to see, but it is important not to allow other readers in the queue to be able to read details about the reader at the desk.

Specific training on OPACs, the use of CD-ROMs, online searching, Internet and so on may also need to be given once the system is installed. Depending on the type of library and on staff availability, a variety of training methods may be used. In their survey of American libraries, Hallmark and Garcia (1992, p. 353) found the provision of handouts and 'tip sheets', giving brief details about the system and how to use it effectively, was most successful. Less successful were group orientation sessions. The use of volunteers from the community was found to be both effective in terms of training and good for public relations, as the volunteers were enthusiastic and felt involved with the library.

USEFUL WWW SITES AND ADDRESSES

SELECTED WEB SITE ADDRESSES: HEALTH AND SAFETY

HEALTH AND SAFETY RISKS AT WORK AND STRATEGIES FOR THEIR PREVENTION

Alternative/Ergonomic Pointing Devices
http://www.ics.com/~diana/mouse.htm
Provided here is a list of products and contact information for alternative pointing devices to the standard mouse.

Computer-Related Ergonomics Information and Products
http://www.distrib.com/ergonomics/homepage.html
An overview of computer-related ergonomic problems and solutions concerning different parts of the body is presented here, along with information on workstation design. Also included is a catalogue of ergonomic products, plus links to ergonomics firms, research and reporting institutions, publications and other resource lists.

An Employer's Guide to the Display Screen Regulations
http://www.system-concepts.com/stds/hse_index.html
Tom Stewart's excellent (if lengthy) guide to the Health and Safety (Display Screen Equipment) Regulations 1992 gives details on how to implement the regulations and the benefits that will be accrued from doing so. It covers breaks, eye tests, user training, assessment of workstations, and ensuring that workstations and the environment are safe and comfortable.

Enviro Village
http://www.envirovillage.com/REnviroVillageHome.htm
Definitions of sick building syndrome and building-related illness are provided, and their causes, solutions and preventions are listed.

Good Health is Good Business: an Introduction to Managing Health Risks at Work
http://www.open.gov.uk/hse/ghigb2.htm
Aimed at managers, this lengthy document forms part of the Health and Safety Executive's campaign to reduce work-related illness. It provides a good overview of work-related illnesses and how they may be managed, particularly respiratory sensitizers, noise at work and musculoskeletal disorders. It includes case studies.

A Guide to Work-Related Stress
http://www.wt.com.au/safetyline/d_pubs/stress2.htm
The nature and causes of work-related stress are discussed in this guide, and information is given on its prevention and management. Although written to help managers comply with the Western Australia Occupational Health and Safety Act 1984, the content is relevant to workers of any country, other than the section on employers' legal position.

Guidelines for Designing Effective and Healthy Learning Environments for Interactive Technologies
http://www.etb.nlm.nih.gov/monograp/ergo/index.html
Weisberg examines VDT-related risks, specifically: potential health hazards from electromagnetic radiation, musculoskeletal disorders, vision complaints, and psychosocial stresses. He also presents, General Workstation Specifications (http://www.etb.nlm.nih.gov/monograp/ergo/specs.html); guidelines on how to design an ergonomically correct workstation.

Macroergonomics, VDT Job Design, and Preventing Work-Related Musculoskeletal Disorders
http://www.dtic.dla.mil/iac/cseriac/g6295vdt.html
This is a synopsis of Hal Hendrick's discussion of a macroergonomic approach that he believes will enable workstations and jobs to be designed that will reduce work-related musculoskeletal disorders, humanize VDT work, enhance productivity, and reduce costs.

RSI–UK Home Page
http://www.demon.co.uk/rsi/
This resource lists articles and other sources which give information on repetitive strain injury (or work-related upper limb disorders), its causes and symptoms, how to treat it, and how to avoid it. It also provides information on ergonomically designed keyboards and other equipment which help prevent RSI.

Sick Building Syndrome
http://www.compulink.co.uk/~gegs/sbs.htm
A summary of the HSE's recently published booklet *How to Deal with Sick Building Syndrome*, this provides a concise list of the main symptoms of Sick Building Syndrome, possible contributory factors, and priorities for the investigation of complaints.

Typing injury FAQ: Keyboard Alternatives
http://www.cs.princeton.edu/~dwallach/tifaq/keyboards.html
This site provides answers to Frequently Asked Questions (FAQ) about keyboard alternatives. Included is a detailed list of products with contact information.

Typing injury FAQ: Software Monitoring Tools
http://www.cs.princeton.edu/~dwallach/tifaq/furniture.html
This site describes tools, primarily software, to help prevent or manage WRULD or RSI.

A Workstation Checklist
http://www.demon.co.uk/rsi/checklst.txt
A comprehensive checklist from *Hazard* magazine. This covers points to check for in your workstation and work environment.

ORGANIZATIONS

Health and Safety Executive
http://www.open.gov.uk/hse/hsehome.htm
This resource provides details on what the HSE does, how to obtain their publications and videos, contacts for enquiries, information on current initiatives and research, and gives pointers to other sources of health and safety information.

Human Factors and Ergonomics Society
http://www.hfes.vt.edu/HFES/
As well as giving details of the society itself, this page offers a facility to purchase human factors publications and other products, links to other ergonomics or human factors pages, and links to Internet search engines.

NEWSGROUPS

news:comp.human-factors
This newsgroup is helpful for keeping up to date with new developments and Web sites related to human factors. (A hint: expand the *Subject* column if you are unable to read the subject of the messages.)

POINTERS TO FURTHER RESOURCES

Ergoweb
http://www.ergoweb.com
Compiled by the University of Utah, this list of ergonomics resources on the Internet contains standards and guidelines, background information, job review and analysis tools and checklists, biomechanical analysis tools, case studies, and much else. Some of the information is US-biased.

Human–Computer Interaction Resources on the Net
http://www.ida.liu.se/labs/aslab/groups/um/hci/
Mikael Ericsson's excellent list of resources is primarily HCI-focused, but there are useful links to more general human factors and ergonomics sources, including those found by search engines such as Yahoo.

SELECTED WEB SITE ADDRESSES: HUMAN–COMPUTER INTERACTION

USER INTERFACE DESIGN

Human–Computer Interface Guidelines
http://groucho.gsfc.nasa.gov/Code_520/Code_522/Documents/HCI_Guidelines/HCI_home.html
Lengthy but comprehensive, this resource presents user interface guidelines addressing graphic/object-oriented interfaces. Guidelines are listed concisely and clearly with pictorial examples of display layouts (which are useful although slow to load). All aspects of screen design are covered.

The Psychology of Menu Selection: Designing Cognitive Control at the Human/Computer Interface
http://www.lap.umd.edu/pomsFolder/pomsHome.html
Kent Norman's online book covers the theory and implementation of menu selection systems, design guidelines, and the implementation and evaluation of menu systems.

Steps to Improving your Interface: a Guide for Users and Non-Experts
http://ourworld.compuserve.com/homepages/chimaera/homepage.htm
This is a good introduction to user interface design (although rather colloquially written). Ten basic guidelines are listed, covering information layout and coding, error messages, toolbars and icons, and screen content, and all linked to further information.

ELECTRONIC PUBLICATIONS

HCI – Electronic Publications
http://www.cs.bgsu.edu/HCI/epub.html
This useful list of electronic publications relevant to HCI is part of the Human–Computer Interaction Virtual Library (http://www.cs.bgsu.edu/HCI/intro.html), which provides links to many other HCI-related resources.

BIBLIOGRAPHIES

The HCI Bibliography
http://www.cibib.org
Gary Perlman's huge HCI Bibliography already contains thousands of records and is still growing. It includes entries on special files, books and reports, conference proceedings, and core journals. Entries consist of title, author, date, pages and publication, with keywords and abstracts when available, and tables of contents of articles in some cases. Records may be accessed by type of publication.

ORGANIZATIONS

British HCI Group
http://www.bcs.org.uk/hci/
Information is given about the group and its aims, and links are provided to HCI news and Web resources, including bibliographies and information on education, publications, laboratories and organizations.

NEWSGROUPS

HCI Communication
http://is.twi.tudelft.nl/hci/communication.html
This is part of the HCI Index (http://is.twi.tudelft.nl/hci/), which provides links to newsgroups and mailing lists related to HCI.

news:comp.human-factors
This newsgroup is helpful for keeping up to date with new developments and Web sites related to human factors. (A hint: expand the *Subject* column if you are unable to read the subject of the messages.)

POINTERS TO FURTHER RESOURCES

HCI Index
http://is.twi.tudelft.nl/hci/
This is a very useful list of resources from Hans de Graaff at Delft University of Technology. It includes information on communication, bibliographies, forthcoming conferences and workshops, publications, reviews of HCI-related books, research groups and organizations and related sources.

Human–Computer Interaction Resources on the Net
http://www.ida.liu.se/labs/aslab/groups/um/hci/
Mikael Ericsson's excellent list of resources provides information on the following: academic and commercial research; companies, organizations and people; jobs; conferences and events; guidelines, styleguides and standards; bibliographies and publications; newsgroups; education; software; and Web resources including links to search engines.

Resources on HCI
http://www.acm.org/~perlman/interactions/resources.html
This page contains pointers to resources collected by Gary Perlman at the University of Ohio. Perlman covers the following HCI-related information: resources available on the Web, frequently asked questions and newsgroups, standards and guidelines, education, bibliographies, organizations, and mailing lists.

PROFESSIONAL SOCIETIES

All-Ukranian Ergonomics Association
Department of Ergonomics
National Research Institution for Design
Post Box 3
252054 KIEV
UKRAINE

Association Español De Ergonomia
Salustiano Olozaga, 5
28001 MADRID
SPAIN

Belgian Ergonomics Society
UCL
Unité Hygiène et Phys. Trav.
Clos Chapelle aux Champs 3038
B-1200 BRUXELLES
BELGIUM

BCS
1 Sanford Street
SWINDON
Wiltshire SN1 1HJ
UK

The British Psychological Society
St Andrew House
Princess Road East
LEICESTER LE1 7DR
UK
Tel: (0116) 2549568

Its members may include workers on the cognitive aspects of human–computer interaction who may be contacted via the society.

The Ergonomics Society
Devonshire House
Devonshire Square
LOUGHBOROUGH
Leicestershire LE11 3DW
UK
Tel: (01509) 234904

The main professional body for ergonomists in the UK.

Chartered Institution of Building Services Engineers
222 Balham High Road
LONDON SW12 9BS
UK
Tel: (0181) 675–5211

Publishers of Technical Memoranda TM6 – 'Lighting for Visual Display Units'.

Croatian Ergonomics Society
Salajeva 3
4100 ZAGREB
CROATIA

Czech Ergonomics Society
Agricultural University – VSZ
Suchdol
160 21 PHAHA 6
CZECH REPUBLIC

Ergonomics Society of Australia and HCI Special Interest Group
Canberra Business Centre
Bradfield Street
DOWNER, ACT, 2602
AUSTRALIA

Ergonomics Society of Taiwan
Department of Industrial Design
Chang Gung College of Medicine and Technology
Kwei-Shan, Tao-Yaun 333
TAIWAN, ROC

Gesellschaft für Arbeitswissenschaft
Institut für Arbeitsphysiologie
Universität Dortmund
Ardeystr. 67
D-44139 DORTMUND H
GERMANY

Hellenic Ergonomics Society
Ergonomia Ltd
3th Septembriou, 77
ATHENS 10434
GREECE

HF Association of Canada/ACE
HFAC/ACE
6519B Mississauga Road
MISSISSAUGA
Ontario L5N 1A6
CANADA

Human Factors and Ergonomics Society
PO Box 1369
SANTA MONICA, CA 90406–1369
USA

Hungarian Ergonomics Society
Totfalusi KM st 24
H-1201 BUDAPEST
HUNGARY

Indian Ergonomics Society
11A Mohan Bagan Lane
CALCUTTA, 70004
INDIA

The Institution of Electrical Engineers
Savoy Place
LONDON WC2R 0BL
UK
Tel: (0171) 240–1871

May offer information on the human aspects of IT

Irish Ergonomics Society
City Engineers
Limerick Corporation
LIMERICK
IRELAND

Israeli Ergonomics Society
Technion-Israel Institute of Technology
Faculty of Industrial Engineering and Management
Technion City
HAIFA 32000
ISRAEL

Japan Ergonomics Research Society
Nihon University School of Medicine
Department of Hygiene
30-1 Ohyaguchi Kamimachi
Itabashi-ku
TOKYO 173,
JAPAN

Nederlandse Vereniging Voor Ergonomie
Dutch Ergonomics Society
WG Plein 564
1054 SJ AMSTERDAM
THE NETHERLANDS

New Zealand Ergonomics and HCI Special Interest Group
33 Monaghan Avenue
Karari
WELLINGTON
NEW ZEALAND

Nordic Ergonomics Society
H&G A/S, PB 5055 Majorstua
0301 OSLO
NORWAY

Polish Ergonomics Society
National Labour Inspectorate
Krucza Street 38/42
00-512 WARSZAWA
POLAND

Portuguese Association of Ergonomics
Estrada da Costa
1499 LISBOA Codex
PORTUGAL

Österreichische Arbeitsgemeinschaft für Ergonomie
c/o Institut für Arbeits-und Betriebswissenschaft der TU-Wein
Theresianumgasse 27
A-1040 WIEN,
AUSTRIA

Russian Ergonomics Association
Voinovastr. 9, ap.1
191187 ST PETERSBURG
RUSSIA

Slovak Ergonomics Association
c/o Faculty of Material Science and Technology
STU Paulinska St. 16
917 24 TRNAVA
SLOVAK REPUBLIC

Société D'Ergonomie De Langue Française
ERGODIN
64 rue d'Ophem
B-1000 BRUXELLES
BELGIUM

Società Italiana Di Ergonomia
Via S. Barnaba, 8
20122 MILANO
ITALY

South-East Ergonomics Society (SEAES)
School of Housing, Building and Planning
Universiti Sains Malaysia
11800 MINDEN
PENANG
MALAYSIA

RESEARCH/CONSULTANCY BODIES

Ergonomics Information Analysis Centre
School of Manufacturing and Mechanical Engineering
University of Birmingham
Edgbaston
BIRMINGHAM B15 2TT
UK
Tel: (0121) 414–3344

Offers consultancy and information on human factors. Produces *Ergonomic Abstracts* and existing and customized bibliographies in selected areas of ergonomics.

Humane Technology
203 Swithland Lane
ROTHLEY
Leicestershire
UK
Tel: (0116) 302321

Offers consultancy, runs seminars, courses and sponsors conferences on alleged health hazards of VDU work.

HUSAT (Human Sciences and Advanced Technology)
The Elms
Elm Grove
LOUGHBOROUGH
Leicestershire LE11 1RG
UK
Tel: (01509) 611088

Researches, publishes and undertakes consultancy in all areas of human–computer interaction.

The Industrial Society
Peter Runge House
3 Carlton House Terrace
LONDON SW1Y 5DG
UK
Tel: (0171) 839–4300

Carries out surveys into working conditions associated with work done on VDUs for organizations.

The National Computing Centre Ltd
Oxford House
Oxford Road
MANCHESTER M1 7ED
UK
Tel: (0161) 228–6333

Advises and publishes on aspects of IT installations.

System Concepts Ltd
2 Savoy Court
Strand
LONDON WC2R 0EZ
UK
Tel: (0171) 240–3388

Independent consultants specializing in systems but also incorporates 'The Ergonomic Training Centre' which runs courses.

MEDICAL ASSOCIATIONS AND SELF-HELP GROUPS

College of Optometrists
10 Knaresborough Place
LONDON SW5 0TG
UK
Tel: (0171) 373–7765

Provides information on the clinical aspects of eyes.

Eye Care Information Service
PO Box 3597
LONDON SE1 6DY
UK
Tel: (0171) 357–7730

Welcomes telephone and written enquiries on all aspects of eye care.

Institution of Occupational Safety and Health
The Grange
Highfield Drive
Wigstone
LEICESTER LE11 1NN
UK
Tel: (0116) 2571399

London Hazards Centre
Interchanges Studies
Dalby Street
LONDON NW5 3NQ
UK
Tel: (0171) 267–3387

Publishes *VDU Hazards Handbook*; *Sick Building Syndrome*; and *Fluorescent Lighting*.

Medical Research Council
20 Park Crescent
LONDON W1N 4AL
UK
Tel: (0171) 636–5422

Government body which can give information on research in medical matters.

INFORMATION/ADVICE FOR VDU OFFICE WORKERS

ACAS – Advisory, Conciliation and Arbitration Service
Brandon House
180 Borough High Street
LONDON SE1 1LW
UK
Tel: (0171) 210–3911

May give advice on work-related issues and publish bibliographies and papers.

British Safety Council
70 Chancellors Road
LONDON W6 9RS
Tel: (0181) 741–1231

Issues publications and runs courses.

Health and Safety Executive
Broad Lane
SHEFFIELD S3 7HQ
Tel: (0114) 291–2300

Offers general advice, publishes books and leaflets and supports HSELINE, a publicly available database.

Eclipse Group Ltd
18–20 Highberry Place
LONDON N5 1QP
Tel: (0171) 354–5858

Produces journals on the subject of industrial relations services.

Labour Research Dept.
78 Blackfriars Road
LONDON SE1 8HF
Tel: (0171) 928–3649

Publishes *VDUs: Health and Safety: A User's Guide*, *Health and Safety Law*, *Stress at work*, and other booklets.

City Centre
4th Floor
Sophia House
32/35 Featherstone Street
LONDON EC1Y 8QX
Tel: (0171) 608–1338

Publishes general advice and and provides information on working with VDUs. They also provide a VDU fact pack for which there is a charge.

TRADE UNIONS OFFERING INFORMATION

BIFU (Banking, Insurance and Finance Union)
Sheffield House
1B Amity Grove
LONDON SW20 0LG
UK
Tel: (0181) 946–9151

Publishes guidelines on work with VDUs.

GMB
22 Worple Road
LONDON SW19 4DF
UK
Tel: (0181) 947–3131

Publishes *Stress at Work, a GMB Guide.*

AUEW, Amalgamated Union of Engineering Workers
Onslow Hall
Little Green
RICHMOND
Surrey TW9 1QN
UK
Tel: (0181) 948–2271

CPSA, Civil and Public Services Association
160 Falcon Road
LONDON SW11 2LN
UK
Tel: (0171) 924–2727

Publishes 'VDU Users Guide' which is available free to its members.

UNISON
1 Mabledon Place
LONDON WC1H 9AJ
UK
Tel: (0171) 388–2366

Publishes 'Display Screen Equipment Regulations, 1992' which is a guide for its members.

EQUIPMENT SUPPLIERS

3M Library Systems
3M House
PO Box 1
Market Place
BRACKNELL
Berkshire RG12 1JU
UK
Tel: (01344) 858162

Suppliers of inset security devices.

Acco UK
Halesowen Industrial Park
Hereward Rise
HALESOWEN
West Midlands B62 8AN
UK
Tel: (0121) 550–8883

Provides a range of computer accessories such as tilt/swivel stands, copyholders, filter screens and acoustic hoods. Also suppliers of system furniture.

AccuCorp Inc
PO Box 66
CHRISTIANSBURG VA 24073
USA
Tel: 703–961–2001

Suppliers of AccuKey.

Apak Systems Ltd
Apak House
Badminton Court
Stations Road
Yate
BRISTOL BS17 5HZ
UK
Tel: (01454) 316086

Omni-directional laser scanners.

Automated Library Systems Ltd
Vector House
Brownfields
WELWYN GARDEN CITY
Hertfordshire AL7 1AN
UK
Tel: (01707) 336251

Self-service library issue, renewals and return service stations.

Balmforth Engineering Ltd
Finway
Library Systems Division
Dallow Road
LUTON LU1 1TE
UK
Tel: (01582) 31171

Suppliers of library furniture.

Blyde-Barton Furniture Ltd
Reepham Road
NORWICH NR6 5LE
UK
Tel: (01603) 789000

Office furniture suppliers.

Carlton Furniture Group Ltd
Monkhill
PONTEFRACT
West Yorkshire WF8 2NS
UK
Tel: (01977) 791191

Suppliers of office furniture.

Comfort Keyboard System
Health Care Keyboard Company
1240-G W. Feerick Street
WAUWATOSA WI 53222
USA
Tel: 414–536–2160

Suppliers of Comfort Keyboard System.

Computer Health Break
Escape Ergonomics Inc
1111 W El Camino Real
Suite 109
Mailstop 403
SUNNYVALE CA
USA
Tel: 408–730–8410

Supplier of software aimed at preventing WRULD.

DataHand Systems Inc
10789 North 90th Street
SCOTTSDALE AR 85260–6727
USA
Tel: 800–875–7171

Suppliers of DataHand; a keyboard system.

Don Gresswell Ltd
Bridge House
Vera Avenue
Grange Park
LONDON N21 1RB
UK
Tel: (0181) 360–6622

Library and office equipment and furniture suppliers. They also supply self-issue systems.

EFG Business Furniture
73 Welbeck Street
LONDON W1M 7HA
UK
Tel: (0171) 486–7051

Suppliers of office furniture.

Electronic Reading Systems Ltd
Unit 1
The Business Park
Clifton Road
SHEFFORD
Bedfordshire SG17 5AB
UK
Tel: (01462) 816213

Barcode scanning equipment including cordless laser and omni-directional readers.

Finmade Ltd
13 Anyards Road
COBHAM
Surrey KT11 2LW
UK
Tel: (01932) 865180

Office furniture suppliers.

HCS Global Computer Products
2 Cartsdyke Avenue
Greenock
INVERCLYDE PA15 1DT
Tel: (0181) 5608585 – London
Tel: (01475) 787878 – Greenock

Provides a range of ergonomic computer products via mail order.

Infogrip Inc
1141 E Main Street
VENTURA CA 93001
USA
Tel: 800–397–0921

Supplies The Bat Personal Keyboard.

Kinesis Corporation
22232 17th Avenue SE
BOTHELL WA 98021–7425
USA
Tel: 206–402–8100

Suppliers of Kinesis Ergonomic Keyboard.

Library Design and Engineering Ltd
CHICHESTER
West Sussex PO19 2TZ
UK
Tel: (01243) 788111

Suppliers of library shelving, study carrels, desks, chairs and other equipment.

Library Furnishing Consultants Ltd
Phoenix House
54 Dennington Road
WELLINGBOROUGH
Northants NN8 2QH
Tel: (01933) 442777
UK

Librex Educational Co.
Colwick Road
NOTTINGHAM NG2 4BG
UK
Tel: (0115) 9504664

Supplies library equipment and furniture.

Luzo UK Ltd
Redbrook Lane
Brereton
RUGELEY
Staffordshire WS15 1QY
UK
Tel: (01889) 582112

Designers and manufacturers of lighting fittings.

PCD Maltron Ltd
15 Orchard Lane
EAST MOLESEY
Surrey KT8 0BN
UK
Tel: (0181) 398–3265 or

Manufacturer and Distributer
TelePrint Systems, Inc
4 Henson Place, Suite No. 5
CHAMPAIGN IL 61820
USA
Tel: 217–359–9690

Suppliers of the Maltron Keyboard.

Point Eight Ltd
Shaw Road
DUDLEY
West Midlands DY2 8TP
UK
Tel: (01384) 238282

Library furniture manufacturers and suppliers offering custom-built solutions and standard products.

President Office Furniture Ltd
Lincoln House, Old Parkbury Lane
Colney Street
ST ALBANS
Hertfordshire AL2 2DX
UK
Tel: (01923) 857211

Furniture suppliers.

Project Office Furniture plc
Showroom and Factories
HAVERHILL
Suffolk CB9 8QJ
UK
Tel: (01440) 705411

Suppliers of office furniture and lighting.

RAN Enterprises
One Woodland Park Drive
HAVERHILL MA 01830
USA
Tel: 800–451–4487

Supplies EyerCise a software program aimed at preventing WRULD.

SB Electronic Systems Ltd
Arden Grove
HARPENDEN
Hertfordshire AL5 4SL
UK
Tel: (01582) 76991

Supplier of hand-held and fixed laser scanners.

Serota
49 Ingram House
Daling Way
LONDON E3 5NL
UK
Tel: (0181) 981–5452

Library furniture suppliers.

Shaun Howard Associates Ltd
Unit 13
Cunliffe Drive
KETTERING
Northants NN16 8LD
UK
Tel: (01536) 310033

Design, manufacture and supply of library furnishings.

Special Systems Industry BV
Lange Voorhout 45a
2514 EC Den Haag
The Netherlands

Manufacturers of the Velotype Keyboard.

Reska Products
Mount House
Bond Avenue
Bletchley
MILTON KEYNES MK1 1SD
UK
Tel: (01908) 371001

Specialize in library issue desks and other custom-made furniture.

MEDIA

CDD Products Ltd
Allen House
Station Road
EGHAM
Surrey PW20 9LB
Tel: (01784) 472234

Markets a package intended to help employers meet their staff training obligations under the Display Screen Regulations Act, 1992. Included is the video 'In your Own Interest'.

Human Focus International Limited
Mill Green Business Park
Mill Green Road
MITCHAM
Surrey CR4 4HT
England
Tel: (0181) 640 3535

Produces a video training package 'In Your Own Internet – VDU Work Techniques'.

TV Choice Ltd
22 Charing Cross Road
LONDON WC2H 0HR
UK
Tel: (0171) 379–0873

Offers videos on various aspects of IT management.

REFERENCES

Åborg, C., Aronsson, G. and Dallner, M. (1993), 'Work organization and health aspects of work with VDUs', in Luczak, H., Çakir, A. and Çakir, G. (eds), *Work With Display Units 92*, pp. 359–63, Amsterdam: Elsevier.

Alderfer, C. P. (1969), 'An empirical test of a new theory of human needs', *Organizational Behaviour and Human Performance*, **4**, 142–75.

Alderfer, C. P. (1972), *Existence, Relatedness and Growth: Human Needs in Organizational Settings*, New York: Free Press.

Allen, A. J. (1991), 'Automation, barriers, and communication: how an integrated library system changes public services–technical services relationships', *Reference Librarian*, **34**, 69–77.

Anderson, J. R. (1980), *Cognitive Psychology and its Implications*, San Francisco, CA: Freeman.

Apex: London and Home Counties Area Technology Sub-committee (1985), *New Technology: A Health and Safety Report*, London: Apex.

Argyle, M. (1972), *The Social Psychology of Work*, Harmondsworth: Penguin.

Astle, D. (1995), 'Special section: library reorganization for the 21st century', Staff involvement – the key to the successful merger of monograph and serial acquisitions functions at Clemson University Library, *Library Acquisitions: Practice and Theory*, **19**(4), 427–30.

Atkinson, H. C. (1984), 'Strategies for change: Part II', *Library Journal*, **109**(5), 556–7.

Baddeley, A. D. and Hitch, G. (1974), *Working Memory. Recent Advances in Learning and Motivation, Vol. 8*, New York: Academic Press.

Baddeley, A. D. and Lieberman, K. (1980), *Spatial Working Memory. Attention and Performance, Vol. 8*, Hillsdale, NJ: Erlbaum.

Baddeley, J. (1988), 'Winning ways with users', *Systems International*, April, 77–9.

Bagley, D. E. and Oyston, E. (1982), *Automation in a Polytechnic Library: Fifteen Years' Development at Hatfield*, London: Library Association.

Ballard, J. (1993), 'RSI on trial', *New Scientist*, 11 September, 24.

Barclay, N. L. (1986), 'Breaking the world speed record', *The Guardian*, 27 March, 25.

Bates, M. J. (1990), 'Where should the person stop and the information search interface start?', *Information Processing and Management*, **26**(5), 575–91.

Beale, R. and Jackson, T. (1990), *Neural Computing: an Introduction*, Bristol: Adam Higer.

Becket, M. and Smith, H. B. (1986), 'Designing a reference station for the information age', *Library Journal*, **111**, 15 April, 42–6.

Beckman, M. (1987), 'Library Equipment in a Changing Environment' in: *53rd IFLA Council and General Conference, Brighton, 16–21 August 1987*, pp. 16–2, 16–3.

Begg, K. (1991), 'Changing systems: putting it all together and making it work – or, how we did it good at Boston College', in Pitkin, G. M. (ed.), *Library Systems Migration: Changing Automated Systems in Libraries and Information Centers*, pp. 102–23, London: Meckler.

Belkin, N. J., Chang, S., Downs, J., Saracevic, T. and Zhao, S. (1990), 'Taking account of user tasks, goals and behaviour for the design of online public access catalogs', in Henderson, D. (ed.), *ASIS '90: Information in the Year 2000: from Research to Applications. Proceedings of the 53rd Annual Meeting of the American Society for Information Science. Toronto, Canada. 4–8 November 1990. Vol. 27*, pp. 69–79, Medford, NJ: Learned Information Inc.

Belkin, N. J., Marchetti, P. G., Albrecht, M., Fusco, L., Skogvold, S., Stokke, H. and Troina, G. (1991), 'User interfaces for information systems', *Journal of Information Science*, **17**(6), 327–44.

Bendix, T. and Jessen, F. (1986), 'Wrist support during typing – a controlled electromyographic study', *Applied Ergonomics*, **17**(3), 162–8.

Bently, C. (1993), *Practical PRINCE: a Guide to Structured Project Management*, Manchester: NCC Blackwell.

Bergqvist, U. (1986), 'The alleged reproductive hazard of VDU work – an overview', in *Allegations of Reproductive Hazards from VDUs: The Second International Scientific Meeting, London, 25 and 26 November 1986*, pp. 3–14, Quorn: Humane Technology.

Bergqvist, U. and Knave, B. (1994), 'Eye discomforts and work with visual display terminals', *Scandinavian Journal of Work Environmental Health*, **20**, 27–33.

Bergqvist, U., Knave, B., Voss, M. and Wilbom, R. (1992), 'A longitudinal study of VDT work and health', *International Journal of Human Computer Interaction*, **4**, 197–219.

Bergqvist, U., Knave, B. and Wilbom, R. (1995), 'Eye discomforts during work with visual display terminals', in Grieco, A., Molteni, G., Piccoli, B., and Occhipinti, E. (eds), *Work With Display Units '94*, pp. 83–8, Oxford: Elsevier.

Bergqvist, U., Wolgast, E., Nilsson, B. and Voss, M. (1995a), 'The influence of VDT work on musculoskeletal disorders', *Ergonomics*, **38**(4), 754–62.

Bergqvist, U., Wolgast, E., Nilsson, B. and Voss, M. (1995b), 'Musculoskeletal disorders among visual display terminal workers: individual, ergonomic, and work organizational factors', *Ergonomics*, **38**(4), 763–76.

Bernstein, B. (1989), 'Developing new interfaces for online services', in Nixon, C. and Padgett, L. (eds), *National Online Meeting, Proceedings 1989, New York, 9–11 May 1989*, pp. 37–47, Medford, NJ: Learned Information Inc.

Bichteler, J. (1986), 'Human aspects of high tech in special libraries', *Special Libraries*, **77**(3), 121–8.

Bichteler, J. (1987), 'Technostress in libraries: Causes, effects and solutions', *Electronic Library*, **5**(5), 282–7.

Biddiscombe, R. (1996), 'The changing role of the reference librarian', in Biddiscombe, R. (ed.), *The End-user Revolution*, pp. 79–95, London: Library Association.

Bjerkedal, T. and Egenaes, J. (1986), 'Video display terminals and birth defects' in Knave, B. and Wideback, P. G. (eds), *Work with display units*, pp. 111–14, Oxford: Elsevier.

Blackwell, R. and Chang, A. (1988), 'Video display terminals and pregnancy. A review', *Journal of Obstetrics and Gynaecology*, **95**, 446–53.

Blunden-Ellis, J. and Graham, M. E. (1994), 'A UK market survey of library automation system vendors (1992–1993)', *Program*, **28**(2), 109–24.

Bonner, P. (1988), 'The next wave: graphical interfaces are taking over the screen', *PC/Computing*, November, 72–82.

Borgman, C. L. (1986a), 'Online catalog screen displays: a human factors critique', in Williams, J. F. (ed.), *Online Catalog Displays: A Series of Discussions, Conference Sponsored by the Council for Library Resources*, pp. 135–56, Washington, DC: Council in Library Resources.

Borgman, C. L. (1986b), 'The user's mental model of an information retrieval system: an experiment on a prototype online catalog', *International Journal of Man–Machine Studies*, **24**, 47–64

Borgman, C. L. (1988), 'Human factors in the use of information systems: research methods and results', in Tudor-Šilovic, N. and Mihel, I. (eds), *Information research: Research Methods in Library and Information Science: Proceedings of an International Seminar on Information Research, Dubrovnic, Yugoslavia, 19–24 May 1986*, pp. 139–65, London: Taylor Graham.

Borgman, C. L., Case, D. O. and Meadow, C. T. (1989a), 'The design and evaluation of a front-end user interface for energy researchers', *Journal of the American Society for Information Science*, **40**(2), March, 99–109.

Borgman, C. L., Chignell, M. H. and Valdez, F. (1989b), 'Designing an information retrieval interface based on children's categorization of knowledge: a pilot study', in Katzer, J. and Newby, G. B. (eds), *ASIS '89: Proceedings of the 52nd Annual Meeting of the American Society for Information Science, Washington, DC, 30 October–2 November 1989, Volume 26*, pp. 81–95, Medford, NJ: Learned Information.

Borgman, C. L., Gallagher, A. L, Krieger, D. and Bower, J. (1990), 'Children's use of an interactive catalog of science materials', in Henderson, D. (ed.), *ASIS '90: Information in the Year 2000: from Research to Applications: Proceedings of the 53rd Annual Meeting of the American Society for Information Science, Toronto, Canada, 4–8 November 1990, Vol. 2*, pp. 55–68, Medford, NJ: Learned Information.

Bosch, V. M. and Hancock-Beaulieu, M. (1995), 'CDROM user interface evaluation: the appropriateness of GUIs', *Online and CDROM Review*, **19**(5), 255–70.

Bower, G. H. and Reitman, J. S. (1972), 'Mnemonic elaboration in multilist learning', *Journal of Verbal Learning and Verbal Behaviour*, **253**, 478–85.

Boyce, P. R. (1981), *Human Factors in Lighting*, London: Applied Science Publishers.

Boyle, H. M. (1986), *VDUs and Health*, A letter from Mr H. M. Boyle, British Embassy, Paris to Mr C. Barrow, Alvey Directorate, 9 April.

Braidwood, P. (1988), 'VDUs: Are they safe in pregnancy?' *Observer*, 7 August, 35.

Bramwell, R. S. and Davidson, M. J. (1993), 'Visual display units and pregnancy outcome: a prospective study', *Journal of Psychosomatic Obstetrics and Gynaecology*, **14**, 197–210.

Brand, J. L. and Judd K. W. (1993), 'Angle of hard copy and text editing performance', *Human Factors*, **35**, 57–69.

Brandt, L. P. A. and Nielsen, C. V. (1990), 'Congenital malformations among children of women working with video display terminals', *Environmental Health*, **16**, 329–33.

Breeding, M. (1994), *PC-based Integrated Library Systems: Essential Guide to the Library IBM PC – Vol. 13*, Westport, CT., London: Meckler.

Brent, R. L., Gordon, W. E., Bennett, W. R. and Beckman, D. A. (1993), 'Reproductive and teratologic effects of electromagnetic fields', *Reproductive Toxicology Review*, **7**, 535–90.

Bridger, R. S. (1995), *Introduction to Ergonomics*, London: McGraw-Hill.

British Standards Institution (1993), *British Standard EN 29241-1*, London: BSI.

Brown, I. D. (1965), 'A comparison of two subsidiary tasks used to measure fatigue in car drivers', *Ergonomics*, **8**, 467–73.

Brown, I. D. and Poulton, E. C. (1961), 'Measuring the "spare mental capacity" of car drivers by a subsidiary task', *Ergonomics*, **5**, 35–40.

Bryant, H. E. and Love, E. J. (1989), 'Video display terminal use and spontaneous abortion risk', *International Journal Epidemiology*, **18**, 132–8.

BS 5266 (1988), *Emergency Lighting, Part 1: Code of Practice for Emergency Lighting of Premises Other Than Cinemas and Certain Other Specified Premises Used for Entertainment*, London: BSI.

BS 7179: Part 4 (1990), *Ergonomics of Design and Use of Visual Display Terminals (VDTs) in Offices, Part 4: Specification for Keyboards*, London: BSI.

BS 7179: Part 5 (1990), *Ergonomics of Design and Use of Visual Display Terminals (VDTs) in Offices, Part 5: Specification for VDT Workstations*, London: BSI.

BS 7179: Part 6 (1990), *Ergonomics of Design and Use of Visual Display Terminals (VDTs) in Offices, Part 6: Code of Practice for the Design of VDT Work Environments,* London: BSI.

BS 8206 (1985), *Lighting for Buildings, Part 1: Code of Practice for Artificial Lighting*, London, BSI.

BUBL (1997), URL: http://bubl.ac.uk/archive/lis/systems/.

Bunge, C. A. (1991), 'CD-ROM stress', *Library Journal*, April 15, 63–4.

Burton, P. F. and Petrie, J. H. (1986), *The Librarian's Guide to Microcomputers for Information Management*, Wokingham: Van Nostrand Reinhold (UK).

Burton, P. (1987), 'Microcomputers as innovation: policies for implementation in libraries and information services', *Electronic Library*, **5**(4), 210–20.

Buxton, A. and Trenner, L. (1987), 'An experiment to assess the friendliness of error messages from interactive information retrieval systems', *Journal of Information Science*, **13**(4), 197–209.

Cakir, A. E. (1995), 'Do split keyboards help to reduce strain?', in Greico, A., Molteni, G., Occhipinto, E. and Piccoli, B. (eds), *Proceedings of 'Work with display units 94'*, pp. 283–8, Oxford: Elsevier.

Cakir, A., Hart, D. J. and Stewart, T. F. M. (1980), *Visual Display Terminals: A Manual Covering Ergonomics, Workspace Design, Health and Safety, Task Organisation*, Chichester: Wiley.

Caputo, J. S. (1991), *Stress and Burnout in Library Service*, Phoenix, AZ: Oryx Press.

Card, S. K. (1982), 'User perception mechanisms in the search of computer command menus', in *Proceedings of the Conference on Human Factors in Computer Systems, Gaithersberg, MD, 15–17 March 1982*, pp. 190–6, Washington, DC: ACM.

Card, S. K. and Moran, T. P. (1980), 'The keystroke-level model for user performance time with interactive systems', *Communications of the ACM*, 205–20.

Carroll, D. (1996), *Alternative/ergonomic Pointing Devices*, http://www.mv/com/users/carroll/mouse.htm.

Carter, J. B. and Banister, E. W. (1994), 'Musculoskeletal problems in VDT work: a review', *Ergonomics*, **37**(10), 1623–48.

Casale, M. (1996), 'Are you sitting comfortably?', *Library Manager*, February, 1996, 26–7.

Chapanis, A. (1960), 'Human engineering', in Flagle, C. D., Higgins, W. H. and Roy, R. N. (eds), *Operations Research and Systems Engineering*, pp. 534–82, Baltimore, MD: Johns Hopkins University Press.

Chartered Institution of Building Services Engineers (CIBSE) (1989), *CIBSE Lighting Guide: Areas for Visual Display Terminals*, London: CIBSE Publications Department.

Chartered Institution of Building Services (CIBS) (1982), *CIBS Lighting Guide for Libraries*, London: CIBS Publications Department.

Cherry, J. M. and Clinton, M. (1989), 'A profile of user background and user satisfaction with the University of Toronto OPAC and the implications for user training and user interfaces', in Katzer, J. and Newby, G. B (eds), *ASIS '89: Proceedings of the 52nd Annual Meeting of the American Society for Information Science, Washington, DC, 30 October–2 November 1989, Vol. 26*, pp. 121–8, Medford, NJ: Learned Information Inc.

Cheung, S. T. Y. (1993), 'A study of the office automation training methodologies adopted by Hong Kong companies', *Journal of Information Science*, **19**(4), 267–78.

Chin, J. P, Diehl, V. A. and Norman, K. L. (1988), 'Development of an instrument measuring user satisfaction of the human–computer interface', in *CHI '88. Proceedings of the conference on Human Factors in Computing Systems*, pp. 213–18, ACM, Reading, MA: Addison-Wesley.

Christ, R. E. (1975), 'Review and analysis of colour coding research for visual displays', *Human Factors*, **17**, 542–70.

Christie, B. and Gardiner, M. M. (1990), 'Evaluation of the human–computer interface', in Wilson, R. J. and Corlett, E. N. (eds), *Evaluation of Human Work: A Practical Ergonomics Methodology*, pp. 271–320, London: Taylor & Francis.

CIE 60 (1984), *Vision and The Visual Display Unit Work Station*, Vienna: Commission International de l'Eclairage.
Civil and Public Services Association Research Department (1983), *New Technology Bulletin*, September.
Clarke, J. (1995), 'Concurrent session III – professional advice on handling change. Understanding transition: the people side of managing change', *The Serials Librarian*, **25**(3/4), 193–202.
Clarke, J. E. (1985), 'The human/IT interface: a librarian's viewpoint', *Library and Information Research News*, **8**(30), 5–9.
Clayton, M. (1987), *Managing Library Automation*, Aldershot: Gower.
Clayton, M. and Batt, C. (1992), *Managing Library Automation*, 2nd edn, Aldershot: Ashgate.
Cleeve, M. (1995), 'The library is for turning: human resource management and re-engineering', *Library Management*, **16**(4), 37–41.
Cluff, E. (1985), 'Examining the health hazards of the VDU', *Computing: The Magazine*, 27 June, 30.
Cole, I., Lansdale, M. and Christie, B. (1985), 'Dialogue design guidelines', in Christie, B. (ed.), *Human Factors of Information Technology in the Office*, pp. 212–42, Chichester: John Wiley.
Coleman, V. (1978), *Stress Control*, London: Pan.
Commings, K. (1991), 'Ongoing training for integrated library automation systems', *Computers in Libraries*, **11**(10), 42–3.
Concise Oxford Dictionary (1990), 8th edn, Oxford: Clarendon Press.
Connor, C. M. (1992), 'Staff training in libraries: the implications of automation', *Library Management*, **13**(6), 15–24.
Conrad, R. (1964), 'Acoustic comparisons in immediate memory', *British Journal of Psychology*, **55**, 75–84.
Corbin, J. (1985), *Managing the Library Automation Project*, Phoenix, AZ: Oryx Press.
Cox, T. (1993), *Stress Research and Stress Management: Putting Theory to Work, HSE Contract Research Report No. 61/1993*, Sheffield: HSE Books.
Croner's Health and Safety at Work Bulletin (1993), 'Work related upper limb disorders', 6.
Cross, R. and Bawden, D. (1987), 'Information technology: human and organizational factors', *Journal of Information Science*, **13**, 277–84.
Czaja, S. J., Cary, J. M., Drury, C. G. and Cohen, B. G. (1987), 'An ergonomic evaluation of traditional and automated office environments', *Office Technology and People*, **3**(3), 231–46.
Dainoff, M. J. (1986), *People and Productivity: A Manager's Guide to Ergonomics in the Electronic Office*, Toronto: Holt, Rinehart & Winston.
Dakshinamurti, G. (1985), 'Automation's effect on library personnel', *Canadian Library Journal*, **42**(6), 343–51.
Damodaran, L., Simpson, A. and Wilson, P. (1980), *Designing Systems for People*, Manchester: NCC Publications.
Daniels, R. J. (1995), 'Effects on non-professional staff of the implementation of

computer-based library systems in college libraries: some case studies', *Program*, **29**(1), 1–13.

Das, B. and Sengupta, A. K. (1996), 'Industrial workstation design: a systematic ergonomic approach', *Applied Ergonomics*, **27**(3), 157–63.

Davis, C. H. and Shaw, D. (1989), 'Comparison of retrieval system interfaces using an objective measure of screen design effectiveness', *Library and Information Science Research*, **11**(4), 325–34.

De Jong, J. R. and Roggeveen, C. (1969), 'The sitting employee and the standing customer', *Ergonomics*, **12**(2), 348–55.

Dean, S. C. (1987), *Durham County Library: Annual report, 1986–87*, Durham: Durham County Library.

Deese, J. (1964), *Principles of Psychology*, Boston, MA: Allyn & Bacon.

Dickson, J. (1984), 'An analysis of user errors in searching an online catalog', *Cataloguing and Classification Quarterly*, **4**(3), 1–17.

Dillon, A. (1992), 'Reading from paper versus screens: a critical review of the empirical literature', *Ergonomics: 3rd special issue on Cognitive Ergonomics*, **35**(10), 1297–326.

Dillon, A. (1994), *Designing Usable Electronic Text*, London: Taylor & Francis.

Dillon, A., Richardson, J. and McKnight, C. (1990), 'Navigation in hypertext: a critical review of the concept', in Diaper, D., Gilmore, D., Cockton, G. and Shackel, B. (eds), *INTERACT '90*, pp. 587–92, North Holland: Amsterdam.

Döckel, H. (1992), 'Managing the impact of automation on library personnel', *Mousaion*, **10**(2), 83–92.

Doerfel, H. R., Graffunder, H. and Piesch, E. (1986), 'Measurement of the effective dose equivalent due to low energy x-rays emitted from video display terminals and colour television sets', *Radiation Protection Dosimetry*, **14**(2), 123–6.

Dressel, D. L. and Francis, J. (1987), 'Office productivity: contributions of the workstation', *Behaviour and Information Technology*, **6**(3), 279–84.

DTI (Department of Trade and Industry) (1990), *A Guide to Usability*, Milton Keynes: Open University.

Dul, J. and Weerdmeester, B. (1993), *Ergonomics for Beginners: A Quick Reference Guide*, London: Taylor & Francis.

Drake, M. A. (1993), Technological innovation and organizational charge, *Journal of Library Administration*, **19**(3–4), 39–53.

Dyer, H., Fossey, D. and McKee, K. (1993), 'The impact of automated library systems on job design and staffing structures', *Program*, **27**(1), 1–16.

Dyson, J. (1986), 'Training for specific skills – automation', in Prytherch, R. (ed.), *Handbook of Library Training Practice*, pp. 277–321, Aldershot: Gower.

Edholm, O. G. (1967), *The Biology of Work*, London: Weidenfeld & Nicolson.

Egan, D. E., Remde, J. R., Landauer, T. K., Lochbaum, C. C. and Gomez, L. M. (1989), 'Behavioural evaluation and analysis of a hypertext browser', in *CHI '89. Proceedings of the conference on Human Factors in Computing Systems*, pp. 93–8, ACM.

Else, L. (1985), 'VDUs: a risk to health?', *Computing: The Newspaper*, 21 March, 25.

Epple, M., Gardner, J. and Warwick, R. T. (1992), 'Staff training and automated systems: 20 tips for success', *The Journal of Academic Librarianship*, **18**(2), 87–9.

Ericson, A. and Kallén, B. (1986a), 'An epidemiological study of work with video screens and pregnancy outcome: I. A registry study', *American Journal of Industrial Medicine*, **9**, 447–57.

Ericson, A. and Kallén, B. (1986b), 'An epidemiological study of work with video screens and pregnancy outcome: II. A case-control study', *American Journal of Industrial Medicine*, **9**, 459–75.

EU (European Union) (1989), 'Council Directive of 12 June 1989 on the introduction of measures to encourage improvements in the safety and health of workers at work 89/391/ECC', *Official Journal of the European Communities*, L183, 32, 29 June 1989, 1–8.

Falzon, P. (1990), 'Human–computer interaction: lessons from human–human communication', in Falzon, P. (ed.), *Cognitive Ergonomics*, pp. 51–66, London: Academic Press.

Fanger, P. O. (1970), *Thermal Comfort*, Copenhagen: Danish Technical Press.

Fayen, E. G. (1983), *The Online Catalog: Improving Public Access to Library Materials*, White Plains, NY: Knowledge Industry Publications.

Fernandez, J. E., Cihangirli, M., Hommertzheim, D. L and Sabuncuoglu, I. (1988), 'The effects of input devices on task performance', in Aghazadeh, F. (ed.), *Trends in Ergonomics/Human Factors V: Proceedings of Annual International Industrial Ergonomics and Safety Conference, New Orleans, Louisiana 8–10 June 1988*, pp. 83–9, Oxford: Elsevier.

Ferris, A. (1987), 'Better by design', *Systems International*, December, 27, 28, 30, 32.

Fine, S. (1982), 'Human factors and human consequences: opening commentary', in Kent, A. and Galvin, T. J. (eds), *Information Technology: Critical Choices for Library Decision Makers*, pp. 209–24, New York: Marcel Dekker.

Fine, S. F. (1986), 'Technological innovation, diffusion and resistance: an historical perspective', *Journal of Library Administration*, **7**(1), 83–108.

Finer, R. (1982), 'The human side of change' in *CLAIM Report no. 14*, Centre for Library and Information Management, pp. 31–49, Loughborough: Loughborough University.

Finnegan, M. J., Pickering, C. A. C. and Burge, P. S. (1984), 'The sick building syndrome: prevalence studies', *British Medical Journal*, **8**, December, 1573–5.

Fleischer, A. G., Rademacher, U. and Windberg, H. J. (1987), 'Individual characteristics of sitting behaviour', *Ergonomics*, **30**(4), 703–9.

Foster, J. and Coles, P. (1977), 'An experimental study of typographic cueing in printed text', *Ergonomics*, **20**, 57–66.

Fox, J. G. (1983), 'Industrial music', in Oborne, D. J. and Gruneberg, M. M. (eds), *The Physical Environment at Work*, pp. 211–26, Chichester: John Wiley.

Gage, N. and Berliner, D. (1979), *Educational Psychology*, Chicago, IL: Rand-McNally.

Galer, I. (1987), *Ergonomics Handbook*, 2nd edn, Sevenoaks: Butterworths.
Galitz, W. O. (1984), *Humanising Office Automation: the Impact of Ergonomics on Productivity*, Wellesley, MA: QED Information Sciences.
Galitz, W. O. (1989), *Handbook of Screen Format Design*, 3rd edn, Wellesley, MA: QED Information Sciences.
Geyser, E. P. and van Brakel, P. A. (1991), 'Man–machine interaction as a factor in the design of computerized information retrieval systems', *South African Journal of Library and Information Science*, **59**(4), 256–60.
German, M. P., Nowak, E. L. and Wilke, J. S. (1987), 'Form design simplifies online search procedures', *Information Technology and Libraries*, **6**(4), 313–16.
Gibson, J. E. (1965), 'Adaptive and Learning Control Systems', in Machol, R. E. (ed.), *System Engineering Handbook*, pp. 30–1 to 30–20, New York: McGraw-Hill.
Gilad, I. and Pollatschek, M. A. (1986), 'Layout simulation for keyboards', *Behaviour and Information Technology*, **5**(3), 273–81.
Gittings, D. (1986), *Query Language Systems*, London: Edward Arnold.
Goldhaber, M. K., Polen, M. R. and Hiatt, R. A. (1988), 'The risk of miscarriage and birth defects among women who use visual display terminals during pregnancy', *American Journal of Industrial Medicine*, **13**, 695–706.
Gould, J. D. and Boies, S. J. (1978), 'Writing, dictating and speaking letters', *Science*, **201**(22), September, 1145–7.
Goulding, A. (1996), *Managing Change for Library Support Staff*, Aldershot: Avebury.
Graham, K. R. (1987), 'The library assistant', in Revill, D. (ed.), *Personnel Management in Polytechnic Libraries*, pp. 142–54, Aldershot: Gower.
Grandjean, E. (1980), *Fitting the Task to the Man: An Ergonomic Approach*, London: Taylor & Francis.
Grandjean, E. (1987), *Ergonomics in Computerized Offices*, London: Taylor & Francis.
Grandjean, E. (1988), *Fitting the Task to the Man*, 4th edn, London: Taylor & Francis.
Gratton, P. D. (1983), *Automation in Derbyshire County Libraries*, London: Library Association.
Green, P. and Wei-Haas, L. (1985), 'The rapid development of user interfaces: experience with the Wizard of Oz Method', in *Proceedings of the Human Factors Society 29th Annual Meeting*, pp. 470–4, Santa Monica, CA: Human Factors Society.
Grey, S. M. and Wilson, J. R. (1984), *The Ergonomics of Library Issue Desks: Report to the British Library Research and Development Department on Project No SI/G/563*, London: British Library.
Grossman, W. (1991), 'Not just a pretty face', *Personal Computer World*, July, 188–91.
Grudin, J. (1989), 'The case against user interface consistency', *Communications of the ACM*, **32**(10), October, 1164–73.
Hagberg, M. (1995), 'The "mouse-arm Syndrome" – concurrence of musculo-skeletal symptoms and possible pathogenesis among VDU operators', in Greico,

A., Molteni, G., Occhipinto, E. and Piccoli, B. (eds), *Proceedings of 'Work with Display Units 94'*, pp. 381–6, Oxford: Elsevier.

Halasz, F. and Moran, T. P. (1982), 'Analogy considered harmful' in *Proceedings of the Conference on Human Factors in Computer Systems, Gaithersburg, MD, 15–17 March 1982, Washington DC*, pp. 383–5, Washington, DC: Association for Computing Machinery.

Hallmark, J. and Garcia, C. R. (1992), 'System migration: experiences from the field', *Information Technology and Libraries*, **11**(4), 345–58.

Halpern, C. A. and Davis, P. J. (1993), 'An evaluation of workstation adjustment and musculoskeletal discomfort: designing for diversity', in *Proceedings of the Human Factors and Ergonomics Society 37th Annual Meeting, Seattle, Washington, October 11–15, 1993, Vol. 2*, pp. 817–21, Santa Monica, CA: Human Factors and Ergonomics Society.

Hamdy, W. R. (1987), 'The Impact of Visual Display Units on Vision', Master's dissertation, Loughborough: Loughborough University.

Hammond, N., Gardiner, M. M., Christie, B. and Marshall, C. (1987), 'The role of cognitive psychology in user–interface design', in Gardiner, M. M. and Christie, B. (eds), *Applying Cognitive Psychology to User Interface Design*, pp. 13–52, Chichester: John Wiley.

Hancock-Beaulieu, M. (1992), 'User friendliness and human–computer interaction in online library catalogues', *Program*, **26**(1), 29–37.

Handy, C. B. (1981), *Understanding Organizations*, 2nd edn, Harmondsworth: Penguin.

Harris, G. (1990), 'The user interface to online information', in Raitt, D. I. (ed.), *Online Information 90: Proceedings of the 14th International Online Information Meeting, London, 11–13 December 1990*, pp. 319–24, Oxford and New Jersey: Learned Information (Europe).

Harley, B. (1995), 'Planning for sustainable automation', *Information Technology and Libraries*, **14**(3), 176–9.

Hartshorn, H. R. and Hix, D. (1989), 'Human–computer interface development: concepts and systems for its management', *ACM Computing Surveys*, **21**(1), March, 5–92.

Harvey, F. (1993), 'The European Community supplies directives: a guide for local authorities', Stamford: Chartered Institute of Processing and Supply.

Hedge, A. (1987), 'Office health hazards: an annotated bibliography', *Ergonomics*, **30**(5), 733–72.

Hedge, A., Burge, P. S., Robertson, A. S., Wilson, S. and Harrisbass, J. (1989), 'Work-related illness in offices: a proposed model of the "sick building syndrome"', *Environment International*, **15**, 143–58.

Henning, R. A., Alteras-Webb, S. M., Jacques, P., Kissel, G. V. and Sullivan, A. B. (1993), 'Frequent short breaks during computer work: the effects on productivity and well-being in a field study', in Luczak, H., Cakir, A. and Cakir, G. (eds), *Work With Display Units '92*, Oxford: Elsevier.

Henry, H. K. (1991), 'Human–computer interfaces and OPACs: introductory thoughts related to INNOPAC', *Library Hi-Tech*, **9**(2), 63–8.

Hepworth, J. B. (1990), 'The training implications of library automation: approaches to staff and end-user training in health libraries', *Health Libraries Review*, **7**(4), 210–3.

Herring, J. E. and Mackenzie, J. A. (1986), *Planning for Library Automation: Aberdeen City Libraries*, London: Library Association.

Herzberg, F., Mausner, B. and Snyderman, B. (1959), *The Motivation to Work*, New York: John Wiley.

Heseltine, R. (1993), 'Choosing in the dark: strategic issues in the selection of library automation systems', *Its News*, **27**, 13–8.

Hicks, R. and Essinger, J. (1991), *Making computers more human – Designing for human–computer interaction*, Oxford: Elsevier.

Hildreth, C. R. (1985), 'The user interface in online catalogues: the telling difference', in Kinsella, J. (ed.), *Online Public Access to Library Files, Conference Proceedings, University of Bath, 3–5 September 1984*, pp. 111–32, Oxford: Elsevier.

Hobday, S. (1985), 'Letting the shape of the keyboard take the strain', *Computing*, 31 October, 33.

Hodges, J. E. (1990), 'Stress in the library', *Library Association Record*, **92**(10), 751, 753–4.

Holmes, S. (1995), 'An Exploration of Potential Stressors Among Library Assistants', Master's dissertation, Loughborough: Loughborough University.

Holmes, T. H. and Rahe, H. (1967), 'The social readjustment rating scale', *Journal of Psychosomatic Research*, **11**, 213–18.

Holroyd, K. (1986), 'Job design and new technology – a trade union initiative', in *Job Design and VDU Work: A One Day Conference, 1 October 1986*, pp. 31–51, Quorn: Humane Technology.

Horberry, T. J., Gale, A. G. and Taylor, S. P. (1995), 'Vision screening for display screen users', in Grieco, A., Molteni, G., Piccoli, B. and Occhipinti, E. (eds), *Work With Diplay Units '94*, pp. 101–6, Oxford: Elsevier.

HSC (Health and Safety Commission) (1992), *A Guide to the Health and Safety at Work etc. Act 1974*, Sheffield: HSE Information Centre.

HSE (Health and Safety Executive) (1987), *Lighting at Work*, Sheffield: HSE Information Centre.

HSE (Health and Safety Executive) (1989), *Noise at Work: The Noise at Work Regulations 1989*, Sheffield: HSE Information Centre.

HSE (Health and Safety Executive) (1990a), *Work Related Upper Limb Disorders – a Guide to Prevention*, Sheffield: HSE Information Centre.

HSE (Health and Safety Executive) (1990b), *Noise at Work, Noise Assessment, Information and Control. Noise Guide 3: Equipment and procedures for noise surveys*, Sheffield: HSE Information Centre.

HSE (Health and Safety Executive) (1992a), *Management of Health and Safety at Work, Approved Code of Practice*, Sheffield: HSE Information Centre.

HSE (Health and Safety Executive) (1992b), *Workplace Health, Safety and Welfare, Approved Code of Practice*, Sheffield: HSE Information Centre.

HSE (Health and Safety Executive) (1992c), *Work Equipment, Guidance on Regulations*, Sheffield: HSE Information Centre.

HSE (Health and Safety Executive) (1992d), *Display Screen Equipment Work, Guidance on Regulations*, Sheffield: HSE Information Centre.

HSE (Health and Safety Executive) (1995), *Stress at Work – a Guide for Employers, HS (G)116*, Sheffield: HSE Information Centre.

Huff Arneson, R. (1993), 'Life on the bleeding edge', in Head, J. W. and McCabe, G. B. (eds), *Insider's Guide to Library Automation: Essays of Practical Experience*, pp. 115–25, Westport, CT: Greenwood Press.

Hughes, G. (1985), 'A large employer's approach to the use of VDUs', in *Alleged Health Hazards of Work at VDUs: Current Industrial Relations Issues: A Conference Held by Humane Technology in London on 26 and 27 November, 1985*, pp. 141–60, Quorn: Humane Technology.

Hulser, R. P. (1993), 'Overview of graphical user interfaces', *Computers in Libraries*, **13**(2), 18–23.

Hunter, I. M. L. (1964), *Memory*, London: Whitefriars Press.

Hunter, P. (1983), 'Vertical lights do the trick', *Computer Weekly*, 24 November, 20.

Hünting, W. (1984), 'Postural loads at VDT work-stations', in Pearce, B. G. (ed.), *Health Hazards of VDTs?*, pp. 119–36, Chichester: John Wiley.

Hutchins, E., Hollan, J. D. and Norman, D. A. (1985), 'Direct manipulation interfaces', in *Human–Computer Interaction, Vol. 1*, pp. 311–38, Lawrence Erlbaum Associates.

Iacono, S. and Kling, R. (1987), 'Changing office technologies and transformations of clerical jobs', in Robert, E. and Kraut, L. (eds), *Technology and the Transformation of White-collar Work*, Hillsdale, NJ: Erlbaum Associates.

Illingworth, V. (ed.) (1991), *Dictionary of Computing*, Oxford: Oxford University Press.

ISO 7730 (1984), *Moderate Thermal Environments – Determination of the PMV and PPD Indices and Specification of the Conditions for Thermal Comfort*, Geneva.

ISO 8995 (1989), *Principles of Visual Ergonomics – the Lighting of Indoor Work Systems*, Geneva.

James, R. (1987), 'Training and the management of automation', *Training and Education*, **4**(3), 53–7.

Jaschinski-Kruza, W. (1990), 'On the preferred viewing distances to screen and document at VDU workplaces', *Ergonomics*, **33**(8), 1055–63.

Jaschinski-Kruza, W. (1991), 'Eyestrain in VDU users: viewing distance and the resting position of ocular muscles', *Human Factors*, **33**(1), 69–83.

Jeffreys, S. (1984), 'VDUs: more than meets the eye', *Computer News*, 26 April, 22.

Johansson, A. and Shahnavaz, H. (1995), 'Comparison of VDT workplaces in Australia and Sweden', in Grieco, A., Molteni, G., Piccoli, B. and Occhipinti, E. (eds), *Work With Display Units '94*, pp. 21–6, Oxford: Elsevier.

Johnson, C. P. (1996), 'The changing nature of jobs: a paraprofessional time series', *College and Research Libraries*, **57**(1), 59–67.

Johnson, G. I., Clegg, C. W. and Ravden, S. J. (1989), 'Towards a practical method of user interface evaluation', *Applied Ergonomics*, **20**(4), 255–60.

Johnson, P. (1991), *Automation and Organizational Change in Libraries*, Boston, MA: G. K. Hall & Co.

Jones, D. E. (1989), 'Library support staff and technology: perceptions and opinions', *Library Trends*, **37**(4), 432–56.

Jones, W. P. (1988), '"As we might think"?: psychological considerations in the design of a personal filing system', in Guindon, R. (ed.), *Cognitive Science and its Applications for Human–computer Interaction*, pp. 235–88, Hillsdale, NJ: Erlbaum.

Jones, N. and Jordan, P. (1987), *Staff Management in Library and Information Work*, 2nd edn, Aldershot: Gower.

Kahn, H. and Cooper, C. L. (1986), 'Computing stress', *Current Psychological Research and Reviews*, Summer, 148–62.

Kavet, R. and Tell, R. A. (1991), 'VDTs: field levels, epidemiology, and laboratory studies, *Health Physiology*, **61**(1), 47–57.

Kayis, B. and Hoang, K. (1995), 'Computerised analysis of prolonged seated posture', in Anzai, Y., Ogawa, K. and Mori, H. (eds), *Symbiosis of Human and Artifact*, pp. 629–34, Oxford: Elsevier.

Kinney, T. and Pasak, P. (1990), 'The "naive subject searcher" model: a tool for evaluating library catalog user interfaces', in Genaway, D. C. (ed.), *IOLS '90. Integrated Online Library Systems: Proceedings of the 5th Integrated Online Library Systems Conference, New York, 2–3 May, 1990*, pp. 103–14, Medford, NJ: Learned Information.

Klatzky, R. L. (1980), *Human Memory: Structures and Processes*, San Francisco, CA: Freeman.

Klobas, J. E. (1990), 'Managing technological change in libraries and information services', *The Electronic Library*, **8**(5), 344–9.

Knave, B. G. (1983), 'The visual display unit', in *Ericsson Information Systems, Ergonomic Principles in Office Automation: State of the Art Reports and Guidelines on Human Factors in the Office Environment*, pp. 11–41, Ericsson Bromma, Sweden: Information Systems.

Kompier, M. and Levi, L. (1994), *Stress at Work: Causes, Effects and Prevention*, Dublin: European Foundation for the Improvement of Living and Working Conditions.

Kroemer, K. H. E. (1995), 'Alternative keyboards and alternatives to keyboards', in Greico, A., Molteni, G., Occhipinto, E. and Piccoli, B. (eds), *Proceedings of 'Work With Display Units 94'*, pp. 277–82, Oxford: Elsevier.

Kurppa, K., Holmberg, P. C., Rantala, K., Nurminen, T. and Saxen, L. (1985), 'Birth defects and exposure to video display terminals during pregnancy', *Environmental Health*, **11**, 353–6.

Kyllonen, P. C. and Alluisi, E. A. (1987), 'Learning and forgetting facts and skills', in Salvendy, G. (ed.), *Handbook of Human Factors*, pp. 45–71, New York: John Wiley.

Lefebvre, C. J. (1985), 'Training for Automation: A Study of the Introduction of Junior Staff to Automated Circulation Systems in Academic Libraries', Master's dissertation, Loughborough: Loughborough University.

Lewis, A. C. (1995), 'The use of PRINCE project management methodology in choosing a new library system at the University of Wales Bangor', *Program*,

29(3), 231–40.

Library and Information Technology Centre (1987), *State of the Art of the Application of New Information Technologies in Libraries and their Impact on Library Functions in the United Kingdom*, London: Library and Information Technology Centre.

Lim, S. and Carayon, P. (1995), 'Psycho-social work factors and upper extremity musculosketal discomfort among office workers', in Grieco, A., Molteni, G., Piccoli, B. and Occhipinti, E. (eds), *Work with Display Units '94*, pp. 57–62, Oxford: Elsevier.

Lindbohm, M-L, Hietanen, M., Kyronen, P., Sallmen, M., von Nandelstadh, P., Taskinen, H., Pekkarinen, M., Ylikoski, M. and Hemminki, K. (1992), 'Magnetic fields of video display terminals and spontaneous abortion', *American Journal of Epidemiology*, **136**, 1041–51.

LITC Report No. 1. (1992), *Planning and Implementing an Automated Library System*, London: Library Information Technology Centre.

LITC Report No. 2. (1992), *Guide to Choosing an Automated Library System*, London: Library Information Technology Centre.

LITC Report No. 3. (1992), *Evaluating Library Systems at a Demonstration*, London: Library Information Technology Centre.

London Hazards Centre (1988), *Repetition Strain Injuries: Hidden Harm from Over-use*, London: London Hazards Centre Trusts Limited.

Lovecy, I. (1984), *Automating Library Procedures: A Survivor's Handbook*, London: Library Association.

Lucas, H. C. (1974), *Toward Creative Systems Design*, Columbia, SC: Columbia University Press.

Lundin, L. (1991), *On Building-related Causes of the Sick Building Syndrome*, Stockholm: Almqvist and Wiksell.

McDonald, A. D., Cherry, N. M., Delorme, C. and McDonald, J. C. (1986), 'Visual display units and pregnancy: evidence from the Montreal survey', *Journal of Occupational Medicine*, **28**, 1226–31.

McDonald, A. D., McDonald, J. C., Armstrong, B., Cherry, N. M., Nolin, A. D. and Robert, D. (1988), 'Work with visual display units in pregnancy', *British Journal of Industrial Medicine*, **45**, 509–15.

McDonald, J. and Schvaneveldt, R. W. (1988), 'The application of user knowledge to interface design', in Guindon, R. (ed.), *Cognitive Science and its Applications for Human–computer Interaction*, pp. 289–335, Hillsdale, NJ: Erlbaum.

Mace, C. A. (1968), *The Psychology of Study*, 3rd edn, Harmondsworth: Penguin.

MacKay, C. (1985), 'Occupational upper limb disorders – causes and prevention', in *Alleged Health Hazards of Work at VDUs: Current Industrial Relation Issues: A Conference Held by Humane Technology in London on 26 and 27 November 1985*, pp. 77–88, Quorn: Humane Technology.

MacKay, C. (1987), 'The alleged reproductive hazards of VDUs and work and stress', *The Practitioner*, **1**(1), 49–57.

MacKay, C. and Cox, T. (1984), 'Occupational stress associated with visual display unit operation', in Pearce, B. G. (ed.), *Health Hazards of VDTs?*, pp. 137–43,

Chichester: John Wiley.

Maddox, M. D. (1989), 'Designing online system interfaces for non-technical people: or "Excuse the dumb question, but what the heck does 'online' mean?"', in Nixon, C. and Padgett, L. (eds), *National Online Meeting: Proceedings 1989, New York, 9–11 May 1989*, pp. 271–8, Medford, NJ: Learned Information Inc.

Malone, T. W. and Lepper, M. (1983), 'Motivation, cognition, and computerized instruction', Paper Presented at the Office of Naval Research Conference on Aptitude, Learning and Instruction: Cognitive and Affective Process Analysis, Stamford, California.

Mandal, A. C. (1986), 'Investigation of the lumbar flexion of office workers', in Corlett, N., Wilson, J. and Manenica, M. (eds), *The Ergonomics of Working Postures: Models, Methods and Cases: The Proceedings of the First International Occupational Ergonomics Symposium, Zadar, Yugoslavia, 15–17 April 1985*, pp. 345–54, London: Taylor & Francis.

Mann, P. (1984), 'Under pressure: how can libraries survive and expand?', *Library Association Record*, **86**(4), 167.

Marchant, M. P. (1982), 'Participative management, job satisfaction, and service', *Library Journal*, **107**(8), 782–4.

Marchionini, G. (1989), 'Making the transition from print to electronic encyclopedias: adaptations of mental models', *International Journal of Man–Machine Studies*, **30**, 591–618.

Marchionini, G. (1992), 'Interfaces for end-user information seeking', *Journal of the American Society for Information Science*, **43**(2), 156–63.

Marek, T., Noworol, C., Gedliczka, A. and Matuszek, L. (1984), 'A study of a modified VDT-stand arrangement', *Behaviour and Information Technology*, **3**(4), 405–9.

Marshall, C., Nelson, C. and Gardiner, M. M. (1987), 'Design guidelines', in Gardiner, M. M, Christie, B. (eds), *Applying Cognitive Psychology to User–Interface Design*, pp. 221–78, Chichester: John Wiley.

Maslow, A. H. (1943), 'A theory of human motivation', *Psychological Review*, **50**, 370–96.

Matthews, J. (1995), 'Moving to the next generation: Aston University's selection and implementation of Galaxy 2000', *Vine*, **101**, December, 42–9.

Matthews, J. R. (1980), *Choosing an Automated Library System: A Planning Guide*, Chicago, IL: American Library Association.

Matthews, J. R. (1987), 'Suggested guidelines for screen layouts and design of online catalogs', *Library Trends*, **35**(4), 3–61.

Meichenbaum, D. (1977), *Cognitive Behaviour Modification: An Integrative Approach*, New York: Plenum.

Mild, K. H. (1986), 'Biological effects of low frequency electromagnetic fields', in *Allegations of Reproductive Hazards from VDUs: The Second International Scientific Meeting, London, 25 and 26 November 1986*, pp. 135–40, Quorn: Humane Technology.

Miller, G. A. (1956), 'The magical number seven plus or minus two: some limits on our capacity for processing information', *Psychological Review*, **63**, 81–97.

Milne, J. R. (1994), 'Research reports: hypertext and its implications for library services', *Library and Information Research News*, **60**, 24–9.

Morris, A. (1992), 'Working with display screen equipment – do you know your rights?', *The Electronic Library*, **10**(4), 209–18.

Morris, A. and Dennison, P. (1995), 'Sick building syndrome: survey findings of libraries in Great Britain', *Library Management*, **16**(3), 34–42.

Msuya, J. (1990), 'Planning the automation of libraries in developing countries: a systems analysis approach', *Library Review*, **39**(3), 29–32.

Munrush, J. L. (1990), 'Options in learning: instructor led and computer based training (CBT)', *Journal of Library Administration*, **2**(2), 47–55.

Murrell, K. F. H. (1971), *Ergonomics: Man in His Working Environment*, London: Chapman & Hall.

Nadler, D. A. (1980), *Concepts for the Management of Organizational Change*, London: O. R. & C. Consultants.

Naqvi, S. A. A. (1994), 'Study of forward sloping seats for VDT workstations', *Journal of Human Ergology*, **23**(1), 41–9.

National Electronics Council (1983), *Human Factors and Information Technology*, London: National Electronics Council.

National Radiological Protection Board (1994), *Health Effects Related to the Use of Visual Display Units, Vol. 5, No. 2*, Oxford: National Radiological Protection Board.

Newmark, M. E. and Penry, J. K. (1979), *Photosensitive Epilepsy: A Review*, New York: Raven Press.

Niedzwetzki, J. (1991), 'Colour and lighting in public libraries', *Australasian Public Libraries and Information Services*, **4**(2), 103–5.

Nielsen, C. V. and Brandt, L. P. A. (1990), 'Spontaneous abortion among women using video display terminals', *Scandinavian Journal of Work Environment and Health*, **16**, 323–8.

Norlin, D. A., Cardman, E. B., Davis, E. B., Dossett, R., Henigman, B., Mischo, W. H. and Troutman, L. (1992), 'Interface design and development: the human factor', *Library Hi-Tech*, **10**(3), 7–24.

Norman, D. A. (1988), *The Psychology of Everyday Things*, New York: Basic Books.

Norman, D. A. and Fisher, D. (1982), 'Why alphabetic keyboards are not easy to use: Keyboard layout doesn't much matter', *Human Factors*, **24**(5), 509–19.

Norman, K. L. and Chin, J. P. (1989), 'The menu metaphor: food for thought', *Behaviour and Information Technology*, **8**(2), 125–34.

Nova Solutions (1993), *VDT Workstation Handbook*, Effingham, IL: Nova Office Furniture.

Novak, G. (1987), 'The forgiving building: a library building consultants' symposium on the design, construction and remodelling of libraries to support a high-tech future', *Library Hi-Tech*, **5**(4), 77–99.

Nurminen, T. and Kurppa, K. (1988), 'Office employment, work with video display terminals, and course of pregnancy', *Environmental Health*, **14**, 293–8.

Oborne, D. J. (1985), *Computers at Work: A Behavioural Approach*, Chichester: John Wiley.

Oborne, D. J. (1995), *Ergonomics at Work*, 3rd edn, Chichester: John Wiley.

Occupational Safety and Health (1992), *Occupational Overuse Syndrome, Treatment and Rehabilitation: a Practitioner's Guide*, Wellington, NZ: Department of Labour.

Oder, N. (1994), 'Multimedia "Crucible" from Penguin and Voyager', *Publishers Weekly*, **241**, 5 September, 25.

Olsgaard, J. N. (1989), 'The physiological and managerial impact of automation on libraries', *Library Trends*, **37**(4), 484–94.

Ong, C. N., Thein, M. M. and Berquist, U. (1990), 'A review of adverse effects on reproduction amongst female computer terminal workers', *Annals Academy of Medicine*, **19**(5), 649–55.

Orna, L. (1992), 'Information design and information services: what information professionals should know about design, in order to deliver value-added information products', *Aslib Proceedings*, **44**(9), 305–8.

Osaka, N. (1985), 'The effect of VDU colour on visual fatigue in the fovea and periphery of the visual field', *Displays, Technology and Applications*, **6**(3), 138–40.

OSHA (Occupational Safety and Health Administration) (1981), 'Occupational noise exposure; hearing conservation amendment', *Federal Register*, **46**, 4078–179.

OSHA (Occupational Safety and Health Administration) (1983), 'Occupational noise exposure; hearing conservation amendment; final rule', *Federal Register*, **48**, 9738–83.

Pachent, G. (1996), 'Network '95: Choosing a third generation automated information system for Suffolk Libraries and Heritage', *Program*, **30**(3), 213–28.

Parker, S. (1996), 'Members' day: a nation connected', Flyer for a meeting organized by the Library Association, 23 October, 1996, Brunei Gallery, University of London.

Parsons, K. (1990), 'Human response to thermal environments: principles and methods', in Wilson, J. R. and Corlett, E. N. (eds), *Evaluation of Human Work: a Practical Ergonomics Methodology*, London: Taylor & Francis.

Paznik, M. J. (1986), 'Ergonomics does pay', *Administrative Management*, **47**(8), 16–24.

Pearce, B. (1986), 'Introduction' in *Job Design and VDU Work: A One Day Conference, 1 October 1986*, pp. 1–3, Quorn: Humane Technology.

Peterson, L. R. and Peterson, M. J. (1959), 'Short-term retention of individual verbal items', *Journal of Experimental Psychology*, **58**, 193–8.

Pheasant, S. (1986), *Bodyspace: Anthropometry, Ergonomics and Design*, London: Taylor & Francis.

Pheasant, S. (1991), *Ergonomics, Work and Health*, London: Macmillan.

Polson, P. (1988), 'The consequences of consistent and inconsistent user interfaces', in Guindon, R. (ed.), *Cognitive Science and Its Applications for Human–computer Interaction*, pp. 59–106, Hillsdale, NJ: Lawrence Erlbaum.

Puttapithakporn, S. (1990), 'Interface design and user problems and errors: a case study of novice searchers', *RQ*, **30**(2), Winter, 195–204.

Quinn, J. B. (1980), *Strategies for Change: Logical Incrementalism*, Homewood, IL: Richard D. Irwin.

Quinlan, T. (1987), 'The management of change', in Work Research Unit (ed.), *Information Technology – Its Profitable Introduction to Business*, pp. 34–41, London: ACAS.

Ralli, T. (1988), Personal communication with the authors, University of Western Australia.

Rantanen, J. (1991), 'User interface management systems and prototyping tools: a state of the art survey and feasibility assessment', in Hammond, J. H., Hall, R. R. and Kaplan, I. (eds), *People before Technology: Proceedings of OZCHI 91, the CHISIG Annual Conference, Sydney, Australia, 28–29 November, 1991, CHISIG*, pp. 39–46, Downer, ACT: Ergonomics Society of Australia.

Ravden, S. J. (1989), *Evaluating Usability of Human–computer Interfaces: A Practical Method*, Hemel Hempstead: Ellis Horwood.

Rea, G. (1987), 'Designing an issue desk', in Rea, G. and Storey, C. (eds), *Ergonomics in Libraries*, Library Association, University Colleges and Research Section South Western Group, Discussion Papers No. 2, Bath.

Reeves, P. E. (1984), *The Challenge of Major Reorganisation: Computers in Essex County Libraries*, London: Library Association.

Rennie, A. (1988), 'The human face of design', *Systems International*, August, 41–2.

Resnik, R. A. (1992), 'Designing an end-user menu interface', in Williams, M. (ed), *Proceedings of the 13th National Online Meeting 1992*, pp. 293–300, Medford, NJ: Learned Information.

Rice, R. E. and Borgman, C. L. (1983), 'The use of computer-monitored data in information science and communications research', *Journal of the American Society for Information Science*, **34**(4), 247–56.

Robertson, M. M. (1992), 'Ergonomic considerations for the human environment', *School Library Media Quarterly*, **20**(4), 211–15.

Robertson, P. J. (1989), *A Guide to Using Color on Alphanumeric Displays: Technical Report*, Hursley Wood, UK: IBM.

Roman, E., Beral, V., Pelerin, M. and Hermon, C. (1992), 'Spontaneous abortion and work with visual display units', *British Journal of Industrial Medicine*, **49**, 507–12.

Rooks, D. C. (1988), *Motivating Today's Library Staff: A Management Guide*, Phoenix, AZ: Oryx Press.

Rowley, J. E. (1990), 'Guidelines on the evaluation and selection of library software packages', *Aslib Proceedings*, **42**(9), 225–35.

RSI Association (1993), *Information Sheet No 3: RSI and the Law*, London: RSI Association.

Rycroft, R. J. G. and Calnan, C. D. (1984), 'Facial rashes among visual display unit (VDU) operators', in Pearce, B. (ed.), *Health Hazards of VDTs?*, pp. 13–15, Chichester: John Wiley.

Sake, J. G. (1996), 'The blind men and the elephant', in Felleman, H. (ed.), *Best Loved Poems of the American people*, US: Doubleday, URL:http://www.wave.net/upg/tqm/blind1.html

Sanders, M. S. and McCormick, E. J. (1987), *Human Factors in Engineering and Design*, 6th edn, London: McGraw-Hill.

Saunders, K. (1993), 'What do people want from a library?', *Library Management*, **14**(2), 8–10.

Saunders, L. M. and Lee Kwon, M. (1991), 'The management of change: minimizing the negative impact on staff and patrons', in Pitkin, G. M. (ed.), *Library Systems Migration: Changing Automated Systems in Libraries and Information Centers*, pp. 69–85, London: Meckler.

Scalet, E. A. (1987), *VDT Health and Safety: Issues and Solutions*, Lawrence, KS: Ergosyst Associates.

Schneider, W. (1982), *Automatic/control Processing Concepts and Their Implications for the Training of Skills – Final Report (Technical Report No. 8101)*, Champaign, IL: University of Illinois, Human Attention Research Laboratory.

Schneider, W. (1984), *Training High Performance Skills: Fallacies and Guidelines (Technical Report No. HARL-ONR-8301)*, Champaign, IL: University of Illinois, Human Attention Research Laboratory.

Schnorr, T. M., Grajewski, B. A., Hornung, R. W., Thun, M. J., Egeland, G. M., Murray, W. E., Conover, D. L. and Halperin, W. E. (1991), 'Video display terminals and the risk of spontaneous abortion', *New England Journal of Medicine*, **324**, 717–33.

Schrenk, L. P. (1969), 'Aiding the decision maker – a decision process model', *Ergonomics*, **12**(4), 543–58.

Sell, R. G. (1986), 'Job design – why bother? The experience of the work research unit', in *Job Design and VDU Work: A One Day Conference, 1 October 1986*, pp. 5–17, Quorn: Humane Technology.

Shackel, B. (ed.) (1984), *Applied Ergonomics Handbook*, London: Butterworth.

Shackel, B. (1989), Personal communication with the authors.

Shackel, B. (1991), 'Usability – context, framework, definition, design and evaluation', in Shackel, B. and Richardson, S. (eds), *Human Factors for Informatics Usability*, pp. 21–39, Cambridge: Cambridge University Press.

Shaw, D. (1986), 'Staff opinions in library automation planning: a case study', *Special Libraries*, **77**(3), 140–51.

Shaw, D. (1991), 'The human–computer interface for information retrieval', *Annual Review of Information Science and Technology (ARIST)*, **26**, 155–95.

Sheedy, J. E. (1993), 'How to eliminate vision symptoms – treat the eyes and fix the environment: a report from the VDT eye clinic', in Luczak, H., Cakir, A. and Cakir, G. (eds), *Work with Display Units '92*, pp. 258–62, London: Elsevier.

Shires, N. L. and Olszak, L. P. (1992), 'What our screens should look like: an introduction to effective OPAC screens', *RQ*, **31**(3), 357–69.

Shneiderman, B. (1991), 'A taxonomy and rule base for the selection of interaction styles', in Shackel, B. and Richardson, S. (eds), *Human Factors for Informatics Usability*, pp. 326–42, Cambridge: Cambridge University Press.

Shneiderman, B. (1992), *Designing the User Interface: Strategies for Effective Human–Computer Interaction*, 2nd edn, Reading, MA: Addison-Wesley.

Shoham, S. (1991), 'The ecology of public library reference rooms', *Public Library*

Quarterly, **11**(4), 43–51.
Sidgreaves, I. (1985), 'Decision-making', in Cowley, J. (ed.), *The Management of Polytechnic Libraries*, pp. 59–83, Aldershot: Gower, in association with COPOL.
SIGCAT (1992), 'CD-ROM consistent interface guidelines: a final report', *CD-ROM Librarian*, February, 18–29.
Slovak, A. J. M. and Trevers, C. (1988), 'Solving workplace problems associated with VDTs', *Applied Ergonomics*, **19**(2), 99–102.
Smith, N. M., Birch, N. E. and Marchant, M. P. (1984), 'Stress, distress, and burnout: a survey of public reference librarians', *Public Libraries*, **23**(3), 83–5.
Smith, S. L. and Mosier, J. N. (1986), *Guidelines for Designing User Interface Software*, Bedford, MA: Mitre Corporation.
Snow, B. in Caduceus column (1986), 'What jargon is really necessary when teaching (and learning) online skills?', *Online*, July, 235–42.
Somerville, A. N., Caren, L., Kaplan, I. and Reiman, D. J. (1986), 'The new science and engineering library at the University of Rochester: blending traditional with electronic', *Science and Technology Libraries*, **7**(2), 71–93.
Sperandio, J. (1987), 'Software ergonomics of interface design', *Behaviour and Information Technology*, **6**(3), 271–8.
Spigai, F. (1987), 'Displays in database search systems', *Library Trends*, **35**(4), 63–115.
Stobie, I. (1986), 'Alternative keyboards: goodbye to QWERTY?', *Practical Computing*, **9**(6), 52–5.
Storey, C. (1992), 'Great expectations: the human aspects of library automation', *Journal of Library and Information Science*, **18**(2), 1–15.
Suter, E. (1985), 'The legal aspects', in *Alleged Health Hazards of Work at VDUs: Current Industrial Relations Issues: A Conference Held by Humane Technology in London on 26 and 27 November 1985*, pp. 185–95, Quorn: Humane Technology.
Sykes, P. (1986), 'Automation and non-professional staff at the Polytechnic of the South Bank', *Training and Education*, **3**(3), 50–6.
Sykes, P. (1991), 'Automation and non-professional staff: the neglected majority', *Serials*, **4**(3), 233–43.
Tedd, L. (1993), *An Introduction to Computer-based Library Systems*, 3rd edn, Chichester: John Wiley.
Tikkanen, J., Heinonen, O. P., Kurppa, K. and Rantala, K. (1990), 'Cardiovascular malformations and maternal exposure to video display terminals during pregnancy', *European Journal of Epidemiology*, **6** (1), 61–6.
Tindall, A. (1993), *Tenosynovitis: A Case of Mistaken Identity*, London: Iron Trades Insurance Co.
Tougas, G. and Nordin, M. C. (1987), 'Seat feature recommendations for workstations', *Applied Ergonomics*, **18**(3), 207–10.
Trades Union Congress (1996), *New Technologies at Work: A Discussion Document*, London: Trades Union Congress.
Tullis, T. S. (1981), 'An evaluation of alphanumeric, graphic and color information displays', *Human Factors*, **23**(5), 541–50.
Turner, N. (1981), 'The attitude of non-systems staff to automated systems', in

CLAIM Report No. 8, pp. 13–22, Loughborough: Centre for Library and Information Management, Loughborough University.

Underwood, G. (1976), *Attention and Memory*, Oxford: Pergamon.

Union of Communication Workers (UCW) (1993), 'VDU questionnaire results', *Safety and Health Practitioner*, **11**(4), 8.

Van Cott, H. P. and Warrick, M. J. (1972), 'Man as a system component', in Van Cott, H. P. and Kincade, R. G. (eds), *Human Engineering Guide to Equipment Design*, pp. 17–40, New York: McGraw-Hill.

Van der Veer, G. C, Wijk, R. and Felt, M. A. M. (1990), 'Metaphors and metacommunication in the development of mental models', in Falzon, P. (ed.), *Cognitive Ergonomics*, pp. 132–49, London: Academic Press.

Van Nes, F. L. (1986), 'Space, colour and typography on visual display terminals', *Behaviour and Information Technology*, **5**(2), 99–118.

Vandierendonck, A., Van Hoe, R. and De Soete, G. (1988), 'Menu search as a function of menu organisation, categorisation and experience', *Acta Psychologica*, **69**(3), 231–48.

VDU Award (1986), *Library and Information News*, November, 6.

Velton, E. (1968), 'A laboratory task for induction of mood states', *Behaviour Research and Therapy*, **6**, 299–342.

Vince, I. (1987), *Sick Building Syndrome*, London: IBC Technical Services.

Wallach, D. S. (1996), *Typing injury FAQ: keyboard alternatives*, http://www.cs.princeton.edu/~dwallach/tifaq/keyboards.html

Warg, L. E. (1988), 'Redundant information as a means of aiding performance in human machine interface', in Aghazadeh, F. (ed.), *Trends in Ergonomics/Human Factors V*, Amsterdam: North-Holland.

Waters, D. (1986), 'Assessing the impact of new technology on library employees', *Library Automated Systems Information Exchange*, **17**(1), 20–7.

Weinschenk, S. and Yeo, S. C. (1995), *Guidelines for Enterprise-wise GUI Design*, Chichester: John Wiley.

Weinstein, N. (1977), 'Noise and intellectual performance: a confirmation and extension', *Journal of Applied Psychology*, **62**, 104–7.

Welsch, E. K. (1990), 'Technology's potential for enhancing library access to information: present trends and future opportunities', *Australian Library Review*, **7**(1), 5–13.

West, B. C. (1987), 'Staff management and communication', in Revill, D. (ed.), *Personnel Management in Polytechnic Libraries*, pp. 76–99, Aldershot: Gower.

Westerholm, P. and Ericson, A. (1987), 'Pregnancy outcome and VDU-work in a cohort of insurance clerks', in Knave, B. and Wildeback, P., *Work With Display Units '86*, pp. 104–7, New York: North Holland.

Westlake, D. (1988), 'Designing a library for information technology', Paper Presented to the Library Association Information Technology Group Annual Conference, 3 July, London: Library Association.

White, H. S. (1985), *Library Personnel Management*, White Plains, NY: Knowledge Industry.

Whitford, I. (1986), 'Mouse motion', *Systems International*, September, 31–2.

Wickens, C. D. (1984), *Engineering Psychology and Human Performance*, Columbus, OH: Charles Merrill.

Wickens, C. D. (1987), 'Information processing, decision-making and cognition', in Salvendy, G. (ed.), *Handbook of Human Factors*, pp. 72–107, New York: John Wiley.

Wickielgren, W. A. (1964), 'Size of rehearsal group in short-term memory', *Journal of Experimental Psychology*, **68**, 413–19.

Wickielgren, W. A. (1981), 'Human learning and memory', *Annual Review of Psychology*, **32**, 21–52.

Wilkins, A. J. (1985), *Visual Display Units and Fluorescent Lighting Affect Movements of the Eyes Across Text (Report No. HF104)*, Winchester: IBM Hursley Human Factors Laboratory.

Williams, J. T. and Leaf, W. A. (1986), 'Subjective and Objective Judgements of Screen Formats', in *Proceedings of the Human Factors Society 30th Annual Meeting, Sept 29–Oct 3 1986, Dayton, Ohio*, pp. 689–93, Medford, NJ: Learned Information.

Wilson, S. and Hedge, A. (1987), *The Office Environment Survey: a Study of Building Sickness*, London: Building Use Studies.

Windham, G. C., Senster, L., Swan, S. H. and Neutra, R. R. (1990), 'Use of video display terminals during pregnancy and the risk of spontaneous abortion, low birth weight, or intrauterine growth retardation', *American Journal of Industrial Medicine*, **18**, 675–88.

Winstead, E. B. (1994), 'Staff reactions to automation', *Computers in Libraries*, **14**(4), 18–21.

World Health Organisation (1983), *Indoor Air Pollutants: Exposure and Health Effects, World Health Organisation EURO Reports and Studies 78*, Rome: World Health Organisation.

World Health Organisation (1990), *Update on Visual Display Terminals, WHO/OCH/90.3,* Rome: World Health Organisation.

Yang, C. L. and Carayon, P. (1995), 'Effect of job demands and social support on worker stress: a study of VDT users', *Behaviour and Information Technology*, **14**(1), 32–40.

Yates, J. W. and Karwowski, W. (1987), 'Maximum acceptable lifting loads during seated and standing work positions', *Applied Ergonomics*, **18**(3), 239–43.

Yee, M. M. (1991), 'System design and cataloguing meet the user: user interfaces to online public access catalogs', *Journal of the American Society of Information Science*, **42**(2), 78–98.

Zaret, M. (1984), 'Cataracts and visual display units', in Pearce, B. G. (ed.), *Health Hazards of VDTs?,* pp. 47–59, Chichester: John Wiley.

Zoltan, E. and Chapanis, A. (1982), 'What do professional persons feel about computers?', *Behaviour and Information Technology*, **1**(1), 55–68.

Zuboff, S. (1988), *In the Age of the Smart Machine: The Future of Work and Power*, Oxford: Heinemann.

SUBJECT INDEX

abbreviations 216
Aberdeen City Libraries 318
absenteeism 74, 284, 323
Accukey keyboard 127–8
achievement, sense of 9
acronyms 216
adaptability of software interfaces 272
adjustable equipment and workstations 105–7, 111, 116–18, 175, 188, 331
age
  abilities, attitudes and susceptibilities related to 13, 43, 50, 76–7, 94, 144, 161, 165, 213 14, 300, 326
  of library buildings 105
air conditioning 165–70 *passim*
air movement 167–8
alphabetical keyboards 124
America *see* United States
America Online 256–7
American Illuminating Engineering Society 142
analogy 218–20
anthropometric data 21–4, 176
anti-static mats and sprays 99
anxiety *see* stress
Apple Mackintosh 212, 218–19
appraisal of new systems *see* evaluation
arousal 6 8
arthritis 43
artificial intelligence 239
artificial light 144–5
'associative experts' 212–14
Aston University 275, 277
attention and attention span 5–7
attitudes to change 297–304, 323
automation
  effects of 66, 286 93 *passim*, 300–301
  in libraries 69–71
  positive reactions to 101, 304, 315
  reasons for undertaking 207–8, 274, 288
  see also 'first time around'; second time around

backaches 32, 34, 67, 95, 107, 197
background information for staff 330–31
backtracking 215
Balans chairs 109–10
barcode readers 137, 201, 223
Bat Personal Keyboard 127, 129
bifocal lenses 52
'Big Bang' implementation 280
'Big Brother' 304, 323
birth abnormalities *see* reproductive hazards
BLCMP Tallis system 240
blinds 144, 149, 166, 171
body temperature 164–5
bold characters 250–51
books, lifting and carrying of 44, 108
Boolean logic 213, 218
boredom 19
Boston College 330–31
'bouncing' through menus 288
Bradley, Lesley 95
brain, human 3–5
  left and right sides of 71
breakdown of computer systems *see* failure
British Library 60
British Standards Institution (BSI) 268
Browne issue system 70, 285
BRS After Dark 236
BRS/SEARCH and BRS/MENTOR 214
buffer area 189
building sickness *see* sick building syndrome

cabling 146, 190–93
Canada 81, 83–4, 96
capacity to perform tasks 10
captions, on-screen 245–6
carpeting 99
'casual experts' 211–13
catalogues *see* online public access
cataloguing 63, 189, 199–201, 228
cataracts 97 8
CD-ROM 70, 180, 207, 211, 213, 217, 230, 238, 268, 276, 319, 340
CD-ROM Consistent Interface Committee (CD-CINC) 268–9
CEN (European Standardization Organization) 268
chairs, design and use of 35, 38–9, 109, 112–14, 119, 180, 199
changeover procedures 319–20
character sets and fonts 58–60
characters
  maximum number per line on screen 245
  strings of 12
  upper and lower case 245
'cheat sheets' 337
children, special requirements of 23
China 168
chording keyboards 127–8
chunking 12–13, 247
citation indexing 239
Civil and Public Services Association (CPSA) 60
CLO values 168
cloakrooms 168
clocks, ticking of 160
clothing 99–100, 152, 167–8
clusters of birth abnormalities 78
coding techniques 250–51
colour
  aesthetic considerations in use of 254
  emotive effects of 253
  of keyboard keys 131
  of lighting 148–9
  of lights on VDUs 120
  personal preferences about 55
  sensitivity to 50
  use in screen displays 251–5
  of workplace environment 152
colour blindness 251–2
Comfort Keyboard System 127
command languages 225–7, 269
Commission Internationale de l'Eclairage 142–3
Common Command Language (CCL) 269
communication
  with decision makers *see* decision makers
  with library users *see* users of libraries
  between people and machines 3, 208
  with staff *see* involvement
complaints from staff 70–71, 154, 323
computer-aided software engineering (CASE) 271–2
computer-assisted learning 333–4
computers, strengths and weaknesses of 19–21
conceptual errors by VDU users 264
confirmatory messages 262–3
Congenital Disabilities Civil Liability Act 95
consistency within software applications 220, 242, 256
consultants, assistance from 308, 317
consultation with staff *see* involvement
'contact' needs 8
control
  of pace of work 288
  VDU users sense of 257
  over windows, lighting and temperature 170–71
'conversational' interfaces 257
copyholders *see* document holders
Council of Civil Service Unions 96
counselling and co-counselling 76–7
crashes *see* failure
critical path techniques 275
The Crucible 238
cumulative trauma disorder 40
curtains 166, 171
custom-made computer systems 278
customization of screen displays 213

daily personal noise exposure ($L_{EP,d}$) 157–8
data, loss or corruption of 303, 330
data entry and data entry staff 36, 149, 177, 200, 257, 265; *see also* form fill-in
data-gloves 128, 224
Data Protection Act 332
database management systems (DBMS) 272
databases 215, 225
DataHand 128, 130
DataStar 215

daylight 144–5
dB(A) scale 155–7
de-bugging 280
de-skilling 286–7, 293, 306
decibels (dB) 155
decision makers 304–5, 313 14
decision-making 14, 17–18
demonstrations of new systems
  313–14, 333
Denmark 81, 84
Derbyshire County Libraries 313
desensitizers 198–9
desk lamps 146–8
desk-top publishing 136
desks, design of 114–17
Dewey Decimal System 13
DHSS (Department of Health and
  Social Security) 60
Dialog database 215
dialogue design and dialogue styles
  208, 222–40
dimmer systems 145
direct manipulation interfaces 229–32,
  257
Disability Working/Living Allowance
  46
disabled people 21–3, 118
disc pressure 107–8, 111, 113, 133
distance between staff and customers
  197
distractions 15–16, 161
document holders 117, 187
document readers 223
documentation of computer systems
  336–8
double glazing 144, 163
dust covers 188
'dustbin' analogy 218–19, 230
duty of care 45
Dvorak keyboard 123

ear protection 158
ease of use of software 256–65
Edwards, Ross 96
ego, the 8
electric shocks *see* static electricity
electromagnetic radiation 96–7
electronic books 239
electrostatic fields 93, 98–9
ELFs (extremely low frequency
  radiation) 78, 92–3
embedded menus 238
emergency lighting 153
Employers Medical Advisory Service
  100
Employment Service 304
enjoyment of work 9; *see also* job
  satisfaction
environmental conditions 11, 68
epilepsy 100
ergonomics 46, 105–6, 111, 126–7
error handling by VDUs 263–4
error messages 263
ESCAPE key 132
Essex County Libraries 314
etched screens 62–3
European Standardization
  Organization (CEN) 268
European Union
  regulations and directives 23, 30, 65,
    73, 141, 157, 284
  tendering process 276
evaluation
  of new systems 322–3; *see also*
    software
  of training 338
exercise 76, 289, 291
experience levels of VDU users,
  specification of 257
eyeball trackers 224
eyes
  problems with *see* vision
  working of 49–53

failure of computer systems 303, 315,
  332, 335, 338
fans 167
fatigue 19
fear of computers and change 298–304,
  314–15, 317
feedback
  on learning 15
  on manuals 337
  positive and negative 18 19
  to and from staff 285, 287, 321
  from users 321–2
  from VDU screens 261, 264
filter panels 61–2
Finland 79, 82–3, 85–6
fire extinguishers and blankets 100
'first time around' automation 297,
  299, 301, 316, 319
FlexPro Keyboard 127
fluorescent tubes 148
footrests 113 14, 117 18
forearm rests 133–4
form fill-in 214, 228–9
French Society of Ophthalmologists 65

frequency of sound 155
frozen shoulder 41
function keys 131, 133, 227
functional measurements 23
functionality of systems 207–8, 270, 275–7

'gambler's fallacy' 18
Gardner keyboard 124, 127
glare 50–51, 60–62, 119, 131, 146–52 *passim*, 331
graphical user interfaces (GUIs) 229 30, 252, 267
graphics *see* visual images
Graphix OCR system 60
grievances of staff *see* complaints
grieving *see* mourning
Grimsby 78

Hambro Life 149
Hatfield Polytechnic 328
Hawthorne effect 149
hazards, identification of 31–3, 331
head, the, position and movement of 185–7
headaches 34, 65, 67, 107, 148– 9, 323
health and safety standards 29–33, 289 90, 303, 306–7, 313; *see also* legal obligations of employers
hearing damage 161
heating and ventilation 164–70
help
  context-sensitive 238, 260
  on-screen 216, 219, 259–60, 338
high pressure discharge lamps 148–9
Highland Regional Council 95
Hillingdon 191–2, 339
Hobday, Stephen 126
home environment, the 96
hoods for VDUs 63
human factor, the xv xvi, 269, 311–12; *see also* social contact
human information processing 3 19
human mind, the 210
humidity and humidifiers 166–7, 171
hypersensitivity 43
hypertext 238–9, 253

icons, use of 229–30, 252
illumination, recommended levels of 142–4
'imagers' and 'non-imagers' 218
in-house development of computer systems 278

industrial tribunals 95
infrared radiation 92
initiative required of computer users 256–7
input devices for computers 222–4, 266
interaction between people 179
interactive tasks with computers 179
International Organization for Standardization (ISO) 267–8
Internet, the 178–80, 238, 266, 276, 285, 303, 340
involvement
  in the learning process 16
  of staff 9, 74–5, 293, 309–23 *passim*
  of users 271–2
ion concentrations 93–4, 98–9
Ireland 96
isolation of staff 70
issue desks 44, 114, 176, 180, 184–5, 195, 285–6
issue systems 70, 76, 101, 285; *see also* self-issue library systems

Japan 40
jargon 261
job design and redesign 9, 68–9, 283–6, 291–3, 307
job enrichment and enlargement 284, 291–4
job losses 71, 300–301, 306, 314–15
job rotation 291
job satisfaction 106–7, 283–7, 292–3, 306, 309, 311, 322–3
Johnstone, Hazel 95

Kata thermometers 167
keyboards 36, 38, 121–33, 222–3
  split and tilt 126–8
Kinesis Ergonomic Keyboard 127
Kingston Polytechnic 309
Klockenberg keyboard 126–7
kneeling chairs 109–10

laptop computers 190
laser printers 100–101
latent psychology 210
leadership 9, 308
leaflets for library users 318
learning 13–17
  active and passive 16
legal obligations of employers 30–32, 45–6, 55, 73–4, 95, 101, 106, 141, 157, 165, 284, 289, 331
legibility of VDU screens 245

$L_{EP,d}$ *see* daily personal noise exposure
Library Information Technology Centre 325
Library Resource Exhibition 275
light pens 137
lighting 50, 63, 141–54
direct and indirect 145–6
for individual tasks 146–8
recommendations about 153–4
linear tungsten halogen lamps 148
literature on computer systems 275
'local experts' 334–5
'locus of control' 256
London Hazards Centre 290
low energy bulbs 149
lumbago 107
lumbar pads 111, 113
luminance 142

Macproject software package 275
macros 212, 225
Malt, Lilian 126
Maltron keyboard 126–9
management information 17
managers of libraries
role of 293–4, 298–301, 305–10 *passim*, 321, 324
stress amongst 69
training of 326, 329, 335
manuals 260, 336–8
MARC database 199–200
matt surfaces 152
medical examinations for staff 66
medication 76
memory, short-term and long-term 5, 11–14, 217
mental capacity 6, 10
mental models 217–19, 237, 260
menu selection 232–8
nesting within 236–7
metabolic heat 167–8
metaphor 219
mice 135–6, 223, 230–31
Micromesh filters 61, 63
microwave emission 92, 98
Miller, Arthur 238
modular furniture 189–90
monitoring of computer use 273, 315
see also surveillance
morale of staff 20, 102, 285, 323–6
motivation 6–10, 283–4, 291
classification of 8
for learning 15
for undertaking automation *see* automation
'mourning' for old working practices 298, 305, 312, 315
mouse-arm syndrome 135, 223
multi-dimensional information 17
multimedia systems 238
muscle pain 40–42, 48, 185
music, 'piped' 8

naive users 211–16, 227, 234, 238–40
National Academy of Sciences, US 91
National Association of Working Women, US 66
National Electronics Council 226, 288
National Radiological Protection Board 91, 98
natural daylight 144–5
natural language 213, 225, 227, 239–40
natural wastage of staff 314–15
needs, hierarchy of 8–9
negligence by employers 45–6
net curtains 144
Netherlands 73
neutral density filters 61
New Zealand 38, 40
noise 7, 155–63, 250
from keyboards 131
from people talking 159
from printers 136
recommended levels for offices 158
normal distribution 23, 176
Norway 80
NOTIS system 331
numbers, strings of 12
numeric keypads 132

objectives, setting of 310–11, 322
occupational cervcrobrachial disorder 40
occupational overuse syndrome 40
Occupational Safety and Health Administration, US 157
OKAPI system 239
on-site training 327
Online (magazine) 220
online information retrieval 177, 201–2, 215, 220, 286, 300, 318
online public access catalogue (OPAC) 134–6, 165, 178, 180, 192, 207, 214–16, 237, 239–40, 247, 258–9, 267, 277–9, 326, 340
Online Public Access Information Retrieval System (OPAIRS) 211–12

online tutorials 259
open plan offices 158
operation requirement (OR) documentation 275
options on screen 14; *see also* menu selection
overload and underload of work 288
'ownership' of information and systems 300, 317
ozone 100–101, 136

palette menus 234–5
palm rests 134
parallel processing by the human brain 5–6
parallel running of new systems 280, 304, 320
paraprofessional staff 302
participative management 309–10; *see also* involvement of staff
passers by, employers duty to 141
passwords 13
pay 9–10, 302, 304, 306
PCBs (polychlorinated biphenyls) 94
perception, process of 10–11, 216–17
  'bottom-up' and 'top-down' 5
periodicals *see* serials
phased implementation 320
physiological needs 8
pictorial messages 13, 16
pilot operation of new systems 320
planning and planning groups 308–13, 321
pointing devices 135
polarization filters 62
pop-up menus 234
positive feedback 15, 18
posture 35–9, 42, 107–12, 167, 179, 289, 331
predicted mean vote (PMV) 168–9
pregnancy 43, 101; *see also* reproductive hazards
PRINCE (Projects IN Controlled Environments) 274–5
print quality 136
printer tables 193
probabilities 18
productivity 105, 114, 152–3, 194, 208–9, 284, 311, 325
professional and non-professional staff 300–302, 305, 310
project management techniques 274–5
promotion prospects 294
prompting messages 262
proof reading 161
prototyping 266, 271–2, 279
psychological closure 215–16
psychological space 198
pull-down menus 232, 243

quarter-wave coatings 62–3
query languages 225
quitting applications 264
QWERTY keyboard 122–4

radiant heat 166
radiation 78, 89–90, 92–6
rashes, facial 98–9
reach, length of 182–4
reaction times 19
reading
  process of 52–3
  from a VDU 64, 247
receptiveness 10–11
recognition
  as distinct from recall 14, 216, 234
  for effort and performance 9, 68
recovery by VDUs 264
red lights 120
redundancies *see* job losses
reference manuals *see* manuals
reflected light 150–52; *see also* glare
reflected noise 159
refresher courses 338
relationships with colleagues 9; *see also* social contact
relative humidity 166–7, 171
reproductive hazards 33, 78–96
response mechanisms, human 18–19
response times of VDUs 265
responsibility, carrying of 9, 286–7
rest breaks 19, 65, 289–91, 331
reversible fatigue syndrome 40
rheumatoid arthritis 43
risks *see* hazards
room layout of libraries 159
RSIs (repetitive strain injuries) 32, 36, 40

safe learning environments 16
SAMMIE package 35
scanning devices 223
screening
  of employees vision 52
  of potential employees 43
screens of VDUs 55–7, 62–4, 119–20
  adjustable arm supports for 189
  displays 215–20, 241–55, 266

dust on 92
positioning of 185–7, 199
users requirements of 241–2
see also filter panels
scrolling 64
'second time around' automation 309–10
security devices 193, 198
sedentary work *see* sitting
selection of computer systems 274–7
selective dissemination of information (SDI) schemes 207
'self-actualization' needs 8
self-issue library systems 180–81, 199
senses, human 3–4
sensitizers 198–9
serials, processing of 200–201
sexual health *see* reproductive hazards
shared workstations 106–7
Sholes layout *see* QWERTY keyboard
short cuts with computers 212, 227
short-term memory *see* working memory
sick building syndrome 169–70
sight tests 51–2
simplicity/complexity of screen displays, measurement of 246
simultaneous performance of tasks by people 6
single channel theory 6
sitting 94–5, 107–8; *see also* chairs
'Six Blind Men of Hindustan' 219
size of people, variations in 21–2, 118
smoking 79
social contact 8, 285–6, 293, 301–2
social readjustment rating scale 80–87
soft function keys 227
software, evaluation and re-evaluation of 270–77, 280
software interface 207–21
definition of 208
solar heat gain 144
sound *see* noise
sound absorbent barriers and materials 163
South Bank Polytechnic 303, 310, 327
special purpose keys on keyboards 131–2
spontaneous abortion 79, 88, 94
staff development opportunities 287, 299
staffing structures in libraries 294; *see also* professional and non-professional staff
standardization in software 219, 265–7
standards-setting bodies 267–8
static electricity 93, 99, 193–4
status messages 262
stillbirths 79
stimuli 6, 11, 250–51
stress 15, 33, 37, 41, 44, 66–77, 99, 114, 161, 170, 208, 284–6, 288, 291, 298–9, 304, 325, 330
causes of 68–71, 93–4
definition of 66–7
individuals reactions to 75–6
in libraries 69–70, 77
measurement of 71 3
Suffolk Libraries and Heritage 274–5
supervision in libraries 286–7; *see also* managers
support for computer users 334–9
surveillance of staff by computers 304; *see also* monitoring
Sweden 40, 73, 81–2, 86, 158
syntax in computer language 225, 234
system furniture 194
system staff, training of 329

talking, noise of 159
task performance messages 262
task analysis 175–80, 202, 278–9
temperature 165–70 passim; *see also* body temperature
tendering process 276
tennis elbow 41
'territorial rights' of staff 293
testing of computer systems 280
thermal indices 168–9
thermometers 167
timing
of automation projects 319
of training sessions 327–8
'tip sheets' 340
touch screens and pads 134–5, 223–4
tracker balls 223
trade unions 306–7
Trades Union Congress 304, 306
training 218, 226, 234, 258–9, 266, 303, 306, 317, 322, 325–40
content of 328–32
factors influencing impact of 15–17
methods of 332–4
transfer of 16
transfer
of staff to other work 301–2
of training 16
troubleshooting 259, 330

trunking *see* cabling
turnover of staff 284
typing errors 161

ultrasound 94
ultraviolet radiation 92
underlining 250
underwear 99–100
undo facility 264
union representation 306–7
Union of Communication Workers (UCW) 29
United States 40, 82, 85, 87
  Department of Education 301
University of Bath 176, 185, 195
University of Tasmania 293
University of Wales, Bangor 274
uplighting 146, 149
user interface management systems 271–2
user-friendliness 256, 270
user-initiated dialogues 256–7
users of computer systems
  definition of 30–31
  groups of 210–14
  requirements of 275–6, 279
users of libraries
  communication with 318–19, 321–2, 339–40
  expectations of 321
  training for 339–40

vapour-deposited coatings 62
variations between people 23–5
VDUs (visual display units)
  as a cause of stress 70–71
  health hazards associated with 78–102
  recommended maximum times for working with 289–90
  technical aspects of 54–63, 119–20
  types of work done on 177–9
  see also screens
ventilation *see* heating and ventilation
viewing angles for VDU screens 185 7, 199
virtual reality 224
vision
  of the library and of new systems 308
  problems with 32, 34, 48–65, 289
visits by staff to other libraries 316, 333
visual images 12–13, 15–16
Vitamin A 92
voice input 240

warnings to staff on potential risks 46–7
welcome screens 261–2
white space on VDU screens 247, 255
WIMPs environment (Windows, Icons, Menus, Pointers) 229–34 *passim*
windowless rooms 144
windows
  coatings for 144
  on screen 229–30
Windows interface 56, 59, 212, 218, 244, 250, 257, 261, 338
'Wizard of Oz' prototyping 271
women
  musculoskeletal problems in 37, 42–3
  susceptibility to sick building syndrome 169
  susceptibility to stress 94
word processing 63, 179
work content *see* job design
work groups 293, 301–2
Work Research Unit 284
working conditions 9, 38
working memory 5, 11–14, 214–17
  limitations on 13
workplaces
  definition of 105
  modelling of 35
workstations
  adjustable to size of users 23
  definition of 105, 175
  design of 46, 106–37, 175–94
  layout of 195–213, 285
World Health Organization 169
World Wide Web (WWW) 179, 238, 264
wrist supports 133–4
writer s cramp 41
WRULDS (work-related upper limb disorders) 36 8, 40–47, 122, 125, 195
  causes of 41–2

X-ray emissions 61, 78, 89–91

Yerkes-Dodson law 7

INDEX OF AUTHORS CITED

Åborg, C. *et al* 289
Albrecht, M. *see* Belkin, N. J. et al (1991)
Alderfer, C. P. 8
Allen, A. J. 309
Alluisi, E. A. 12, 14
Alteras-Webb, S. M. *see* Henning, R. A. et al
Anderson, J. R. 14–15
Argyle, M. 291, 301
Armstrong, B. *see* McDonald, A. D. et al
Aronsson, G. *see* Åborg, C. et al
Astle, D. 317
Atkinson, H. C. 315

Baddeley, J. 12, 309, 338
Bagley, D. E. 328
Ballard, J. 42–3
Banister, E. W. 34
Barclay, N. L. 124
Bates, M. J. 208
Batt, C. 307, 313
Bawden, D. 303
Beale, R. 4
Becket, M. 193
Beckman, D. A. *see* Brent, R. L. *et al*
Beckman, M. 188, 191
Begg, K. 330–1
Belkin, N. J.
et al (1990) 214, 279
et al (1991) 228
Bendix, T. 133–4
Bennett, W. R. *see* Brent, R. L. *et al*
Bently, C. 275
Beral, V. *see* Roman, E. et al
Bergqvist, U. 48, 78, 91–2, 94
et al (1992) 48
et al (1995a) 36
et al (1995b) 36 7
see also Ong, C. N. et al
Berliner, D. 14
Bernstein, B. 208, 213, 263, 278
Bichteler, J. 300, 308, 326
Biddiscome, R. 302
Birch, N. E. *see* Smith, N. M. *et al*
Bjerkedal, T. 80
Blunden-Ellis, J. 275
Boies, S. J. 231
Bonner, P. 231
Borgman, C. L. 218, 227, 245, 260, 264, 273
et al (1989a) 245
et al (1989b) 210
et al (1990) 213, 236
Bosch, V. M. 230
Bower, G. H. 12
Bower, J. *see* Borgman, C. L. *et al* (1990)
Boyce, P. R. 145
Boyle, H. M. 65
Braidwood, P. 96
Bramwell, R. S. 80, 88
Brand, J. L. 187
Brandt, L. P. A. 81, 84, 88
Breeding, M. 275
Brent, R. L. *et al* 93
Bridger, R. S. 114, 166, 284, 292
Brown, I. D. 10
Bryant, H. E. 81, 88
Bunge, C. A. 70
Burge, P. S. *see* Finnegan, M. J. *et al*; Hedge, A. *et al*
Burton, P. 298, 328, 337
Buxton, A. 263

Cakir, A. 127
et al 61, 75, 100, 111, 166–7, 286
Calnan, C. D. 99
Caputo, J. S. 70, 292
Carayon, P. 37, 71, 288
Card, S. K. 210, 237

Cardman, E. B. *see* Norlin, D. A. *et al*
Caren, L. *see* Somerville, A. N. *et al*
Carroll, D. 135
Carter, J. B. 34
Casale, M. 191
Case, D. O. *see* Borgman, C. L. *et al* (1989a)
Chang, S. *see* Belkin *et al* (1990)
Chapanis, A. 19, 300
Cherry, J. M. 258
Cherry, N. M. *see* McDonald, A. D. *et al*
Cheung, S. T. Y. 332
Chignell, M. H. *see* Borgman, C. L. *et al* (1989a)
Chin, J. P. 228, 238
et al 236
Christ, R. E. 251
Christie, B. 273; *see also* Cole, I. *et al*; Hammond, N. *et al*
Cihangirli, M. *see* Fernandez, J. E. *et al*
Clarke, J. 297–8, 317, 319, 339
Clayton, M. 300, 307, 313, 323, 337
Cleeve, M. 302, 308–9, 317, 324
Clegg, C. W. *see* Johnson, G. I. *et al*
Clinton, M. 258
Cluff, E. 29
Cole, I. *et al* 210–12, 225, 227, 229–30, 243, 247, 255, 257
Coles, P. 250
Commings, K. 330, 336
Connor, C. M. 326–7, 331
Conover, D. L. *see* Schnorr, T. M. *et al*
Conrad, R. 12
Cooper, C. L. 288
Corbin, J. 293
Cox, T. 67, 70
Cross, R. 303

Dainoff, M. J. 36
Dakshinamurti, G. 309
Dallner, M. *see* èborg, C. *et al*
Damodaran, L. *et al* 149–50, 153, 165, 183, 187, 293, 312, 322, 336
Daniels, R. J. 303, 315, 321, 338
Das, B. 183
Davidson, M. J. 80, 88
Davis, C. H. 250
Davis, E. B. *see* Norlin, D. A. *et al*
Davis, P. J. 105, 194
De Jong, J. R. 197
De Soete, G. *see* Vandierendonck, A. *et al*
Dean, S. C. 307
Deese, J. 12–13
Dennison, P. 105, 170
Dickson, J. 264
Diehl, V. A. *see* Chin, J. P. *et al*
Dillon, A. 64, 247
et al 64
Döckel, H. 298, 309, 311
Doerfel, H. R. *et al* 91
Dossett, R. *see* Norlin, D. A. *et al*
Downs, J. *see* Belkin *et al* (1990)
Dressel, D. L. 105, 194
Dul, J. 117, 157
Dyer, H. *et al* 294
Dyson, J. 328

Edholm, O. G. 142
Egan, D. E. *et al* 238
Egeland, G. M. *see* Schnorr, T. M. *et al*
Egenaes, J. 80
Else, L. 71
Epple, M. *et al* 327, 333, 338
Ericson, A. 81–2, 86
Essinger, J. 208, 232, 268

Falzon, P. 240
Fanger, P. O. 168
Fayen, E. G. 192
Felt, M. A. M. *see* Van der Veer, G. C. *et al*
Ferguson, Professor 122
Fernandez, J. E. *et al* 224
Ferris, A. 131, 134
Fine, S. xv, 298, 300–1, 304, 313
Finer, R. 293
Finnegan, M. J. *et al* 169
Fisher, D. 124 5
Fleischer, A. G. *et al* 111
Fossey, D. *see* Dyer, H. *et al*
Foster, J. 250
Francis, J. 105, 194
Fusco, L. *see* Belkin, N. J. *et al* (1991)

Gage, N. 14
Galer, I. 142, 144, 150, 164–5, 167
Galitz, W. O. 243, 245–7, 252–3, 255, 261–2, 267
Gallagher, A. L. *see* Borgman, C. L. *et al* (1990)
Garcia, C. R. 319, 321, 326–7, 340
Gardiner, M. M. 273; *see also* Hammond, N. *et al*; Marek, T. *et al*
Gardner, J. *see* Epple, M. *et al*
Gedliczka, A. *see* Marek, T. *et al*
German, M. P. *et al* 286
Geyser, E. P. 217–18

Gibson, J. E. 12
Gilad, I. 122
Gittings, D. 225
Goldhaber, M. K. *et al* 79, 82, 88
Gomez, L. M. *see* Egan, D. E. *et al*
Gordon, W. E. *see* Brent, R. L. *et al*
Gould, J. D. 231
Goulding, A. 301–2, 321
Graffunder, H. *see* Doerfel, H. R. *et al*
Graham, K. R. 339
Graham, M. E. 275
Grajewski, B. A. *see* Schnorr, T. M. *et al*
Grandjean, E. 56, 59–63, 91–3, 97–8, 107–8, 111–13, 117, 121, 131, 133, 142–3, 148, 150, 152, 155, 161–2, 177–9, 182, 185
Gratton, P. D. 313, 334–5
Green, P. 271
Grey, S. M. 197–8
Grossman, W. 214, 261
Grudin, J. 220

Hagberg, M. 135
Halasz, F. 219
Hallmark, J. 319, 321, 326–7, 340
Halperin, W. E. *see* Schnorr, T. M. *et al*
Halpern, C. A. 105, 194
Hamdy, W. R. 64, 97
Hammond, N. *et al* 272
Hancock-Bealieu, M. 230, 237, 239
Handy, C. B. 314
Harley, B. 311
Harris, G. 252
Harrisbass, J. *see* Hedge, A. *et al*
Hart, D. J. *see* Cakir, A. *et al*
Hartshorn, H. R. 271
Harvey, F. 276
Hedge, A. 101, 169
et al 170
Heinonen, O. P. *see* Tikkanen, J. *et al*
Hemminki, K. *see* Kurppa, K. *et al*
Henigman, B. *see* Norlin, D. A. *et al*
Henning, R. A. *et al* 291
Henry, H. K. 207, 231, 236
Hepworth, J. B. 325
Hermon, C. *see* Roman, E. *et al*
Herring, J. E. 318
Herzberg, F. *et al* 8–9
Heseltine, R. 276
Hiatt, R. A. *see* Goldhaber, M. K. *et al*
Hicks, R. 208, 232, 268
Hietanen, M. *see* Kurppa, K. *et al*
Hildreth, C. R. 247, 277
Hitch, G. 12
Hix, D. 271
Hoang, K. 35, 111
Hobday, S. 122
Hodges, J. E. 69
Hollan, J. D. *see* Hutchins, E. *et al*
Holmberg, P. C. *see* Kurppa, K. *et al*
Holmes, S. 69
Holmes, T. H. 71–2, 80
Holroyd, K. 292
Hommertzheim, D. L. *see* Fernandez, J. E. *et al*
Hornung, R. W. *see* Schnorr, T. M. *et al*
Huff Arneson, R. 335
Hughes, G. 43
Hulser, R. P. 208, 214
Hunter, I. M. L. 11
Hunter, P. 149
Hünting, W. 36
Hutchins, E. *et al* 257

Iacono, S. 286
Illingworth, V. 229

Jackson, T. 4
Jacques, P. *see* Henning, R. A. *et al*
James, R. 309 10
Jaschinski-Kruza, W. 186
Jeffreys, S. 93
Jessen, F. 133–4
Johansson, A. 34
Johnson, C. P. 287
Johnson, G. I. *et al* 208, 277
Johnson, P. 301–2
Jones, D. E. 302, 304
Jones, N. 306–7
Jones, W. P. 208
Jordan, P. 306–7
Judd, K. W. 187

Kahn, H. 288
Kallén, B. 81–2
Kaplan, I. *see* Somerville, A. N. *et al*
Karwowski, W. 108
Kayis, B. 35, 111
Kinney, T. 211, 213
Kissel, G. V. *see* Henning, R. A. *et al*
Klatzky, R. L. 12
Kling, R. 286
Klobas, J. E. 297, 308, 314
Knave, B. 48, 59; *see also* Bergqvist, U. *et al* (1992)
Kolers, ? *et al* 64
Kompier, M. 68, 73–5, 77

Krieger, D. *see* Borgman, C. L. *et al* (1990)
Kroemer, K. H. E. 127–8, 130
Kruppa, K. 85
Kurppa, K. *et al* 82; *see also* Tikkanen, J. *et al*
Kyllonen, P. C. 12, 14
Kyronen, P. *see* Kurppa, K. *et al*

Landauer, T. K. *see* Egan, D. E. *et al*
Lansdale, M. *see* Cole, I. *et al*
Leaf, W. A. 272
Lee Kwon, M. 337
Lefebvre, C. J. 333
Lepper, M. 15
Levi, L. 68, 73–5, 77
Lewis, A. C. 274–5
Lieberman, K. 12
Lim, S. 37
Lindbohm, M.-L. *et al* 79, 83
Lochbaum, C. C. *see* Egan, D. E. *et al*
Love, E. J. 81, 88
Lovecy, I. 289, 301–3, 315–16, 322, 328, 337
Lucas, H. C. 310
Lundin, L. 170

McCormick, E. J. 168
McDonald, A. D. *et al* 79, 83–4, 88
McDonald, J. 237; *see also* McDonald, A. D. *et al*
Mace, C. A. 14–15, 18
MacKay, C. 36–7, 70, 94
McKee, K. *see* Dyer, H. *et al*
Mackenzie, J. A. 318
McKnight, C. *see* Dillon, A. *et al*
Maddox, M. D. 224, 261, 279–80
Malone, T. W. 15
Mandal, A. C. 111
Mann, P. 300
Marchant, M. P. 309–10, 323; see also Smith, N. M. *et al*
Marchetti, P. G. *see* Belkin, N. J. *et al* (1991)
Marchionini, G. 208, 215, 217, 219
Marek, T. *et al* 178
Marshall, C.
et al 256, 262
see also Hammond, N. *et al*
Matthews, J. 220, 274–5, 277, 332
Matuszek, L. *see* Marek, T. *et al*
Mausner, B. *see* Herzberg, F. *et al*
Meadow, C, T. *see* Borgman, C. L. *et al* (1989a)

Meichenbaum, D. 15
Mild, K. H. 93
Miller, G. A. 12
Milne, J. R. 238
Mischo, W. H. *see* Norlin, D. A. *et al*
Moran, T. P. 210, 219
Morris, A. 29, 105–6, 170
Mosier, J. N. 232
Msuya, J. 270
Munrush, J. L. 333
Murray, W. E. *see* Schnorr, T. M. *et al*
Murrell, K. F. H. 19

Nadler, D. A. 312
Naqvi, S. A. A. 109
Nelson, C. *see* Marek, T. *et al*
Neutra, R. R. *see* Windham, G. C. *et al*
Newmark, M. E. 100
Niedzwetzki, J. 152
Nielsen, C. V. 81, 84, 88
Nilsson, B. *see* Bergqvist, U. *et al* (1995a and 1995b)
Nolin, A. D. *see* McDonald, A. D. *et al*
Nordin, M. C. 109
Norlin, D. A. *et al* 214, 255, 260, 265
Norman, D. A. 124 5, 210; *see also* Hutchins, E. *et al*
Norman, K. L. 228, 238; *see also* Chin, J. P. *et al*
Novak, G. 136
Nowak, E. L. *see* German, M. P. *et al*
Noworol, C. *see* Marek, T. *et al*
Nurminent, T. 851 *see also* Kurppa, K. *et al*

Oborne, D. J. 250–1
Oder, N. 238
Olsgaard, J. N. 307
Olszak, L. P. 241, 262–3, 267
Ong, C. N. *et al* 78–9
Orna, L. 214
Osaka, N. 55
Oyston, E. 328

Pachent, G. 274–5
Parker, S. 285
Parsons, K. 169
Pasak, P. 211, 213
Paznik, M. J. 114, 152
Pearce, B. 288
Pekkarinen, M. *see* Kurppa, K. *et al*
Pelerin, M. *see* Roman, E. *et al*
Penry, J. K. 100
Peterson, L. R. and M. J. 11

Petrie, J. H. 328, 337
Pheasant, S. 22, 24, 109, 117, 121, 142, 144, 183
Pickering, C. A. C. *see* Finnegan, M. J. *et al*
Piesch, E. *see* Doerfel, H. R. *et al*
Polen, M. R. *see* Goldhaber, M. K. *et al*
Pollatschek, M. A. 122
Polson, P. 220
Poulton, E. C. 10
Puttapithakporn, S. 236–7

Quinlan, T. 314, 316, 323
Quinn, J. B. 317

Rademacher, U. *see* Fleischer, A. G. *et al*
Rahe, H. 71–2, 80
Ralli, T. 38, 44, 188
Rantala, K. *see* Kurppa, K. *et al*; Tikkanen, J. *et al*
Rantanen, J. 272
Ravden, S. J. 277; *see also* Johnson, G. I. *et al*
Rea, G. 176, 185, 188, 195, 197
Reeves, P. E. 314
Reiman, D. J. *see* Somerville, A. N. *et al*
Reitman, J. S. 12
Remde, J. R. *see* Egan, D. E. *et al*
Rennie, A. 208
Resnik, R. A. 213, 225
Rice, R. E. 273
Richardson, J. *see* Dillon, A. *et al*
Roberson, M. M. 152
Robert, D. *see* McDonald, A. D. *et al*
Robertson, A. S. *see* Hedge, A. *et al*
Robertson, P. J. 254
Roggeveen, C. 197
Roman, E. *et al* 85, 88
Rooks, D. C. 297, 299–300
Rowley, J. E. 270–1
Rycroft, R. J. G. 99

Sabuncuoglu, I. *see* Fernandez, J. E. *et al*
Sake, J. G. 219
Sallmen, M. *see* Kurppa, K. *et al*
Sanders, M. S. 168
Saracevic, T. *see* Belkin *et al* (1990)
Saunders, K. 159
Saunders, L. M. 337
Saxen, L. *see* Kurppa, K. *et al*
Scalet, E. A. 101
Schneider, W. 16
Schnorr, T. M. *et al* 85, 88
Schrenk, L. P. 17
Schvaneveldt, R. W. 237
Sekey, ? 64
Sell, R. G. 284, 292
Sengupta, A. K. 183
Senster, L. *see* Windham, G. C. *et al*
Shackel, B. 20, 128, 270
Shahnavaz, H. 34
Shaw, D. 213, 232, 234, 236, 238, 250, 301
Sheedy, J. E. 55, 60
Shires, N. L. 241, 262–3, 267
Shneiderman, B. 211–12, 225, 231, 256, 263, 266
Shoham, S. 155
Simpson, A. *see* Damodaran, L. *et al*
Skogvold, S. *see* Belkin, N. J. *et al* (1991)
Slovak, A. J. M. 106
Smith, H. B. 193
Smith, N. M. *et al* 76
Smith, S. L. 232
Snow, B. 220
Snyderman, B. *see* Herzberg, F. *et al*
Somerville, A. N. *et al* 145, 191
Sperandio, J. 243
Spigai, F. 220, 236
Springer, T.J. 114
Stewart, T. F. M. *see* Cakir, A. *et al*
Stobie, I. 123
Stokke, H. *see* Belkin, N. J. *et al* (1991)
Storey, C. 299
Sullivan, A. B. *see* Henning, R. A. *et al*
Suter, E. 95
Swan, S. H. *see* Windham, G. C. *et al*
Sykes, P. 297, 303, 305, 309–10, 316–17, 327, 333

Taskinen, H. *see* Kurppa, K. *et al*
Tedd, L. 320
Thein, M. M. *see* Ong, C. N. *et al*
Thun, M. J. *see* Schnorr, T. M. *et al*
Tietz, ? 64
Tikkanen, J. *et al* 86
Tindall, A. 42, 45
Tougas, G. 109
Trenner, L. 263
Trevers, C. 106
Troina, G. *see* Belkin, N. J. *et al* (1991)
Troutman, L. *see* Norlin, D. A. *et al*
Turner, N. 298, 323

Underwood, G. 12

Valdez, F. *see* Borgman, C. L. *et al* (1989a)
van Brakel, P. A. 217–18
Van Cott, H. P. 19
Van Hoe, R. *see* Vandierendonck, A. *et al*
Van der Veer, G. C. *et al* 218
Vandierendonck, A. *et al* 213
Velton, E. 15
Vince, I. 169–70
von Nandelstadh, P. *see* Kurppa, K. *et al*
Voss, M. *see* Bergqvist, U. *et al* (1992, 1995a and 1995b)

Wallach, D. S. 125, 127–8, 135
Warg, L. E. 217
Warrick, M. J. 19
Warwick, R. T. *see* Epple, M. *et al*
Waters, D. 293
Weerdmeester, B. 117, 157
Wei-Haas, L. 271
Weinschenk, S. 232, 267
Weinstein, N. 161
Welsch, E. K. 325
West, B. C. 309
Westerholm, P. 86
Westlake, D. 191–2
White, H. S. 299, 315
Whitford, I. 135
Wickens, C. D. 6, 11–12
Wickielgren, W. A. 12, 14
Wijk, R. *see* Van der Veer, G. C. *et al*
Wilbom, R. *see* Bergqvist, U. *et al* (1992)
Wilke, J. S. *see* Germann, M. P. *et al*
Wilkins, A. J. 148
Williams, J. T. 272
Wilson, J. R. 197–8
Wilson, P. *see* Damodaran, L. *et al*
Wilson, S. 169; see also Hedge, A. *et al*
Windberg, H. J. *see* Fleischer, A. G. *et al*
Windham, G. C. *et al* 87
Winstead, E. B. 301, 303–4, 317
Wolgast, E. *see* Bergqvist, U. *et al* (1995a and 1995b)
Worringly, ? 288

Yang, C. L. 71, 288
Yates, J. W. 108
Yee, M. M. 278
Yeo, S. C. 232, 267
Ylikoski, M. *see* Kurppa, K. *et al*

Zaret, M. 98
Zhao, S. *see* Belkin *et al* (1990)
Zingrone, Frank 71
Zoltan, E. 300
Zuboff, S. 285